STECK-VAUGHN

CONNECTIONS

Basic Skills in Mathematics

REVIEWERS

Jim Barlow
Retired Vice-Principal
of Adult Education
Waterloo Region District
School Board
Educational Consultant
and Author
Kitchener, Ontario

William Burns
Instructor
San Mateo County
Office of Education
Palo Alto, California

Sherri Claiborne
Literacy Coordinator
Claiborne County
Adult Reading Experience
(CCARE)
Tazewell, Tennessee

Bill Freeland
Almonte Adult School
Almonte, California

Joanie Griffin-Rethlake
Adult Education Division
Harris County Department
of Education
Houston, Texas

Jim Scheil
Jersey City Adult Education Center
Jersey City, New Jersey

STECK-VAUGHN
COMPANY

A Division of Harcourt Brace & Company

www.steck-vaughn.com

Acknowledgments

Executive Editor: Ellen Northcutt

Project Editor: Julie Higgins

Design Manager: Jim Cauthron

Media Researchers: Claudette Landry, Christina Berry

Cover Design: Donna Neal

Cover Production: Donna Neal, Alan Klemp

Electronic Production: PC&F, Inc.

Photograph Credits: Cover and title page: all images © PhotoDisc, except (20,1) © T. Kevin Smyth/The Stock Market; (4, 6, 7, 8) © Chris Tomaidis/ Tony Stone Images; pp. 9, 62, 69, 106, 115, 148 © PhotoDisc; p. 152 © Tim Defrisco/Allsport; p. 156 © Michael Newman/PhotoEdit; p. 228 © CORBIS/ Owen Franken; p. 232 © Werner Bertsch/Bruce Coleman

Illustration Credits: G&S Typesetters, Inc. pages 1, 3, 4a, 5–6, 22, 28, 30, 33–35, 48, 52–54, 56–57, 59–61, 70b, 72–74a, 75–76, 80, 82–83, 90–96, 102, 109, 113, 117, 122–123, 126, 132–133, 135, 142–143, 151b, 155, 159, 164–171, 172b–e, 176, 178, 180, 182–183, 186–187, 189–191, 193–203, 205–207, 209–210, 212, 216–221, 222b–223, 224a, 226–227, 231, 235–236, 238b, 239, 240

PC&F, Inc. pages 4b, 10, 12–13, 15–16, 23, 29, 31, 43, 49, 63, 65, 70a, 74b, 79, 103, 107, 116, 127, 130, 137, 139, 146, 149, 151a, 158, 160, 162, 172a, 213–214, 222a, 222c, 224b, 225, 238a, 242, 245–246, 248–249, 250–251, 254, and calculator icon used throughout

Contents

UNIT 1

UNIT 2

UNIT 3

To the Student

How to Use This Book

This book presents basic mathematics skills and concepts in the context of everyday, real-life applications. The four units are divided into sections, which begin with a real-life activity presented as set-up problems requiring no computation. You first practice identifying the correct operation, equation, or expression that would solve the problem. Then you learn the computation skills needed to solve similar problems. Each section ends with a two-page practice review. The **Answers and Explanations** section at the back of the book is useful for checking your computation skills as well as the answers.

UNITS

Unit 1: Whole Numbers. This unit covers whole number concepts such as place value, order of operations, rounding, equations, and expressions, as well as addition, subtraction, multiplication, and division of whole numbers.

Unit 2: Fractions. Fraction concepts and using the four operations with fractions are presented in this unit.

Unit 3: Decimals. This unit covers decimal concepts and using the four operations with decimals.

Unit 4: Ratios, Proportions, and Percents and Special Topics. In this unit you will learn the concept of ratios and how to solve proportions. You will also use the percent and interest formulas. This unit also covers two special topics: the concepts of probability and finding the circumference and area of circles.

INVENTORY AND POSTTEST

The Inventory is a self-check to see which skills you already know. When you complete all the items in the Inventory, check your work in the Answers and Explanations section in the back of the book. Then fill out the Inventory Correlation Chart. This chart tells you where each skill is taught in this book. When you complete this book, you will take a Posttest. Compare your Posttest score to your Inventory score to see how much your math skills have improved.

REVIEWS

Each unit includes a Unit Review that lets you see how well you have learned all the skills and concepts in that unit. Each Unit Review includes a **Math Extension** activity that provides an opportunity for further practice with the unit skills. Units Two, Three, and Four also include Cumulative Reviews combining all the skills and concepts that have been presented up to that point.

REAL-LIFE APPLICATIONS

All the math skills and concepts are applied to problems in real-life applications in the following four areas:

- Workplace Math—math used on the job, in the office, and in business situations, with such items as order forms, timecards, and pay stubs; and in computing sales discounts, commissions, and sale prices.
- Measurement and Geometry—working with both standard and metric units to find perimeter, area, volume, and circumference; reading maps and scales, and applying percent, ratios, and rates.
- Data Analysis—working with graphs, charts, and tables, and computing the mean and median and probability.
- Algebra Topics—writing and solving equations and using formulas.

PROBLEM-SOLVING STRATEGIES

You will learn many helpful strategies that are used to solve math problems. As you work with these strategies, use the following four-step plan to solve word problems.

Read Read the problem to determine what you need to find out. Identify the information you need to solve the problem.

Plan Plan how you will solve the problem. Decide which operation or operations you will use to solve the problem.

Solve Solve the problem by doing all the computations necessary to find the answer.

Check Check your answer by reading the problem again. Ask yourself, "Does my answer make sense?" Also check your computations to make sure you did them correctly.

CALCULATOR HANDBOOK

A Calculator Handbook is included after Unit 4. It presents step-by-step instructions for applying the four operations—addition, subtraction, multiplication, and division—to whole numbers and decimals, and for solving exponent, square root, and percent problems.

MATH AT WORK

Math at Work is a two-page feature included in each unit. Each Math at Work feature introduces a specific job, describes the math skills the job requires, and includes a math activity related to it. It also gives information about other jobs in the same career area.

MATH CONNECTION

Math Connection is an interdisciplinary feature included in each unit that shows how mathematics is related to another content area. It provides information about the relationship and an exercise to check comprehension.

Inventory

Use this Inventory before you begin Section 1. Don't worry if you can't easily answer all the questions. The Inventory will help you determine which skills you are already strong in and which skills you need to practice.

Read and answer the questions that follow. Check your answers on pages 255–257. Then enter your scores on the chart on page 7. Use the chart to figure out which skills to work on and where to find those skills in this book.

Write the value of the underlined digit in words.

1. 6,302,450 _____

2. 5.3479 _____

Compare each pair of numbers. Write >, <, or = between the two numbers.

3. 46,023 _____ 46,203   4. 0.76 _____ 0.456

Write your answers in the blanks.

5. Round 534,103 to the nearest ten thousand. _____

6. Round 3.725 to the nearest hundredth. _____

7. Write the mixed number that names the shaded portion. _____

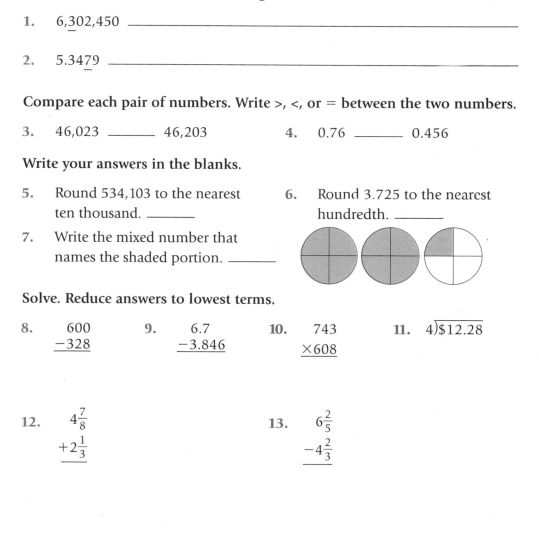

Solve. Reduce answers to lowest terms.

8. $\begin{array}{r} 600 \\ -328 \\ \hline \end{array}$

9. $\begin{array}{r} 6.7 \\ -3.846 \\ \hline \end{array}$

10. $\begin{array}{r} 743 \\ \times 608 \\ \hline \end{array}$

11. $4\overline{)\$12.28}$

12. $\begin{array}{r} 4\frac{7}{8} \\ +2\frac{1}{3} \\ \hline \end{array}$

13. $\begin{array}{r} 6\frac{2}{5} \\ -4\frac{2}{3} \\ \hline \end{array}$

14. $12 \div (2 + 1) \times 6 - 4 =$

15. $42 + 376 + 57 =$

16. $3.27 + 26.4 + 8.531 =$

17. $\$5.08 - \$2.99 =$

18. $0.054 \times 0.3 =$

19. $4\frac{2}{5} \times 3\frac{3}{4} =$

20. $7{,}658 \div 42 =$

21. $31.2 \div 0.06 =$

22. $3\frac{1}{2} \div 4\frac{3}{8} =$

23. $7^2 =$

24. Change 375% to a decimal.

25. Change 0.07 to a percent.

26. Change 80% to a fraction.

27. Change $\frac{3}{4}$ to a percent.

28. What is 8% of 150?

29. What percent of $72 is $18?

30. 54 is 150% of what number?

31. $\frac{8}{20} = \frac{?}{35}$

Circle the best answer for each item.

32. Handy Hardware had 144 gallons of interior paint and 96 gallons of exterior paint in stock. During the week, 48 gallons of interior paint were sold. Which is the correct expression to find the number of gallons of interior paint still in stock?

    (1) 96 − 48
    (2) 144 + 48
    (3) 144 − 48
    (4) 144 − 96
    (5) 144 + 96 − 48

33. Carol is building a rectangular dog kennel that is 12 feet long and 6 feet wide. What is the perimeter of the kennel in feet?

    (1) 18
    (2) 36
    (3) 72
    (4) 144
    (5) 180

34. Darius wants to carpet a living room that measures 20 feet on each side. Which is the correct expression to find the area of the room in square feet?

    (1) 20 × 9
    (2) 20 ÷ 9
    (3) 20^2
    (4) 20^2 ÷ 9
    (5) 20 + 20 + 20 + 20

35. On Friday, Village Deli sold 79 sandwiches on rye, 35 sandwiches on white, and 52 sandwiches on whole wheat bread. Estimate the total number of sandwiches sold on Friday.

    (1) 150
    (2) 160
    (3) 170
    (4) 180
    (5) Not enough information is given.

36. A package contains 8 hamburger buns. Maria bought 16 packages of buns for a picnic. Which is the correct expression to find the number of buns she bought?

    (1) 8 ÷ 16
    (2) 8 + 16
    (3) 16 − 8
    (4) 16 × 8
    (5) 16 ÷ 8

37. The spinner shown has 6 equal sections. What is the probability that the wheel will stop on a 2?

    (1) $\frac{1}{6}$ or $16\frac{2}{3}$%
    (2) $\frac{1}{3}$ or $33\frac{1}{3}$%
    (3) $\frac{1}{2}$ or 50%
    (4) 1 or 100%
    (5) 2 or 200%

38. Jesse has a recipe that calls for $2\frac{1}{4}$ cups of chicken broth. Jesse wants to make one third of this recipe. How many cups of broth should he use?

    (1) $\frac{4}{27}$
    (2) $\frac{3}{4}$
    (3) $2\frac{1}{4}$
    (4) $5\frac{1}{4}$
    (5) $6\frac{3}{4}$

39. Joan's Market has onions on sale at 3 pounds for $1.56. A customer bought 5 pounds. Which is the correct expression to find the cost of 5 pounds of onions?

    (1) $\frac{3}{5} = \frac{?}{\$1.56}$
    (2) $\frac{3}{\$1.56} = \frac{5}{?}$
    (3) $\frac{3}{\$1.56} = \frac{?}{5}$
    (4) $\frac{5}{3} = \frac{\$1.56}{?}$
    (5) $\frac{5}{\$1.56} = \frac{3}{?}$

40. Jermaine works 40 hours per week as a clerk at Central Auto Supply. He spends 15% of his time restocking shelves. Which is the correct expression to find the number of hours he spends restocking shelves each week?

(1) 15 × 40%
(2) 15 ÷ 40%
(3) 15% ÷ 40
(4) 40 × 15%
(5) 40 ÷ 15%

41. The vet told Andy to put his dog on a diet. During the first week of the diet, Andy's dog lost 0.2 kilogram. How many grams did the dog lose?

(1)   0.002
(2)   0.02
(3)   2
(4)   200
(5)   2,000

42. The Food Mart had ground beef on sale for $2.08 per pound. Joy bought 4 pounds of ground beef. Which is the correct expression to find the total cost of the ground beef?

(1) $2.08 × 0.4
(2) $2.08 ÷ 0.4
(3) 4 ÷ $2.08
(4) $2.08 × 4
(5) $2.08 ÷ 4

43. Ricardo's gross pay is $420.00 each week. He had $25.20 deducted for his credit union savings account. What percent of Ricardo's gross pay is deducted for savings?

(1)   0.6%
(2)   6%
(3) $16\frac{2}{3}$%
(4)   20%
(5)   25%

44. Maggie bought a jacket at Discount Fashions. The price tag is shown here. Which is the correct expression to find the original price of the jacket?

(1) $18.60 × 30%
(2) $18.60 × 70%
(3) $18.60 ÷ 30%
(4) 30% × $18.60
(5) 30% ÷ $18.60

45. Which expression best describes how to find the width of the rectangle?

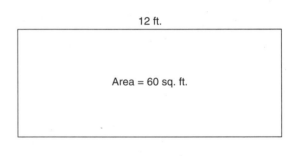

(1) 12 + 60
(2) 60 − 12
(3) 12 × 60
(4) 60 ÷ 12
(5) Not enough information is given.

46. Abdul just started work at a fast-food restaurant. His current wage is $5.25 per hour. After three months, his new wage will be $5.46 per hour. What is the percent of increase in his wages?

(1)   3%
(2)   4%
(3)   20%
(4)   21%
(5)   96%

Write your answers in the space provided.
Show your work.

47. What is the volume in cubic inches of the shipping carton shown?

16 in.

10 in.   12 in.

48. Marcus drives a delivery truck for a bread company. His total route is $36\frac{9}{10}$ miles long. He drives $19\frac{2}{5}$ miles in the morning and finishes the route in the afternoon. Estimate the number of miles Marcus drives in the afternoon.

Item 49 refers to the following table.

Fuel Economy in Miles per Gallon

	Car A	Car B	Car C
Highway	32	23	27
In Town	21	14	18

49. How many miles can car B travel on the highway on 15 gallons of gas?

50. Miranda needs a board that is $4\frac{1}{2}$ feet long for a shelf. She has a board that is $56\frac{1}{2}$ inches long. How many feet are in $56\frac{1}{2}$ inches?

Item 51 refers to the following timecard.

NAME: MARIA RODRIGUEZ
WEEK OF : 5/16–5/22

IN	MON. 5/16	8:43 A.M.
OUT	MON. 5/16	12:36 P.M.
IN	MON. 5/16	1:12 P.M.
OUT	MON. 5/16	5:04 P.M.

51. Approximately how many hours did Maria work before taking a lunch break?

52. Find the area of this circle to the nearest square foot.

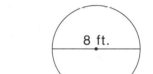

8 ft.

53. Tawanna works at a gift shop. Her customer is buying a statue for $32.50 and a card for $1.75. Sales tax on this purchase is $1.71. The customer gives Tawanna $40.00 in cash. How much change should Tawanna give her customer?

Item 54 refers to the following graph.

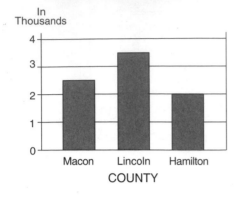

**County of Residence of
Woodland Mall Customers, 6/23**

54. What is the approximate difference in the number of mall customers who live in Lincoln County and the number who live in Hamilton County?

55. The Smiths want to put a fence around a circular flower garden. The diameter of the garden is 9 feet. To the nearest foot, how many feet of fencing will they need?

Items 56 and 57 refer to the following data.

The points scored by a football team during 4 preseason games were: 37, 18, 21, and 24.

56. What is the mean of these data?

57. What is the median of these data?

58. Of the 36 employees at Home Products, Inc., 20 work in the factory. What is the ratio in lowest terms of those who work in the factory compared to all employees?

59. Mai borrowed $3,000 from her parents to buy a new car. She will pay this back in 5 years at 5% interest. How much interest will she pay?

Item 60 refers to the following number line.

$$\begin{array}{c}\text{-10 -9 -8 -7 -6 -5 -4 -3 -2 -1 \ 0 \ +1 +2 +3 +4 +5 +6 +7 +8 +9 +10}\end{array}$$

60. Before going to bed, Joe read that the temperature was 2 degrees. The next morning he heard that the temperature had dropped 7 degrees overnight. What was the temperature in the morning?

Check your answers on pages 255–257.

Inventory Correlation Chart

Circle the number of each item that you answered correctly on the Inventory. Count the number of items in each row that you answered correctly. Write the number in the Total Correct space in each row. (For example, in the Number Sense row, write the number correct in the blank before *out of 5*). Complete this process for the remaining rows. Then add the 16 totals to get your total correct for the whole 60-item Inventory.

Section/Skill	Item Numbers	Total Correct	Pages
1. Number Sense	1, 3, 5, 14, 60	____ out of 5	10–21
2. Adding and Subtracting Whole Numbers	8, 15, 32, 33, 35, 54	____ out of 6	22–33
3. Multiplying and Dividing Whole Numbers	10, 20, 36, 45, 49, 56, 57	____ out of 7	34–51
4. Squares, Cubes, and Square Roots	23, 34, 47	____ out of 3	52–67
5. Fraction Basics	7	____ out of 1	68–81
6. Adding and Subtracting Fractions	12, 13, 48, 51	____ out of 4	82–95
7. Multiplying and Dividing Fractions	19, 22, 38, 50	____ out of 4	96–113
8. Decimal Basics	2, 4, 6	____ out of 3	114–125
9. Adding and Subtracting Decimals	9, 16, 17, 53	____ out of 4	126–135
10. Multiplying and Dividing Decimals	11, 18, 21, 41, 42	____ out of 5	136–155
11. Ratio and Proportions	31, 39, 58	____ out of 3	156–169
12. Percent Basics	24, 25, 26, 27	____ out of 4	170–175
13. Solving for the Part (*p*)	28, 40, 59	____ out of 3	176–186
14. Solving for the Rate (*r*)	29, 43	____ out of 2	187–198
15. Solving for the Base (*b*) and Percent of Change	30, 44, 46	____ out of 3	199–212
16. Special Topics	37, 52, 55	____ out of 3	213–235

TOTAL CORRECT FOR INVENTORY _____ out of 60

Whole Numbers

You may not realize how often you say, read, or think about numbers as you go about your daily activities. You use numbers to compare prices, fill out order forms, or keep score in a sport. Take a few moments to think about the whole numbers you have used today.

◉ Did you pay cash for a purchase? How much change did you get back? Did you estimate the total cost, including tax, ahead of time to see if you had enough money with you?

◉ What numbers have you used at home today? When paying bills, do you write your account numbers or your Social Security number on your checks?

◉ How do you use numbers at work? Do you handle order forms? Do you fill out a time sheet? Do you add your hours and multiply by your hourly wage?

SECTIONS

1 **Number Sense**

2 **Adding and Subtracting Whole Numbers**

3 **Multiplying and Dividing Whole Numbers**

4 **Squares, Cubes, and Square Roots**

1

Number Sense

Using Order Forms

Catalogs and order forms are commonly found in workplaces across the country. Everything from parts to office supplies is purchased using an order form. When workers fill out order forms, numbers are needed to communicate information such as addresses, item codes, quantities, and prices.

EXAMPLE Kelly is an office manager at a factory. The company orders all of its office equipment and cleaning supplies from mail-order companies. Kelly is placing an order for new computer accessories and supplies for one of the departments. Since she will need to get approval for the purchase, Kelly needs to figure out the total amount.

Item Number	Qty.	Description	Item Price	Total
4 1 4 – 5 2 7	1	Color computer monitor	299.99	299.99
3 0 6 – 2 9 1	5	3.5" diskettes	10.99	54.95
4 1 4 – 2 7 1	1	Color scanner	199.99	199.99
4 1 4 – 6 4 4	1	Laser printer	699.99	699.99
3 0 6 – 7 0 2	5	Laser paper	6.99	34.95
–				

Shipping and Handling						
Merchandise Total	Add	Merchandise Total	Add
Up to $25	$4.95	$100.01 to $200	$10.95
$25.01 to $50	$6.95	$200.01 to $300	$12.95
$50.01 to $100	$8.95	$300.01 and over	$14.95

Merchandise Total	1,289.87
Shipping/Handling	
TOTAL	

Based on the merchandise total, how much will Kelly's company need to pay in shipping and handling charges for this order?

(1) $6.95

(2) $8.95

(3) $10.95

(4) $12.95

(5) $14.95

Answer **(5)** is correct. Use the shipping and handling chart to determine which merchandise total range this order falls into. The merchandise total is $1,289.87. For any order over $300.01, the shipping and handling charge is **$14.95.**

Circle the best answer for each item based on the order form on page 10.

1 Which item matches the item number 414-644?

(1) monitor
(2) diskettes
(3) scanner
(4) laser paper
(5) laser printer

2 Mail-order companies often put out different versions of their catalog. The first three digits of an item number lets a company know in which catalog the item was listed. Which three items listed on the order form are from the same catalog?

(1) monitor, diskettes, scanner
(2) diskettes, scanner, printer
(3) monitor, scanner, printer
(4) diskettes, scanner, printer
(5) monitor, scanner, laser paper

3 The total for the laser paper is $34.95. Which expression shows how Kelly found that amount?

(1) 5 × $10.99
(2) 5 × $6.99
(3) 1 × $34.95
(4) 5 × $4.95
(5) 3 × $10.99

4 Which of the following represents the best estimate of the cost of the diskettes and laser paper?

(1) $11 + $7
(2) $55 + $7
(3) $55 + $200
(4) $55 + $35
(5) $11 + $35

5 When Kelly receives the order, she will need to make sure that everything she ordered was sent. The column labeled *Quantity* lists the number of items ordered. Which expression shows the number of items that should be delivered?

(1) 1 + 5 + 5
(2) 1 + 5 + 1 + 5
(3) 1 + 5 + 1 + 1
(4) 1 + 5 + 1 + 1 + 5
(5) 1 + 5 + 1

6 Kelly compared prices with another mail-order company. She finds that she can purchase a similar laser printer for $459.99. Which expression shows how much Kelly could save if she bought the less expensive printer?

(1) $699.99 ÷ $459.99
(2) $699.99 + $459.99
(3) $699.99 × $459.99
(4) $459.99 ÷ $699.99
(5) $699.99 − $459.99

7 **Reason** Write a sentence to explain the following: If diskettes cost $10.99 each, why is the total amount for the diskettes $54.95?

Place Value

Which cash prize would you rather win: $50 or $500? You would pick $500 because you know that the value of $500 is greater than the value of $50. You know this from the placement of the numbers or **digits**.

Our number system uses ten digits: 0, 1, 2, 3, 4, 5, 6, 7, 8, and 9. Writing one or more digits in a row forms a **whole number**. The number 7 is a one-digit number, and the number 154 is a three-digit number. The number 5,000 is a four-digit number even though three of the digits are zeros.

The **place-value** chart below shows the first ten place values in our whole number system. The value of a digit depends on its place in the number. The value of the places increases as you move to the left.

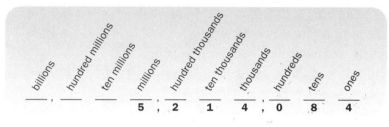

The number 5,214,084 is written on the place-value chart. From the chart, you can see that this whole number has 5 millions, 2 hundred thousands, 1 ten thousand, 4 thousands, 0 (or no) hundreds, 8 tens, and 4 ones. The 4 in the thousands place has a greater value than the 4 in the ones place.

EXAMPLE Write the place value for the 4 in 54,201.

Step 1: Find the number being discussed: the 4 in 54,201.

Step 2: Find the place value of the digit being discussed. Look at the place-value chart if necessary. The 4 is the fourth digit from the right. Thus, the 4 is in the thousands place.

TIP

When you write numbers, place commas every three digits, counting from the right. For example, 32,548,901.

PRACTICE

Write the place value of each underlined digit. Refer to the chart above if necessary.

1 9,650 _____ tens _____

2 97<u>2</u> _____

3 2<u>8</u>,730 _____

4 826,1<u>1</u>0 _____

5 40,<u>0</u>59 _____

6 926,4<u>2</u>0 _____

7 2,1<u>0</u>6,920 _____

8 <u>7</u>,535,000 _____

9 57,4<u>2</u>5 _____

10 9,<u>8</u>53,483 _____

11 <u>3</u>6 _____

12 7<u>6</u>5,999 _____

13 28,7<u>8</u>1 _____

14 <u>3</u>7,654,321 _____

15 56,8<u>3</u>9 _____

16 348,<u>6</u>53 _____

Reading and Writing Whole Numbers

To write a check, you must write the amount in both words and digits.

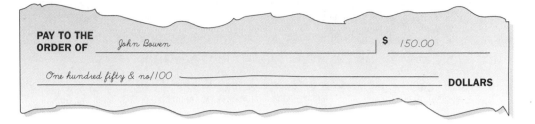

PAY TO THE ORDER OF _John Bowen_ $ _150.00_

One hundred fifty & no/100 _____ **DOLLARS**

Use the following rules to read and write whole numbers:

Rule 1: When writing a number using digits, use zero as a placeholder.

EXAMPLE "Two hundred four" is written **204**. The zero in the tens place has no value, but it is needed to hold the tens place.

Rule 2: To read a number, read each group of digits from left to right, stopping at each comma. Then say the name of the group to the left of the comma.

EXAMPLE 4,819,520 is read **"four million, eight hundred nineteen thousand, five hundred twenty."**

Rule 3: When writing big numbers in words, place a comma after the word *million* and the word *thousand*. Do not place a comma after the word *hundred*. This puts a comma in the same place with the words as with the digits.

EXAMPLE 5,150,425 is written **"five million, one hundred fifty thousand, four hundred twenty-five."**

TIP

Do not say the word *and* when reading whole numbers.

Read 1,523 as "one thousand, five hundred twenty-three."

PRACTICE

Write each number in words.

1 756 _____ seven hundred fifty-six _____

2 43,018 _____

3 115,200 _____

4 5,400,012 _____

Write each number using digits. Then enter it in a calculator.

5 one thousand, seven hundred eight _____ 1,708 _____

6 two hundred fifty thousand, nine hundred eleven _____

7 twelve thousand, sixteen _____

8 nine million, fourteen thousand, five hundred sixty _____

For more practice, see page 242 of the Calculator Handbook.

Comparing Whole Numbers

Would you rather work for $9 or $7 an hour? To answer the question, you would compare the two numbers and pick $9, the greater amount.

The symbols =, >, and < are used to compare numbers. You can write these symbols between two numbers to show how the numbers compare.

100 = 100	100 **equals** 100
$9 > $7	$9 **is greater than** $7
10 < 50	10 **is less than** 50

TIP

Think of the symbols > and < as arrows that always point to the smaller number.

12 < 20 and
20 > 12

Use these rules to compare whole numbers:

Rule 1: A number with more digits is greater than a number with fewer digits.

EXAMPLE Compare 8,500 and 920.

Step 1: Count the number of digits in one number. There are four digits in 8,500.

Step 2: Count the number of digits in the other number. There are three digits in 920.

8,500 > 920 because 8,500 has more digits than 920. We read this as "8,500 is greater than 920."

Rule 2: If the numbers have the same number of digits, work from left to right and compare each place value until the digits in a place value are different.

EXAMPLE Compare 6,410 and 6,481.

Step 1: Compare the digits in the thousands place. 6,410   6,481 → same
Step 2: Compare the digits in the hundreds place. 6,410   6,481 → same
Step 3: Compare the digits in the tens place. 6,410   6,481 → different

6,410 < 6,481 because 1 is less than 8. We read this as "6,410 is less than 6,481."

PRACTICE

Compare each pair of numbers. Write >, <, or =.

1 4,700 __>__ 740

2 38,000 _____ 38,500

3 179 _____ 179

4 210,580 _____ 210,480

5 1,000,000 _____ 10,000,000

6 496 _____ 4,690

7 13,415 _____ 13,415

8 802,165 _____ 803,980

9 5,000 _____ 50,000

10 1,345 _____ 1,435

11 10,334 _____ 10,334

12 479 _____ 476

13 340,635 _____ 340,835

14 5,010 _____ 5,001

15 682,489 _____ 682,489

16 3,800 _____ 3,850

Check your answers on page 258.

ALGEBRA TOPIC Number Line

All whole numbers—positive numbers, negative numbers, and zero—are called **integers.**

Integers that are greater than zero are called **positive numbers.** Integers that are less than zero are called **negative numbers.** Zero is neither positive nor negative. A positive number can be written with or without a positive sign ($+$). Negative numbers must always have a minus sign ($-$) in front of the number.

A **number line** shows the relationship of the positive and negative numbers and zero. On a number line, positive numbers are to the right of zero and negative numbers are to the left of zero.

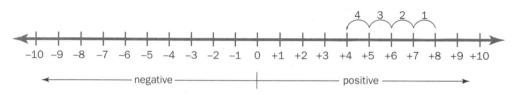

> **TIP**
>
> To solve problems using a number line, count to the right to find an increase or gain. Count to the left to find a decrease or loss.

EXAMPLE Darius is a farmer. During cold weather he checks the temperature hourly. Before dinner, he saw that the temperature was 8 degrees. During the next hour, the temperature fell 4 degrees. What was the temperature at that time?

Step 1: Find the number on the number line: $+8$.

Step 2: Move 4 units to the left since the temperature fell (decreased) 4 degrees.

The temperature at that time was **+4 degrees.**

PRACTICE

Write your answers in the blanks. Refer to the number lines above if necessary.

❶ 3 degrees with a drop of 12 degrees ___ −9 ___

❷ 0 yards with a gain of 4 yards and a loss of 10 yards _____

❸ $3 and $5 more _____

❹ $0 with a loss of $8 and a loss of $3 _____

❺ deduct 10 points from 12 points

_____

❻ −7 percent with an increase of 10 percent

_____

❼ When Elena left for work in the morning, the temperature was −6 degrees. By lunchtime the temperature had increased 5 degrees. What was the temperature when Elena left work to go to lunch?

_____

❽ In a high school football game, the home team lost 4 yards on its first play and gained 6 yards on its second play. In all, how many yards has the home team lost or gained so far?

_____

Check your answers on page 258.

Calculator Skills

Calculators have become an important tool in our society. We use them at home, at work, in class, and even when we go shopping. Throughout the lessons, you will be directed to the Calculator Handbook, which can be found at the back of this book. Refer to the handbook to learn how to use a calculator for the skill being taught.

Key Reference

The illustration below is probably similar to a calculator you have seen or one that you have used.

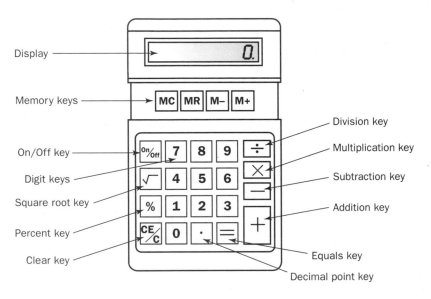

EXAMPLE Enter 1,965 on your calculator.

Step 1: Press the clear key **CE/C** to erase any previous entries.

Step 2: Press the appropriate digit keys   1965

Your calculator display should look like this: 1965.

Adding and Subtracting on a Calculator

The **+** key is used to add numbers. Use the **−** key to subtract numbers. To display the answer, press the **=** key.

EXAMPLE Ariel brings home $1,500 a month. She pays $435 a month for mortgage and $450 for other bills. What is the difference in Ariel's monthly take-home pay and her monthly payments?

Step 1: Find the total of Ariel's monthly payments.

CE/C 435 **+** 450 **=** 885

Step 2: Subtract Ariel's monthly payments from her salary.

CE/C 1500 **−** 885 **=** 615 or **$615**

TIP

Commas are not used with calculators.

1,200 is entered as 1200.

Refer to pages 243–244 in the Calculator Handbook for more information on adding and subtracting with calculators.

Multiplying and Dividing on a Calculator

The (x) key is used to multiply numbers. Use the (÷) key to divide numbers. To clear the display, press the (CE/C) key.

EXAMPLE Michael bought a package of laser paper for $7.99. How much would four packages cost?

Step 1: Press (CE/C) to erase any previous entries.

Step 2: Multiply the cost for each package by the number of packages. Be sure to enter the decimal point.

7.99 (x) 4 (=) 31.96, which is **$31.96**.

PRACTICE

A. **For each number, write what you would see on the calculator display.**

❶ 780 _____ *780* _____

❷ 1,653 _____

❸ 10,241 _____

❹ $18.92 _____

B. **Show how to calculate each problem. Then, use your calculator to find the answer.**

❺ 15 + 89 (CE/C) *15 + 89 = 104*

❻ 73 × 46 _____

❼ 2,187 ÷ 3 _____

❽ 1,406.19 − $94.37 _____

C. **Use your calculator to solve these problems.**

❾ Stephanie ordered the following holiday gifts for company clients:

    Five-box fruit gift $31.95
    Fruit basket $43.95
    Chocolate truffles $18.95

What is the total cost of the order? _____

❿ Roberto said he would bring soda pop to the class holiday party. One 2-liter bottle costs $1.05. How much will Roberto spend if he buys 6 bottles?

_____

Rounding

Rounded numbers are easier to remember and use than exact amounts. They can be used to estimate approximate answers or to check your calculations. Whenever you see the word *about*, you know the amounts are estimates.

EXAMPLE   Miwa earns $34,965 a year as a computer programmer. About how much does she earn a year?

Step 1:	Round the number (34,965) to the nearest thousand. Underline the digit in that place: in this case, the 4 in 34,965	$34,965
Step 2:	Look at the digit to the right of the underlined digit: 9	$34,965
Step 3:	If this digit is 5 or more, add 1 to the underlined digit. If the digit is less than 5, do not change the underlined digit. Since 9 is greater than 5, add 1: 4 + 1 = 5	$35,_ _ _
Step 4:	Change *all* the digits to the right of the underlined digit to zeros.	$35,000

TIP

Rounded numbers always end in one or more zeros.

40; 500; and 10,000 are all rounded numbers.

EXAMPLE   Round 4,514 to the nearest hundred.

Step 1:	Underline the digit in the place you want to round the number: the 5 is in the hundreds place	4,514
Step 2:	Look at the digit to the right of the underlined digit.	4,514
Step 3:	Decide if the underlined digit needs to change. Since 1 is less than 5, the underlined digit does not change.	4,5_ _
Step 4:	Change digits to the right of the underlined digit to zeros.	4,500

PRACTICE

Round each number to the given place value.

1 Round 58 to the nearest ten. _____ 60 _____

2 Round 1,723 to the nearest hundred. _____

3 Round 6,509 to the nearest thousand. _____

4 Round 861 to the nearest hundred. _____

5 Round 19,580 to the nearest thousand. _____

6 Round 209,320 to the nearest ten thousand. _____

7 Round 64,299 to the nearest thousand. _____

8 Round 5,256,000 to the nearest hundred thousand. _____

Check your answers on pages 258–259.

Estimation

Estimation is a useful tool when you work with numbers. An *estimate* gives you a general idea of a value. You can estimate to find an approximate answer or to make sure an answer is reasonable. Rounding is one method of estimation.

EXAMPLE Samuel works at an office supply store. A customer wants to know if 15 packages of pastel-colored copy paper will cost less than $60. Each package sells for $4.99. Excluding tax, is $60 enough money?

Step 1: Ask "Is an exact answer needed?" No, the customer wants a general idea of how much the paper will cost.

Step 2: To estimate, round to the nearest dollar: $4.99 rounded to the nearest dollar is $5.00.

Step 3: Perform the needed operations. Multiply the number of packages by the estimated cost per package: $15 \times \$5 = \$75$. **No, $60 is not enough money.**

EXAMPLE Bridget is a cashier at a department store. A customer is buying a vest for $24, a cotton turtleneck for $16, a denim skirt for $19, and a blouse for $29. How much do the items cost before tax?

Step 1: Is an exact answer needed? Yes. Since Bridget is a cashier, she needs to find the exact total so that she can add the correct tax and the customer can pay for the purchases.

Step 2: Find the total cost before tax.
$24 + $16 + $19 + $29 = $88 **The items total $88.**

TIP

Some situations need only an approximate answer or estimate. Other situations, like how much money to charge someone, need an exact answer.

PRACTICE

Find an estimate *and* an exact answer for each problem.

		Estimate	Exact
❶	$11 + $15 + $23 (round to the nearest $5)	_____	_____
❷	58 feet + 13 feet + 28 feet	_____	_____

Based on the problem, decide to find either an estimate *or* an exact answer.

❸ Chin recommended that his boss buy the following equipment to protect the office computer and to back up information. He listed a surge protector for $19, a back-up disk drive for $119, and an extra back-up disk for $16. About how much money would the items on Chin's list cost?

❹ Elory has a pet-sitting service. She charges $15 for one hour, per pet. Last week she sat 1 hour each with 18 cats and 1 hour each with 22 dogs. How much money did Elory charge customers last week?

Section 1 Review

Write the place value of each underlined digit.

1 2,834 _____

2 328,954 _____

3 7,927,480 _____

4 8,051,939 _____

Write each number in words.

5 28,302 _____

6 1,076,500 _____

Write each number using digits.

7 forty-two thousand, fifty-seven _____

8 three million, four hundred thousand, five hundred ninety

_____

Compare each pair of numbers. Write >, <, or = to make each expression true.

9 5,680 _____ 856

10 32,457 _____ 32,457

11 82,346 _____ 82,546

12 790,300 _____ 709,300

Round each number to the given place value.

13 Round 43 to the nearest ten _____

14 Round 2,453 to the nearest hundred _____

15 Round 307,216 to the nearest ten thousand _____

16 Round 4,293,785 to the nearest million _____

Use your calculator to solve these problems.

17 1,023 ➖ 645 ➕ 270

18 36,450 ➗ 3 ✖ 12

19 2,010 ➕ 45 ➖ 70

20 5,388 ✖ 2 ➖ 375

Circle the best answer for each item.

Items 21–23 refer to the following table.

Weekly Production of Parts

Week 1	12,435
Week 2	14,526
Week 3	12,345
Week 4	14,814
Week 5	13,706

21 Which statement is false?

(1) Fewer parts were produced during week 5 than during week 4.

(2) Fewer parts were produced during week 1 than during week 3.

(3) More parts were produced during week 4 than during week 2.

(4) More parts were produced during week 5 than during week 1.

(5) Fewer parts were produced during week 2 than during week 4.

22 During which week was the greatest number of parts produced?

(1) week 1
(2) week 2
(3) week 3
(4) week 4
(5) week 5

23 The manager stated that about 14,000 parts were produced during one week. He had rounded the number to the nearest thousand. Which week was he talking about?

(1) week 1
(2) week 2
(3) week 3
(4) week 4
(5) week 5

24 The number of cars sold in the state was 1,030,402. Select the expression below that shows in words the number of cars sold.

(1) one million, three thousand, four hundred two

(2) one million, thirty thousand, four hundred two

(3) one hundred thirty thousand, forty-two

(4) one hundred thirty thousand, four hundred two

(5) one million, three thousand, four hundred twenty

25 Marsha spent $2,264 last year for day care. Rounded to the nearest hundred dollars, how much did she spend for day care?

(1) $2,000
(2) $2,200
(3) $2,260
(4) $2,300
(5) $3,000

26 When Kevin left work, the temperature was 8 degrees. By the time he went to sleep, the temperature had dropped 10 degrees. What was the temperature when Kevin went to sleep?

(1) 18 degrees
(2) 2 degrees
(3) 8 degrees
(4) −6 degrees
(5) −2 degrees

Adding and Subtracting Whole Numbers

MEASUREMENT AND GEOMETRY

Perimeter

EXAMPLE   Pete and Bev Ramos want to find out how many feet of fencing they will need to fence in their yard to keep in their dog.

○ **The distance around the edge of something is called its perimeter. This means the perimeter is the sum of the measurements of all the sides.**

The Ramos's yard is a rectangle. They know that the distance across the front of the yard is 200 feet and one side of the yard measures 120 feet. To find the perimeter of the yard, they need to know the measurement of each side. Without measuring, how can they find the remaining two measurements?

○ **In a rectangle, opposite sides are equal in length.**

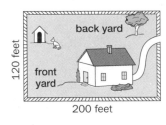

Which is the correct expression to find the perimeter of the yard?
(1)   200 + 200 + 120 + 200
(2)   200 + 200 + 120 + 120
(3)   200 + 120 + 120 + 200 + 120
(4)   200 + 200 + 200 + 200
(5)   200 + 120

Answer **(2)** is the correct choice. Since opposite sides of a rectangle are equal, each length is 200 feet and each width is 120 feet. Pete and Bev need to find the total, or sum, of all the sides, so they will **add all four lengths** together:
200 + 200 + 120 + 120 = 640 feet, so the perimeter of the yard is **640 feet.**

EXAMPLE   Pete and Bev want a 10-foot gate in the side of the yard, which is 120 feet wide. Which expression shows the amount of fencing needed to fence in the yard if they install the gate?
(1)   640 − 10
(2)   640 + 10
(3)   10 − 640
(4)   120 + 120 + 10 + 10
(5)   120 + 120 + 120 − 10

Answer **(1)** is correct. Pete and Bev need to deduct, or **subtract,** the width of the gate (10 feet) from the perimeter of the yard.

640 − 10 = 630 feet, so they need **630 feet of fencing** for the entire yard.

Circle the best answer for each item.

1 Joe is planning a new exhibit at the zoo for a rectangular space that is 40 feet long and 30 feet wide. Which is the correct expression to find the perimeter of the exhibit?

(1) 40 − 30
(2) 40 + 30
(3) 40 + 30 + 30
(4) 40 + 40 + 30 + 30
(5) 40 + 40 + 40 + 40

2 Shaunte works at The Frame Place. She is framing a rectangular painting for a customer. The size of the painting is shown below.

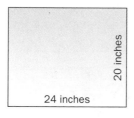

Which expression shows how to find the amount of framing she needs?

(1) 24 − 24
(2) 24 − 20
(3) 20 + 20 + 24
(4) 20 + 24 + 24
(5) 20 + 20 + 24 + 24

3 The landscaping service where Sandy works is putting a fence across a front yard. The yard is 80 feet wide. Sandy needs to leave a 3-foot opening for the sidewalk. Which is the correct expression to find how much fencing she needs on that side?

(1) 80 + 3
(2) 80 − 3
(3) 80 + 80 + 3
(4) 80 + 80 − 3
(5) 80 + 80 + 3 + 3

4 Lee is building a rectangular dog run that is 30 feet long and 6 feet wide. Which is the correct expression to find the perimeter of the dog run?

(1) 30 + 6
(2) 30 − 6
(3) 30 + 30 + 6
(4) 30 + 30 + 6 + 6
(5) 30 + 30 − 6 − 6

5 Jackie is planning to add a border to the walls of her bedroom. The room is 10 feet wide and 14 feet long. To find how much border she needs, Jackie has to figure out the perimeter of her room. Which is the correct expression to find the perimeter of the bedroom?

(1) 10 + 10 + 14 + 14
(2) 10 + 10 − 14
(3) 10 + 14 + 14
(4) 10 + 14
(5) 10 − 14

6 Roberto is building a railing around the edges of a deck. The deck is 10 feet wide and 18 feet long. Which is the correct expression to find the length of the railing?

(1) 10 + 10 + 18
(2) 10 + 10 + 18 + 18
(3) 10 + 18
(4) 10 + 18 + 18
(5) 10 + 10 − 18

7 **Communicate** Explain why you can find the perimeter of a rectangular room by measuring only one length and one width. On a separate piece of paper, write your explanation in one or two sentences. Make a drawing to illustrate your point.

Check your answers on pages 259–260.

Adding Whole Numbers

Addition means putting numbers together to find a total. The total is called the **sum.** You write an addition problem with a plus sign (+). Addition problems can be written in rows or in columns.

If the numbers in a column add up to more than 10, you regroup the number of tens to the tens column. Do the same with hundreds and so on.

EXAMPLE Add: 291 + 12 + 125

Step 1: Line up the numbers so that the digits in each column have the same place value.

```
  291
   12
+125
```

Step 2: Start with the ones column and add the numbers. Working from right to left, add the numbers in each column.

```
  291
   12
+125
    8
```

Step 3: If the numbers in a column add up to more than 10, regroup the number of tens to the tens column. Do the same with hundreds and so on.

```
  1        1
  291      291
   12       12
+125     +125
   28      428
```

TIP

To check addition, add again from bottom to top.

```
  125
   12
+291
  428
```

Many times you will see addition problems written in rows rather than columns. It is often helpful to rewrite the addition problem in a column to help you line up the digits.

EXAMPLE 153 + 148

Rewrite:
```
  153
+148
```

Add:
```
   11
  153
+148
  301
```

To add numbers on a calculator, enter each number, use the plus key between each number, then use the equals key to get the answer.

EXAMPLE Sal adds the number of boxes that both of his teams unpacked on Thursday. He uses his calculator as shown below to add 4,643 and 7,982.

CE/C 4643 **+** 7982 = **12625.**

For more practice adding with the calculator, see Calculator Handbook, page 243.

Sal checks his work by adding the numbers again, in reverse order.

A. Add the numbers below. Check your answers.

①
```
        Check
    254   325
   +325  +254
    579   579
```

②
```
    476
   + 18
```

③
```
    387
   +264
```

④
```
    473
   +182
```

⑤
```
    148
    327
   +232
```

⑥
```
    135
    456
   +784
```

⑦
```
    542
    125
   + 68
```

⑧
```
    817
     76
   +453
```

B. Rewrite the problems below. Line up the numbers. Solve and check. The first one is started for you.

⑨ 38 + 157 =
```
     38
   +157
```

⑩ 450 + 326 + 188 =

⑪ 3,947 + 18,889 + 232 =

⑫ 142 + 7 + 802 + 3 =

C. Solve the problems below with a calculator. Refer to page 243 in the Calculator Handbook for additional information.

⑬ Rachel ordered computer components for her department. She ordered the following:

Cable (1)	$ 39
Monitor (1)	229
Disks (1 box)	25
Software package	499

What was the total cost of the components?

Calculator Answer: _____

Checked Answer: _____

When working with calculators, be careful to enter digits in the correct order.

⑭ Sandra's son is in the third grade. The teacher is encouraging the 26 students in the class to read more. When the students have read 100 books, they win a class party. By which week did they meet their goal? _____

Week 1	13 books
Week 2	12 books
Week 3	21 books
Week 4	19 books
Week 5	16 books
Week 6	18 books
Week 7	17 books
Week 8	15 books
Week 9	16 books
Week 10	16 books

Subtracting Whole Numbers

Subtraction is taking an amount away from another amount. Subtract when you need to find the **difference** or to make a comparison. The minus sign (−) tells you to subtract. Like addition problems, subtraction problems can be written in rows or in columns.

EXAMPLE   Subtract 62 from 145.

Step 1:   Line up the numbers. Make sure the smaller number is on the bottom.

$$\begin{array}{r} 145 \\ -\ 62 \\ \hline \end{array}$$

Step 2:   Start with the ones column and subtract. Working to the left, subtract the numbers in each column.

$$\begin{array}{r} 14\mathbf{5} \\ -\ 6\mathbf{2} \\ \hline 3 \end{array}$$

Step 3:   Regroup whenever the digit being subtracted is greater than the digit above it.

$$\begin{array}{r} {}^{0\ 14} \\ \cancel{14}5 \\ -\ \mathbf{62} \\ \hline 83 \end{array}$$

If a subtraction problem has one or more zeros in it, you can regroup from the first digit to the left that is not a zero.

EXAMPLE   Subtract: 300 − 148

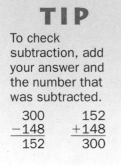

TIP

To check subtraction, add your answer and the number that was subtracted.

$$\begin{array}{r} 300 \\ -148 \\ \hline 152 \end{array} \qquad \begin{array}{r} 152 \\ +148 \\ \hline 300 \end{array}$$

Step 1:   Start with the ones column. You need to regroup from the hundreds column.

$$\begin{array}{r} {}^{2\ 10} \\ \cancel{3}00 \\ -148 \\ \hline \end{array}$$

Step 2:   Next, regroup from the tens column.

$$\begin{array}{r} {}^{9\ 10} \\ {}^{2\ \cancel{10}} \\ \cancel{3}\cancel{0}0 \\ -148 \\ \hline \end{array}$$

Step 3:   Work to the left and subtract each column.

$$\begin{array}{r} {}^{2\ 9\ 10} \\ \cancel{3}\cancel{0}\cancel{0} \\ -148 \\ \hline \mathbf{152} \end{array}$$

Subtracting on a Calculator

To subtract numbers on a calculator, enter each number, use the minus key between each number, then use the equals key to get the answer.

Subtract: 408 − 299

[CE/C] 408 [−] 299 [=] 109

A. Subtract the numbers below. Check your answers using addition.

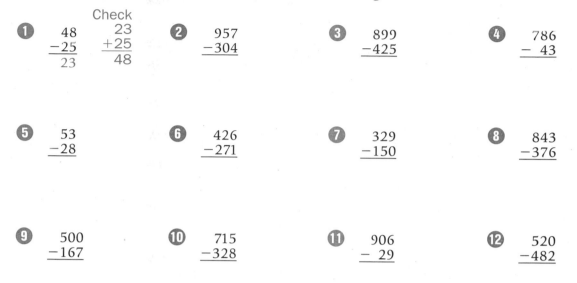

Check
① 48   23
  −25 +25
  23   48

② 957
  −304

③ 899
  −425

④ 786
  − 43

⑤ 53
  −28

⑥ 426
  −271

⑦ 329
  −150

⑧ 843
  −376

⑨ 500
  −167

⑩ 715
  −328

⑪ 906
  − 29

⑫ 520
  −482

B. Rewrite the problems below. Line up the numbers. Solve and check.
The first one is started for you.

⑬ 827 − 254 =   827
                 −254

⑭ 817 − 499 =

⑮ 700 − 543 =

⑯ 684 − 486 =

C. Solve the problems below with a calculator. Refer to page 244 in the
Calculator Handbook for additional information.

⑰ Jaime was balancing his checkbook.
His starting balance was $650. He
wrote checks for $125, $89, $57, and
$21. What was Jaime's ending balance?

Calculator Answer: _____

Checked Answer: _____

⑱ Gayle works at a bookstore. She needs to
order a total of 120 books from different
publishers. She ordered 43 books on
Monday, 54 books on Tuesday, and 12
books on Wednesday. How many more
books does she need to order?

_____

Extra/Not Enough Information

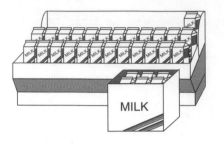

EXAMPLE Lila is in charge of keeping the dairy cases filled at Jons Grocery. The milk case holds 60 half-gallon cartons. When the store opens, Lila takes out any cartons that are out-of-date. Then she adds new cartons to fill the case.

On Monday Lila pulled 14 outdated cartons at 8:00 A.M. and added 38 new cartons. At 10:00 A.M. she added 12 cartons. At noon she put out 19 new cartons, and at 2:00 P.M. she added 18 more. At the end of the day, Lila has to figure out how many new cartons she put in the case.

This problem contains a lot of **extra information.** You don't need all of the details to solve the problem.

Think about what the question asks. Which is the correct expression to find the number of new cartons?
(1) 60 + 14 + 38 + 12 + 19 + 18
(2) 60 + 38 + 12 + 19 + 18
(3) 38 + 12 + 19 + 18
(4) 14 + 38 + 12 + 19 + 18
(5) 14 + 38 + 12 + 19

Answer (3) is the correct choice. Lila needs to find the **total number of new cartons** she put in the case: **38 + 12 + 19 + 18 = 87 cartons,** so Lila put **87 new cartons** in the case. You need only the numbers of **new cartons** to solve the problem. You don't need to know how many cartons the case will hold, 60, or how many cartons were outdated, 14.

Sometimes **not enough information** is given to solve a problem. These types of questions are similar to some situations in real life. Sometimes, you just don't have all the information you need to solve a problem.

EXAMPLE At 2:00 P.M. Lila refilled the milk case by adding 18 more cartons. If the milk case holds 60 cartons, how many new cartons will Lila need to add on Tuesday morning?
(1) 18
(2) 42
(3) 60
(4) 78
(5) Not enough information is given.

Answer **(5)** is correct. You don't know how much milk was sold by the end of the day on Monday, so you can't answer the question.

Circle the best answer for each item.

1 Paula's take-home pay is $1,200 a month. She pays $450 a month for rent and $225 a month for food. Which is the correct expression to find how much more she pays for rent than for food a month?

(1) $1,200 − $450
(2) $1,200 − $225
(3) $450 + $225
(4) $450 − $225
(5) Not enough information is given.

2 Rasheed delivers bottled water three times a week to an office building. He delivers 132 bottles on Wednesdays and 76 bottles on Fridays. What is the total number of bottles Rasheed delivers in a week?

(1) 56
(2) 208
(3) 211
(4) 624
(5) Not enough information is given.

3 Danielle drove 45 miles in 1 hour. Her car gets 18 miles to a gallon of gasoline. Which expression shows how many miles Danielle could drive in 5 hours at that rate?

(1) 45 × 18
(2) 45 × 5
(3) 45 ÷ 5
(4) 45 + 18
(5) Not enough information is given.

4 Kunio is a driver for a recycling company. He collected 56 boxes of newspapers Monday morning and 17 more boxes that afternoon. On Tuesday morning Kunio collected 49 boxes of newspapers and 21 more that afternoon. Which is the correct expression to find the total number of boxes Kunio picked up on Monday?

(1) 49 + 21
(2) 56 + 49
(3) 56 + 17
(4) 17 + 21
(5) Not enough information is given.

5 The shoe store where Eva works had a 3-day sale. By the end of the sale, there were 27 pairs of sandals and 38 pairs of tennis shoes left in stock. How many pairs of sandals were sold during the sale?

(1) 27
(2) 38
(3) 65
(4) 81
(5) Not enough information is given.

6 Sandra is a records clerk. This week she handled requests for 67 marriage licenses, 73 birth certificates, and 81 death certificates. Estimate the total number of requests she handled this week.

(1) 200
(2) 220
(3) 240
(4) 260
(5) 280

7 **Communicate** Think of a situation where you would purchase more than one of a certain item. Write a sentence or two describing what information you would need to find the total cost of this purchase.

Using a Bar Graph

A **bar graph** displays data in horizontal or vertical bars. Bar graphs are used to compare data when only one aspect of the data is numbers. In the example, the data being compared is the number of subscribers in the different sections of town. There are three parts to a bar graph: the **title,** the horizontal and vertical **axis lines,** and a **scale** that is used to show number values.

TIP

Read across from the top of the bar to the mark on the scale. The number by the mark is the value of the bar.

EXAMPLE Tatsu is a clerk at the *Tilden Times,* a daily newspaper. Her boss has asked her to divide the town into six sections so that she could quickly answer questions for him such as the following:

1. Which section of the town has the most subscribers?

2. About how many more subscribers live in the biggest section than in the smallest section?

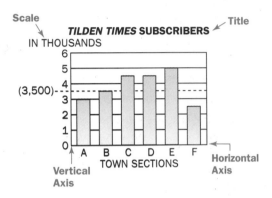

To answer question **1,** find the tallest bar, which represents the greatest number of subscribers. The bar for Section E is the tallest. It reaches the 5 mark and represents **5,000 subscribers.**

For question **2,** find the biggest and smallest section. The biggest section is Section E with 5,000 subscribers. The bar for Section F is the smallest. It represents about 2,500 subscribers. Subtract to find the *difference.* **Section E, the largest section, has about 2,500 more subscribers than Section F, the smallest.**

Drawing a Bar Graph

EXAMPLE Use the information in the table to begin building the bar graph in Part B on the next page.

Tall Buildings in the United States	
Building	Height (Stories)
Amoco Building	80
Chrysler Building	77
Empire State Building	102
First Interstate World Center	75
One World Trade Center	110

Step 1: Based on the information you will be graphing, write a title for the graph.

Step 2: Think about the information that is being listed on the vertical axis. Write a label for the vertical axis.

Step 3: Observe the pattern used for the scale on the vertical axis. Complete the scale following the same pattern.

A. **Use the bar graph at the right to answer the following questions.**

1 Which type of car sold the least? _____4-door_____

2 What was the most popular car sold? _____

3 How many minivans were sold? _____

4 What is the total of the two least popular types of cars sold? _____

5 How many more Sports Vehicles were sold than 2-door cars? _____

6 How many total sales did Country Motors have in 1998? _____

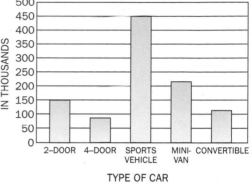

CARS SOLD BY COUNTRY MOTORS, 1998

B. **Complete the bar graph below using the information from the table in the second example on page 30.**

7

AMOCO BUILDING

C. **Use the bar graph you drew in Part B to answer these questions.**

8 What label did you use for the vertical axis? _____

9 What pattern did you use for the vertical axis scale? _____

10 What title did you give your bar graph? _____

11 Which building is the tallest? How many stories does it have? _____

12 Which building has the fewest stories? _____

13 How many more stories does the One World Trade Center have than the Amoco Building?

_____

14 How many stories does the Empire State Building have? _____

15 List the three tallest buildings shown on the graph. _____

_____

Section 2 Review

Add or subtract.

1   237
   +496

2   547
   −385

3 $295
   + 487

4   600
   −321

5 $607
   − 295

6   503
    58
   +367

7 $227
   317
   + 145

8 $993
   − 294

9   762
   −684

10   836
   +414

11 $3,342 + $467 =

12 52,304 − 7,099 =

Solve each word problem. Show your work.

13 Ed works in the mailroom of a large company. This morning he needs to make deliveries to 238 people. He makes deliveries to 149 people before his coffee break. How many people does he have to make deliveries to after his coffee break?

14 On Saturday the Perez family drove to Tall Pines Park where they will camp. They drove 243 miles before lunch and 179 miles after lunch. How many miles did the Perez family drive on Saturday?

15 Gloria delivers newspapers. She has 153 customers who receive the paper every day, 26 who get it on weekends only, and 42 who receive it on Sundays only. How many total customers does Gloria have?

16 Carolyn works in the theater box office. She sold 83 tickets for the 5:00 show and 127 tickets for the 7:30 show. How many more tickets did she sell for the 7:30 show?

Check your answers on pages 262–263.

Circle the best answer for each item.

17 Sue is putting a wallpaper border around her rectangular living room. The living room is 20 feet long and 14 feet wide. Which is the correct expression to find the amount of border she will need?

(1) 20 + 14
(2) 20 − 14
(3) 20 + 20
(4) 20 + 20 + 20 + 14
(5) 20 + 20 + 14 + 14

18 Last month Mark's utility bills included $28 for electricity, $53 for the telephone, and $11 for water. What is the total of Mark's utility bills last month?

(1) $39
(2) $70
(3) $81
(4) $92
(5) $102

19 Bill weighs 208 pounds. His doctor said he should weigh only 185 pounds. Which is the correct expression to find how much weight Bill should lose?

(1) 185 − 208
(2) 208 − 185
(3) 208 + 185
(4) 208 + 208 − 185
(5) 208 − 185 + 185

20 Joshua is taking a course on office procedures. Tuition is $348, and the cost of books and supplies is $104. What is the total cost for Joshua to take this course?

(1) $244
(2) $342
(3) $418
(4) $452
(5) $888

Items 21–23 refer to the following graph.

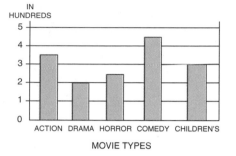

RENTALS AT VIDEOLAND, OCT. 15

21 Which type of movie had the greatest number of rentals?

(1) action
(2) drama
(3) comedy
(4) horror
(5) children's

22 What is the approximate difference between the movie type that had the most rentals and the movie type that had the fewest rentals?

(1) 100
(2) 150
(3) 200
(4) 250
(5) 650

23 How many total movie rentals were there on October 15?

(1) 1,600
(2) 1,550
(3) 1,500
(4) 1,150
(5) 1,000

Check your answers on page 263.

Multiplying and Dividing Whole Numbers

Sorting and Packing

EXAMPLE   Dean waits on customers at a bakery. This morning he will be packing cookies in boxes to sell. Each box holds one dozen cookies.

1 dozen = 12

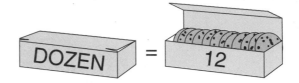

The baker tells Dean that there are 6 trays of cookies ready to be packed. Dean knows each tray holds 100 cookies.

Which is the correct expression to find the total number of cookies?

```
100
100        100
100      ×   6
100        600
100
+100
600
```

(1) $100 \div 6$

(2) 100×6

(3) $100 + 6$

(4) $100 - 6$

(5) $6 - 100$

Answer **(2)** is correct. Dean needs to find a total. He could add 100 six times. He could also get the same result if he multiplies 100 by 6. **100 × 6 = 600**, so Dean needs to pack **600 cookies.**

EXAMPLE   Next, Dean has to figure out how many boxes he will need to pack 600 cookies. Remember, each box holds one dozen (12) cookies.

Which is the correct expression to find how many dozens of cookies he needs to pack?

(1) 600×12

(2) $600 + 12$

(3) $600 \div 12$

(4) 12×12

(5) $600 - 12$

Answer **(3)** is correct. Dean needs to separate an amount into equal groups. He could repeatedly subtract 12 cookies until there are no more cookies left. But subtracting in this way is very cumbersome and time-consuming. It is more likely that Dean will divide to find how many boxes he needs. **600 ÷ 12 = 50**, so Dean needs **50 boxes** to pack 600 cookies.

Circle the best answer for each item.

1 Ms. Loe, a teacher, placed an order. She needs 144 cookies. Dean needs to know how many dozens are in 144. Which expression finds how many dozens of cookies are in 144?

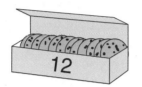

(1) 144 ÷ 12
(2) 12 + 144
(3) 12 ÷ 144
(4) 144 − 144
(5) 144 × 12

2 George bought a sofa that cost $849, including tax. He will pay for it in 3 equal payments. Which is the correct expression to find how much each payment will be?

(1) $849 × 3
(2) $849 − 3
(3) $849 ÷ 3
(4) 3 − $849
(5) $849 + 3

3 Marlowe bought 4 rolls of edging for his garden. Each roll is 10 feet long. Which is the correct expression to find how many feet of edging he bought?

(1) 4 + 10
(2) 10 ÷ 4
(3) 4 × 10
(4) 4 − 10
(5) 10 − 4

4 Leon works in a toy store. He packs 18 model airplanes in each shipping carton. Today he packed 24 cartons of model airplanes. Which is the correct expression to find how many airplanes he packed?

(1) 18 + 24
(2) 24 − 18
(3) 24 ÷ 18
(4) 18 × 24
(5) 18 ÷ 24

5 Valerie used 10 gallons of gas to drive 200 miles in her pickup truck. Which is the correct expression to find how far she drove on each gallon of gas?

(1) 10 × 200
(2) 10 + 200
(3) 10 ÷ 200
(4) 200 − 10
(5) 200 ÷ 10

6 Carl manages a service station. Before a new shipment of tires arrives, he counts 18 tires on hand. Carl then received a shipment of 30 tires. Which is the correct expression to find how many tires are on hand after the shipment arrives?

(1) 18 − 30
(2) 18 + 30
(3) 30 × 18
(4) 30 − 18
(5) 30 ÷ 18

7 **Communicate** Explain why dividing is a shortcut for subtraction and multiplication is a shortcut for addition. Give an example to support your explanations. (Hint: Refer to Workplace Math.)

Check your answers on page 263.

Multiplying Whole Numbers

Multiplying, like adding, helps you find a total. When you need to add the same number many times, use **multiplication.** The times sign, ($\times$), tells you to multiply. Multiplication is also shown by using parentheses: (3)(5) means 3×5.

EXAMPLE   Multiply 62 by 15.

Step 1:   Multiply each digit in the top number by each digit in the bottom number. Work from right to left. Be sure to line up the digits according to place value. If a column is 10 or more, you need to regroup. Since the 1 in the number 15 is in the tens place, you are really multiplying by 10. So, put a 0 in the ones place. (One times any number equals that number.)

$$
\begin{array}{r}
1 \\
62 \\
\times\ \mathbf{15} \\
\hline
310 \\
\end{array}
\qquad
\begin{array}{r}
62 \\
\times\ \mathbf{15} \\
\hline
310 \\
+62\mathbf{0} \\
\end{array}
$$

Step 2.   Add the results.

$$
\begin{array}{r}
62 \\
\times\ 15 \\
\hline
310 \\
+620 \\
\hline
\mathbf{930} \\
\end{array}
$$

TIP

To check multiplication, you can multiply the numbers in the reverse order.

$$
\begin{array}{r}
15 \\
\times 62 \\
\end{array}
$$

EXAMPLE   Multiply 169 by 205.

Step 1:   Multiply as normal. Multiply each digit in 169 by the 5 in 205.

$$
\begin{array}{r}
169 \\
\times 205 \\
\hline
845 \\
\end{array}
$$

Step 2:   Multiply by 0. Since 0 times any number is 0, write 0 directly below the 0 in 205. You do not need to write a row of zeros.

$$
\begin{array}{r}
169 \\
\times 205 \\
\hline
845 \\
0 \\
\end{array}
$$

Step 3:   Multiply by 2. Begin directly under the 2, to the left of the 0 you just wrote. Notice how the answer is the same as if you had written the row of zeros.

$$
\begin{array}{r}
169 \\
\times 205 \\
\hline
845 \\
3380\ \\
\hline
34645 \\
\end{array}
\qquad
\begin{array}{r}
169 \\
\times 205 \\
\hline
845 \\
00 \\
338\ \ \\
\hline
34645 \\
\end{array}
$$

For more practice multiplying with the calculator, see Calculator Handbook, page 245.

EXAMPLE   Keesha is a salesperson at Frank's Furniture. Yesterday she sold 3 dining room sets at \$1,129 each. Use a calculator to find the total sales for the 3 sets.

CE/C 1129 **×** 3 **=** **3387**

A. Multiply. Show your work. The first one is done for you.

1
$$
\begin{array}{r}
\overset{2}{172} \\
\times\ \ 4 \\
\hline
688
\end{array}
$$

2
$$
\begin{array}{r}
234 \\
\times\ \ 2 \\
\hline
\end{array}
$$

3
$$
\begin{array}{r}
384 \\
\times\ \ 6 \\
\hline
\end{array}
$$

4
$$
\begin{array}{r}
73 \\
\times 18 \\
\hline
\end{array}
$$

5
$$
\begin{array}{r}
83 \\
\times 24 \\
\hline
\end{array}
$$

6
$$
\begin{array}{r}
259 \\
\times\ 47 \\
\hline
\end{array}
$$

7
$$
\begin{array}{r}
\$475 \\
\times\ \ 9 \\
\hline
\end{array}
$$

8
$$
\begin{array}{r}
64 \\
\times 300 \\
\hline
\end{array}
$$

9
$$
\begin{array}{r}
837 \\
\times 402 \\
\hline
\end{array}
$$

10
$$
\begin{array}{r}
126 \\
\times 280 \\
\hline
\end{array}
$$

B. Rewrite the problems below. Line up the numbers. Solve and check. The first one is started for you.

11 $\$389 \times 7 =$
$$
\begin{array}{r}
\$389 \\
\times\ \ \ 7 \\
\hline
\end{array}
$$

12 $48 \times 40 =$

13 $527 \times 63 =$

14 $936 \times 508 =$

 C. Solve the problems below with a calculator. Refer to page 245 in the Calculator Handbook for additional information.

15 Javier works in a warehouse. He knows that 1 carton can hold 55 of Product 1 or 40 of Product 2. Javier counted the number of cartons holding each product.

Product 1   198 cartons
Product 2   57 cartons

What was the total number of both products in the packed cartons?

Calculator Answer: _____

Checked Answer: _____

16 Natina makes wreaths to sell at craft fairs. She uses 8 yards of ribbon for each wreath. If she makes 45 wreaths for one craft fair and 48 wreaths for another, how many yards of ribbon does Natina need?

17 Greenhall Corp. held an awards banquet for 125 people at Westside Cafe. If dinner for each person costs $12, how much did the banquet cost?

18 Leal Furniture Manufacturers can make 280 chairs each week. If the factory operates 51 weeks this year, how many chairs can be made?

19 Lobodo Publishers shipped 84 boxes of a new auto repair manual today. Each box holds 24 books. How many books were shipped today?

20 Robert drives a gasoline tank truck. Today he delivered gas to 14 different gas stations. The tank at each station holds 350 gallons. How many gallons of gasoline did Robert deliver?

✔ Check your work by entering the numbers again.

Dividing Whole Numbers

Use **division** to figure out how many times one number goes into or divides another number. Division problems can be written using the division sign, ($\div$). To solve the problem, you can write it using a division bracket, $\overline{)}$.

$$35 \div 7 = 5$$

$$\begin{array}{r} 5 \\ 7\overline{)35} \\ -35 \\ \hline 0 \end{array}$$

EXAMPLE   Divide 2,352 by 12.

Step 1:   How many times does 12 **divide into** 23? $12 \times 1 = 12$ and $12 \times 2 = 24$. Choose the answer that is closer to 23 without going over 23. Write **1** in the answer space over 23. **Multiply** $12 \times 1 = 12$. Write 12 under 23 and **subtract. Bring down** the next digit: 5.

Step 2:   How many times does 12 divide into 115? $12 \times 9 = 108$. Write **9** in the answer space over the 5. Multiply and subtract. Bring down the next digit: 2.

Step 3:   How many times does 12 divide into 72? $12 \times 6 = 72$. Write **6** in the answer space over the 2. Multiply and subtract.

Step 4:   Check your answer. Multiply 196 by 12. The result is the number you divided, 2,352.

$$\begin{array}{r} \mathbf{196} \\ 12\overline{)2,352} \\ -12 \\ \hline 115 \\ -108 \\ \hline 72 \\ -72 \\ \hline 0 \end{array}$$

$$\begin{array}{r} 196 \\ \times\ 12 \\ \hline 392 \\ +196 \\ \hline 2,352 \end{array}$$

> **TIP**
>
> To check division, multiply your answer by the number you divided by. If you divided correctly, the result should be the same as the number you divided.

Dividing Money

Divide as you would with whole numbers. Then, put the decimal point and dollar sign in your answer above the decimal and dollar sign in the problem.

Divide: $3\overline{)\$1,260}$

Solve:
$$\begin{array}{r} \$420 \\ 3\overline{)\$1,260} \\ -12 \\ \hline 6 \\ -60 \\ \hline 00 \end{array}$$

Check:
$$\begin{array}{r} \$420 \\ \times\ \ 3 \\ \hline \$1,260 \end{array}$$

PRACTICE

Divide and check. Show your work. The first one is done for you.

1
$$\begin{array}{r} 54 \\ 8\overline{)432} \\ -40 \\ \hline 32 \\ -32 \\ \hline 0 \end{array} \qquad \begin{array}{r} 54 \\ \times\ \ 8 \\ \hline 432 \end{array}$$

2 $7\overline{)\$1,505}$

3 $6\overline{)5,478}$

4 $36\overline{)5,940}$

Check your answers on page 264.

Remainders

EXAMPLE Divide 475 by 8.

Step 1: How many times does 8 divide into 47? 5 times. Write **5** in the answer space above 47. Multiply, subtract, and bring down the next digit.

Step 2: How many times does 8 divide into 75? 9 times. Write **9** in the answer space above 5. Multiply and subtract.

$$\begin{array}{r} 59\ r3 \\ 8\overline{)475} \\ -40 \\ \hline 75 \\ -72 \\ \hline 3 \end{array}$$

> **TIP**
>
> The letter *r* means remainder.
>
> 59 r 3 means 59 with a remainder of 3.

When there are no more numbers to bring down and you have an amount left over, you have a **remainder.**

Step 3: Write the remainder in the answer space.

Step 4: Check your answer. Multiply 59 by 8, then add the remainder 3.

$$\begin{array}{r} 7 \\ 59 \\ \times\ \ 8 \\ \hline 472 \\ +\ \ 3 \\ \hline 475 \end{array}$$

To divide numbers on a calculator, enter each number. Use the **division key**, ➗, between numbers and the **equals key** to get the answer.

EXAMPLE Divide $2,596 by 15. Be sure to enter the number being divided first.

CE/C 2596 **÷** 15 **=** **173.0667**

Notice in your answer there are digits to the right of the decimal point. The digits 0667 are your remainder. When doing division with money, often you are asked to round your answer to the nearest dollar. The answer rounded to the nearest dollar is $173.

PRACTICE

Divide and check. Show your work. For money amounts, round your answer to the nearest cent. The first one is done for you.

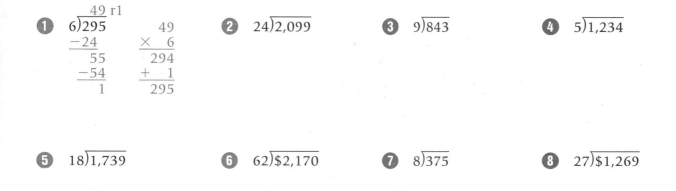

1
$$\begin{array}{r} 49\ r1 \\ 6\overline{)295} \\ -24 \\ \hline 55 \\ -54 \\ \hline 1 \end{array}$$
$$\begin{array}{r} 49 \\ \times\ 6 \\ \hline 294 \\ +\ 1 \\ \hline 295 \end{array}$$

2 24)2,099

3 9)843

4 5)1,234

5 18)1,739

6 62)$2,170

7 8)375

8 27)$1,269

Zeros in the Answer

EXAMPLE   Divide 490 by 7.

Step 1:   How many times does 7 divide into 49? 7 times. Write **7** in the answer space over 49. Multiply, subtract, and bring down the next digit.

Step 2:   How many times does 7 divide into 00 (0)? $0 \div 7 = 0$, or 0 times. Write **0** in the answer space over 0.

Step 3:   Check your answer.

$$\begin{array}{r} 70 \\ 7)\overline{490} \\ -49 \\ \hline 00 \end{array}$$

$$\begin{array}{r} 70 \\ \times\ \ 7 \\ \hline 490 \end{array}$$

TIP

$0 \div$ any number $= 0$

EXAMPLE   Divide 4,872 by 24.

Step 1:   How many times does 24 divide into 48? 2 times. Write **2** in the answer space over 48. Multiply, subtract, and bring down the next digit.

Step 2.   How many times does 24 divide into 7? 0 times. The number 24 cannot divide into 7, so write **0** in the answer space over 7. Bring down the next digit.

Step 3:   How many times does 24 divide into 72? 3 times. Write **3** in the answer space over 2. Multiply and subtract.

Step 4:   Check your answer.

$$\begin{array}{r} 203 \\ 24)\overline{4,872} \\ -48 \\ \hline 72 \\ -72 \\ \hline 0 \end{array}$$

$$\begin{array}{r} 203 \\ \times\ \ 24 \\ \hline 812 \\ +406 \\ \hline 4,872 \end{array}$$

PRACTICE

A.   Divide. Watch for zeros in the answers. Check your answers. The first one is done for you.

①
$$\begin{array}{r} 50 \\ 9)\overline{450} \\ -45 \\ \hline 0 \end{array} \qquad \begin{array}{r} 50 \\ \times\ \ 9 \\ \hline 450 \end{array}$$

② $26)\overline{2,626}$

③ $8)\overline{\$320}$

④ $6)\overline{1,824}$

Check your answers on page 265.

B. **Divide. Watch for zeros. Check your answers. The first one is done for you.**

5 $5\overline{)280}$
56
-25
30
-30
0

Check
56
$\times\ \ 5$
280

6 $27\overline{)5,103}$

7 $19\overline{)5,852}$

8 $8\overline{)\$2,208}$

9 $6\overline{)4,800}$

10 $13\overline{)6,695}$

11 $4\overline{)\$2,320}$

12 $10\overline{)4,550}$

C. **Divide. Watch for remainders. Check your answers. The first one is done for you.**

13 $6\overline{)465}$
77 r3
-42
45
-42
3

Check
77
$\times\ \ 6$
462
$+\ \ 3$
465

14 $7\overline{)3,429}$

15 $9\overline{)5,688}$

16 $21\overline{)1,953}$

17 $8\overline{)1,252}$

18 $46\overline{)4,955}$

19 $4\overline{)\$3,668}$

20 $5\overline{)1,734}$

D. **Solve the problems below with a calculator. Refer to page 246 in the Calculator Handbook for additional information.**

21 Brian works at Frank's Foods. He is packing cans of corn in boxes. The cans fit 24 to a box. How many boxes does Brian need for 1,632 cans?

Calculator Answer: _____

Checked Answer: _____

22 Melanie earns $17,460 a year. How much does she make a month? (Hint: Think how many months in a year.)

23 Adrienne can enter 80 words a minute on her computer. How long will it take her to do an 18,480-word report?

24 Sara works for a music library. One week she listened to 840 minutes of music. How many hours of music did she hear? (Hint: 60 min = 1 hr)

Writing and Solving Equations

An **equation** is two mathematical expressions, or statements, with an equals sign ($=$) in between. Some equations are all numerals, such as $6 + 4 = 10$. But some equations contain unknown numbers, or **variables**, represented by letters such as $y + 4 = 10$. (Any letter can be used.)

Writing Equations

EXAMPLE Write an equation for the following statement: A number increased by nine equals eighteen.

Step 1: Identify the key words that tell you what operation to use. The words *increased by* tell you to *add*.

Step 2: Identify the unknown amount. *A number* is the unknown.

Step 3: Write the equation.

Unknown	add	9	=	18
y	$+$	9	$=$	18

EXAMPLE Write an equation for the following question: How many dozen produces sixty cookies? or the product of twelve and a number equals sixty.

Step 1: Identify the operation. The word *product* means to multiply.

Step 2: Identify the unknown amount. *A number* is the unknown.

Step 3: Write the equation. Notice that multiplication can be shown without using multiplication signs.

Multiply	12 times the unknown	=	60
	(12) (y)	$=$	60

Solving Equations

When solving equations, you want to get the unknown number (variable) alone on one side of the equation. To do this, perform the **inverse** (opposite) operations on *both* sides of the equation.

EXAMPLE Solve $z + 18 = 34$.

Step 1: The inverse of addition is subtraction. Subtract 18 from both sides.

$$\begin{array}{r} z + 18 = 34 \\ -18 = -18 \\ \hline z + 0 = 16 \\ z = 16 \end{array}$$

(You do not write the zero in your final answer.)

Step 2: Check your answer using substitution, or putting in the value for z into the equation: $16 + 18 = 34$

EXAMPLE Solve $w \div 5 = 16$.

Step 1: The inverse of division is multiplication. Since w is being divided by 5, multiply both sides of the equation by 5 to solve for w.

$$\begin{array}{r} w \div 5 = 16 \\ \times 5 = \times 5 \\ \hline w = 80 \end{array}$$

Step 2. Check the answer using substitution.

$80 \div 5 = 16$

TIP

- The inverse of addition is subtraction.
- The inverse of subtraction is addition.
- The inverse of multiplication is division.
- The inverse of division is multiplication.

A. Write an equation for each statement. Do not solve. The first one is done for you.

1 A number divided by eight equals four. _____ $n \div 8 = 4$ _____

2 Sixteen more than a number equals twenty. _____

3 Five times a number equals fifty. _____

4 Twenty-seven decreased by a number equals twelve. _____

5 The quotient of a number divided by six equals five. _____

B. Solve and check. The first one is started for you.

6 $a + 12 = 32$ $a + 12 = 32$
$\underline{-12 = -12}$

7 $d \div 7 = 8$

8 $n - 18 = 56$

9 $d - 47 = 24$

10 $n \div 4 = 15$

11 $9n = 108$

12 $3z = 54$

13 $c + 13 = 13$

14 $y + 19 = 45$

15 $10c = 100$

C. Write and solve the equation for each problem.

16 Anne earns d dollars an hour. She worked 45 hours last week and received $540. How much does she earn an hour?

17 Monica weighs p pounds. After she loses 15 pounds, Monica will weigh 125 pounds. How much does Monica weigh?

18 Van is making a fruit salad for his office party. He bought p pounds of melon and 6 pounds of berries. He bought a total of 18 pounds of fruit for the salad. How many pounds of melon did he buy?

19 Luis was taking inventory at the video rental store where he works. He noted that the total number of videos divided by 26 people who rented videos equals an average of three videos per person. What was the total number of rented videos?

20 At work Andrea received a $25 bonus added to her check. The total amount of the check was $560. How much was her check before the bonus?

Solving Word Problems

Calculating the answer is only part of the problem-solving process. First you have to figure out how to approach and solve the problem.

Choosing the Correct Operation

Look for the meaning behind the words in a problem. Ask yourself, "How should I approach the problem?" The chart below will help you choose the correct operation.

You should . . .	When you need to . . .
Add	Find a **total,** find **how many in all,** put amounts together, or find the sum.
Subtract	Find a **difference,** find **how many more, how many less,** or **how much is left.**
Multiply	Add the same number more than once to find a total or to find a product.
Divide	Break or separate an amount into equal parts; find **how many** or **how much** for **each.**

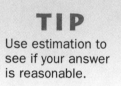

TIP
Use estimation to see if your answer is reasonable.

EXAMPLE Yuri is a cashier at the Empire Theater. He counted ticket stubs for one show and found out 56 adults and 39 children or senior citizens were at the show. What was the total number of ticket sales for the one show?

There are five steps to solving word problems:

Step 1: Read the problem carefully. Make sure you understand what you are being asked to find. *What was the total number of tickets sold?*

Step 2: Think about what you are trying to find.
You are finding *the total number of ticket sales for one show.*

Step 3: Ask yourself how the facts in the problem can help you find the answer. Determine which numbers you need to use.
There are 56 adults and 39 children or senior citizens at the show.

Step 4: Look for clue words that tell you which operation to use.
Total: use addition. **56 + 39 = 95**

Step 5. Make sure your answer is reasonable. Estimate: 60 + 40 = 100. The total number of tickets is about 100. The answer, 95, is reasonable.

Using the Calculator to Solve Word Problems

EXAMPLE Spring has a population of 1,437,306. Denton has a population of 731,523. How many more people live in Spring than in Denton?

The clue words are *how many more.* Therefore, you subtract.

CE/C 1437306 **—** 731523 **=** 705783.

A. Underline clue words in each problem. Write which operation to use. Do not solve the problem. The first one is done for you.

1 Marcus drove 156 miles the first day, 202 miles the second day, and 175 miles the third day. How many miles in all did Marcus drive?
Addition

2 In a 40-hour work week, Carol makes $542. How much does she earn for each hour she works?

3 Kevin is in a jazz band. Sixty-four people paid $4 to hear him play. What is the total amount of money paid to hear the band?

4 Marilyn is an office manager. She has a budget of $1,000 to buy a new computer for the office. The computer sells for $1,399. How much more money does Marilyn need in the budget?

B. Write the operation needed and solve. Use estimation to check the reasonableness of your answers.

5 Midori is making a fruit basket for a party. She buys 12 pounds of melon and 6 pounds of berries. How many pounds of fruit does she buy?

6 Tastee Burger sold 198 hamburgers on Friday. Of the hamburgers sold, 72 were sold after 7:00 P.M. How many hamburgers sold during the rest of the day?

C. Use the five-step problem solving approach to solve the word problems below. You may use a calculator.

7 Marcy is a cashier at Best Video. Each video costs $17. If her customer has $50 to spend, how many videos can the customer buy?

8 Terrence was reviewing his utility bills. He had the following bills: Electricity $53, water $16, telephone $27, trash pickup $5, gas $11. What was the total amount Terrence owes for utilities?

9 Lelia put on layaway a coat that cost $371. She put $25 down to hold the coat. How much money does Lelia still owe for the coat?

10 Carmen is a records clerk. This week she handled requests for 67 marriage licenses, 73 birth certificates, and 81 death certificates. What was the total number of requests she handled this week?

Finding Averages

TIP

The average is also called the **mean**.

Numbers that people group and study to make decisions are called **data**. Finding the **average** is one way to figure out the usual value of a set of data.

EXAMPLE Stan works at a department store that is open 6 days a week. Stan's boss wants him to find the average number of customers who shop between 4 P.M. and 6 P.M. What would be the best way for Stan to get this information?

Stan decides to count the people who go through the checkout lines within the 2-hour period every day for 1 week. Stan collects the following data:

Monday	106	Thursday	85
Tuesday	94	Friday	100
Wednesday	75	Saturday	104

Step 1: To find an average, first find the total of the numbers in the set of data. $106 + 94 + 75 + 85 + 100 + 104 = 564$

Step 2: Next, divide the total by how many numbers are in the data set. There are six numbers in the data set. Divide 564 by 6. The usual (or average) number of shoppers between 4 P.M. and 6 P.M. is **94**.

$$\begin{array}{r} 94 \\ 6\overline{)564} \\ -54 \\ \hline 24 \\ -24 \\ \hline 0 \end{array}$$

Finding the Median

The **median** is another kind of typical value. The *median* is the middle number of a set of data.

EXAMPLE Women's Shoes Unlimited had nine types of shoes on sale in these sizes: 5, 7, 8, 10, 6, 9, 4, 9, 6. What is the median for this set of data?

Step 1: Arrange the data in numerical order, from least to greatest.

4 5 6 6 <u>7</u> 8 9 9 10

Step 2: Since there are nine values in all, the middle value would be the fifth number: four numbers to the left and four numbers to the right. The median, or middle number, is **7**.

To find an average on a calculator, enter each number to find a total. Use the ÷ key to divide the total by how many numbers are in the set of data.

EXAMPLE Find the average for this set of data: 1, 2, 3, 4, 5, 6, 7, 8, 9, 10, 11

CE/C 1 **+** 2 **+** 3 **+** 4 **+** 5 **+** 6 **+** 7 **+** 8 **+** 9 **+** 10 **+** 11 **=** 66

66 **÷** 11 **=** 6

A. **Find the mean and median for each set of numbers. The first one is done for you.**

① Hours worked Monday by a group of employees: 8, 8, 7, 9, 10, 6, 8

Mean:

8 + 8 + 7 + 9 + 10 + 6 + 8 = 56

56 ÷ 7 = 8

Median: 6, 7, 8, 8, 8, 9, 10

Mean ___8___   Median ___8___

② Number of hot dogs sold for three days: 410, 350, 368

Mean _____   Median _____

③ Bowling scores for five games: 261, 254, 105, 280, 165

Mean _____   Median _____

④ Calories eaten each day for a week: 2,450; 2,100; 1,970; 2,430; 2,840; 1,800; 2,860

Mean _____   Median _____

B. **When there is an even amount of numbers in the set of data, there will be two middle numbers. Find the average of the two middle numbers to find the median. The first one is done for you.**

⑤ 7, 8, 10, 10, 8, 7, 12, 15   Median ___9___

7, 7, 8, 8, 10, 10, 12, 15

8 + 10 = 18 $2\overline{)18}$ → 9
                    -18
                    0

⑥ 146, 162, 234, 212, 184, 198, 523, 206   Median _____

⑦ 81, 46, 73, 32, 54, 81, 26, 41   Median _____

⑧ 1,903; 7,418; 5,651; 4,342; 2,950; 3,420   Median _____

C. **Solve each problem below with a calculator.**

⑨ On Saturday Angela kept a record of the number of customers she had each hour during the 9 hours her fruit stand was open: 25, 42, 45, 37, 102, 86, 46, 38, 27. What is the average number of customers per hour?

⑩ John is a manager at a movie theater. He recorded the attendance for one Saturday evening. What is the average attendance for that evening?

Movie Attendance on Saturday Evening, Nov. 1	
Cinema 1 502	Cinema 2 147
Cinema 3 425	Cinema 4 454
Cinema 5 518	Cinema 6 504

Area of a Rectangle

EXAMPLE Donna is putting wood tiling in a rectangular office. The office is 16 feet wide and 20 feet long. She wants to use tiles that measure 1 foot on each side. To find out how many tiles she needs, Donna first needs to find the area of the office.

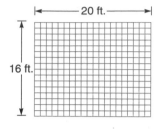

Look at the diagram of the office. There are 16 rows with 20 squares in a row. You can find the area by counting all the squares, or you can add 20 sixteen times. However, the easiest way to find the area is to multiply.

Which is the correct expression to find the area of the office?

(1) 16 + 16 + 20 + 20
(2) 20 + 20 + 20 + 20
(3) 20 × 20
(4) 16 × 20
(5) 16 × 16

Answer **(4)** is the correct choice. Donna uses the area formula to find the area of the office. She knows the length, 20 feet, and the width, 16 feet, of the office. She multiplies the length by the width to find the area. **16 × 20 = 320 square feet,** so the area of the office is **320 square feet.** Because the tiles measure 1 square foot, Donna needs **320 tiles** to cover the floor.

 Area is the amount of surface something takes up.

The formula for the area of a rectangle or a square is written:

$A = (l)(w)$, which means *Area = (length) (width)*.

 Area is measured in square units.

A square foot measures 1 foot on each side.

EXAMPLE Jason works at Picture Frames. The picture he is framing measures 12 inches by 16 inches. He needs to put glass down on the picture before he frames it.

Which is the correct expression to find the surface area of the picture?

(1) 12 + 16
(2) 12 × 16
(3) 12 + 12 + 16 + 16
(4) 12 × 12
(5) 16 × 16

Answer **(2)** is the correct choice. Jason needs to find the surface area of the picture. **12 × 16 = 192 square inches,** so Jason needs **192 square inches of glass.**

Circle the best answer for each item.

1 Yakov is planning a new exhibit at the zoo for a rectangular space that is 40 feet long and 30 feet wide. Which is the correct expression to find the area of the exhibit?

(1) 40×30
(2) $40 + 30$
(3) $40 + 40 + 30$
(4) $40 \times 40 \times 30 \times 30$
(5) $40 + 40 + 30 + 30$

2 The Juarez family is converting their garage into a family room. They are going to install the carpet themselves. The size of the family room is shown below. Which expression shows how to find the amount of carpeting they need?

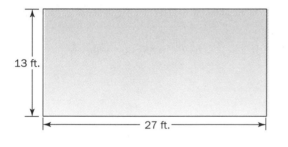

13 ft.

27 ft.

(1) $13 \times 13 \times 27$
(2) $27 + 27 + 13 + 13$
(3) $27 \times 13 \div 2$
(4) 13×27
(5) $27 \div 3$

3 Mrs. Juarez is making four curtain panels. She needs 6 feet of fabric for each curtain panel. There are 3 feet in a yard. Which is the correct expression to find the number of yards of fabric Mrs. Juarez needs?

(1) $6 \times 4 \times 3$
(2) $6 + 4 + 3$
(3) $6 \times 4 \div 3$
(4) $6 + 4$
(5) $6 \div 3$

4 Akeo is putting new vinyl flooring in the basement of her house. The basement is 42 feet long and 32 feet wide. Which expression shows the amount of vinyl flooring Akeo will need?

(1) $42 + 32$
(2) $42 + 32 + 42 + 32$
(3) 42×32
(4) $42 - 32$
(5) $42 \div 32$

5 Nancy is going to put lime on her rectangular garden to fertilize the soil. When deciding how much lime she needs, Nancy needs to know the size of the area to be covered. She knows the length of the garden is 12 feet. What is the area of her garden?

(1)   12 square feet
(2)   24 square feet
(3)   48 square feet
(4) 144 square feet
(5) Not enough information is given.

6 Maria wants to cover the surface of an old coffee table with three-inch square tiles. She measured the tabletop and found that it is 20 inches wide and 40 inches long. What is the area of the tabletop?

(1)   60 square inches
(2)   63 square inches
(3) 120 square inches
(4) 800 square inches
(5) Not enough information is given.

7 **Reason** Given the formula for the area of a rectangle is $A = (l)\,(w)$, develop the formula for finding the area of a square. (Hint: Let s = one side of the square.)

Check your answers on page 268.

Section 3 Review

Multiply or divide. Show your work.

① 456
× 3

② 83
×49

③ $297
× 6

④ 178
×702

⑤ 653
×305

⑥ 4)3,294 **⑦** 7)$6,440 **⑧** 18)5,794 **⑨** 43)8,657 **⑩** 27)4,131

⑪ 39 × 800 =

⑫ $842 × 5 =

⑬ $1,640 ÷ 8 =

⑭ 7,463 ÷ 52 =

Items **15–16** refer to the following information.

Freddie drives a van for a delivery service. The number of miles he drove on each of the 5 days he worked last week are 62, 28, 57, 65, and 63.

⑮ What is the *mean* number of miles Freddie drove last week?

⑯ What is the *median* number of miles Freddie drove last week?

Solve the following:

⑰ Solve for y in this equation: $y - 13 = 29$.

⑱ The Auto Lube Shop charges $22 for an oil change. This week the shop collected $330 for oil changes. How many customers had an oil change?

Use a calculator to solve the following problems.

⑲ 723 ⊗ 25 ⊜

⑳ 40 ⊗ 300 ⊜

㉑ 5,520 ÷ 12 ⊜

㉒ 3,145 ÷ 5 ⊜

Check your answers on pages 268–270.

PRACTICE

Circle the best answer for each item.

23 A professional basketball court is 94 feet long and 50 feet wide. Which is the correct expression for finding the area of the court?

(1) 94 + 50
(2) 94 × 50
(3) 94 + 50 + 50
(4) 94 + 94 + 50 + 50
(5) 94 × 94 × 50 × 50

24 Lena earns $8 an hour making deliveries for a pharmacy. She worked 7 hours on Friday, 8 hours on Saturday, and 10 hours on Sunday. How much did Lena earn on Friday?

(1) $56
(2) $64
(3) $80
(4) $200
(5) Not enough information is given.

25 Felicia bought 6 quarts of oil for her car. The total cost was $12. Which is the correct expression for finding the price of one quart of oil?

(1) 6 × $12
(2) 6 ÷ $12
(3) $12 ÷ 6
(4) $12 + 6
(5) $12 × 6

26 Kelvin dug a rectangular garden in his backyard. The garden is 40 feet long. What is the area of the garden in square feet?

(1) 15
(2) 65
(3) 130
(4) 1,000
(5) Not enough information is given.

27 A shoe store has children's shoes on sale for $5 a pair. Martin bought *p* pairs of shoes for his children. He paid $20. Which is the correct expression to find how many pairs of shoes Martin bought?

(1) 5p$ = $20
(2) $5 + p = $20
(3) $5 − p = $20
(4) $20 − p = $5
(5) $20 − $5 = p

Use your calculator for items 28–29.

28 Perry is a salesperson. The odometer on his car reads 126,000. If his car is 3 years old, about how many miles is Perry driving a year?

(1) 2,000
(2) 4,200
(3) 12,600
(4) 32,000
(5) 42,000

29 The Santos family is taking a driving vacation. Below is a list of cities and the road mileage. How many total miles are they planning on driving this vacation?

Location	Miles
Cincinnati, OH to Chicago, IL	244
Chicago, IL to Denver, CO	996
Denver, CO to Los Angeles, CA	1,059

(1) 1,303
(2) 2,055
(3) 2,289
(4) 2,299
(5) 2,300

4 Squares, Cubes, and Square Roots

MEASUREMENT AND GEOMETRY

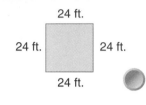

24 ft.

24 ft. | | 24 ft.

24 ft.

Area of a Square

EXAMPLE John works for Culver Carpets. On Tuesday John's boss sent him to meet Sara Vega. She wants to carpet her office. John measured her office. He found that each side is 24 feet long. John knows her office is a square.

○ **A square is a rectangle that has four sides of equal length.**

Recall in Section 3 you used the formula *Area = length × width* to find the area of a rectangle. You can also multiply the length by the width to find the area of a square: $A = s \times s$. But there is another way to write the formula.

○ **$A = s^2$, where s is one side of the square.**

The raised 2 (s^2) tells you how many times to multiply the side by itself. So $A = s^2$ is the same as $A = s \times s$.

Which is the correct expression to find the area of Sara Vega's office?

(1) 24 + 24
(2) 24 + 24 + 24 + 24
(3) 24 × 24
(4) 24 × 2
(5) 24 × 4

Answer **(3)** is correct. John needs to find the area of the office, so he multiplies the side by itself. **24 × 24 = 24^2 = 576 square feet,** so the office area is **576 square feet.**

1 yd.

1 yd. | | 1 yd.

1 yd.

1 square yard

EXAMPLE Carpeting is sold by the square yard. John knows the number of square feet needed (576 square feet). He needs to change the square feet to square yards.

A **square yard** is a square that measures 1 yard on each side. Also, 1 yard = 3 feet. A square that measures 3 feet on each side has 9 square feet, so 1 square yard = 9 square feet.

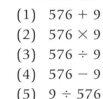

3 ft.

3 ft. | | 3 ft.

3 ft.

9 square feet
1 sq. yd. = 9 sq. ft.

Which is the correct expression to find the area of the office in square yards?

(1) 576 + 9
(2) 576 × 9
(3) 576 ÷ 9
(4) 576 − 9
(5) 9 ÷ 576

Answer **(3)** is correct. John needs to divide the number of square feet by 9. **576 ÷ 9 = 64 square yards,** so the area of the office is **64 square yards.**

PRACTICE

Circle the best answer for each item.

1 Ms. Vega wants to carpet another room that measures 15 feet on each side. John needs to find the area of the room. Which is the correct expression to find the area of the room in square feet?

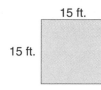

15 ft.

15 ft.

(1) 15×9
(2) $15 + 15 + 15 + 15$
(3) $9 \div 5$
(4) 15^2
(5) $15 \div 9$

2 John figures out that the area of the room is 225 square feet. He needs to find the area in square yards. Which is the correct expression to change 225 square feet to square yards?

(1) 225×9
(2) $9 \div 225$
(3) 9^2
(4) 225^2
(5) $225 \div 9$

3 Stanley wants to use a tarp to cover an area that measures 18 feet on each side. In square feet, which is the correct expression to find the minimum area that the tarp needs to cover?

(1) 18^2
(2) $18 + 18$
(3) 18×9
(4) $18 \div 9$
(5) $18 + 18 + 18 + 18$

4 Silvia has been hired to cover the floor of a kitchen and dining area with tile. The floor measures 12 feet on each side. Which is the correct expression to find the area of the floor in square feet?

(1) $12^2 \div 9$
(2) $12^2 \times 9$
(3) $12 + 12 + 12 + 12$
(4) 12^2
(5) $12 + 12 \div 9$

5 Ramon wants to order bathroom carpet from a catalog. The carpet is sold by the square yard. He finds that the area of his bathroom is 36 square feet. Which is the correct expression to change 36 square feet to square yards?

(1) $36 \div 4$
(2) $36 \div 9$
(3) 36×9
(4) $36 \div 3$
(5) Not enough information is given.

6 Irene wants to tile a hallway. The area of the hallway is 18 square feet. Four tiles are needed to cover one square foot. Which is the correct expression to find how many tiles Irene needs?

(1) $4 \div 18$
(2) $18 + 4$
(3) $18 \div 4$
(4) 4×18
(5) $18 + 18$

18 sq. ft.

7 **Communicate** Explain why you can find the area of a square room by measuring only one side. On a separate piece of paper, write your explanation in a sentence or two. Include a drawing with your explanation.

Check your answers on page 270.

Exponents and Roots

Finding Squares and Cubes

exponent

7^2 means 7 x 7

base
number

An **exponent** tells you how many times a number is to be multiplied by itself. In the expression 7×7, you multiply 7 times itself. Another way to write this expression is 7^2. The 7 is the **base** number of the expression. The 2 is the **exponent.**

The expression 7^2 can be read "seven to the second power" or "seven squared." The square of 7 is 49 because $7 \times 7 = 49$. The **value** of the expression 7^2 is 49.

A number is **cubed** when the number is multiplied by itself three times. Using an exponent, "4 cubed" is written 4^3. The expression 4^3 means $4 \times 4 \times 4$.

EXAMPLE Find the value of the expression 4^3.

Step 1: Multiply 4×4.

$4 \times 4 = 16$

Step 2: Multiply the result from Step 1 by 4.

$16 \times 4 = 64$

The value of the expression 4^3 **is 64.**

> **TIP**
>
> To find the value of a number with an exponent, write the expression as a multiplication problem and multiply.
> $4^3 = 4 \times 4 \times 4$
> $= 64$

Finding Square Roots

To find the square of a number, multiply that number by itself. To find the **square root** of a number, think: "What number squared equals this number?" The symbol for square root is $\sqrt{}$.

EXAMPLE What is the square root of 36?

Step 1: Think of what number squared equals 36.

$6^2 = 6 \times 6 = 36$

Step 2: Write in square root notation.

$\sqrt{36} = 6$

Square Root on a Calculator

To find the square root of a number on a calculator, enter the number and press the **square root** key $\boxed{\sqrt{}}$ to get the answer.

EXAMPLE Find the value of $\sqrt{3,136}$ using a calculator.

$\boxed{\text{CE/C}}$ 3136 $\boxed{\sqrt{}}$ 56

For more practice finding the square root with a calculator, see Calculator Handbook page 247.

A. Write each expression as a multiplication problem and multiply. The first one is done for you.

1 6^3 $\underbrace{\underbrace{6 \times 6}_{36 \times 6} \times 6}_{216}$ **2** 5^2 **3** 1^2 **4** 7^3

B. Refer to the table below to find the value of each expression. The first one is started for you.

TABLE OF SQUARES

1^2	=	1		11^2	=	121
2^2	=	4		12^2	=	144
3^2	=	9		13^2	=	169
4^2	=	16		14^2	=	196
5^2	=	25		15^2	=	225
6^2	=	36		16^2	=	256
7^2	=	49		17^2	=	289
8^2	=	64		18^2	=	324
9^2	=	81		19^2	=	361
10^2	=	100		20^2	=	400

5 8^2 8×8 **6** 11^2 **7** 18^2 **8** 16^2
9 5^3 **10** 8^3 **11** 4^3 **12** 10^3
13 $\sqrt{121}$ **14** $\sqrt{16}$ **15** $\sqrt{289}$ **16** $\sqrt{196}$

C. Find the value of the expressions below with a calculator. Refer to page 247 in the Calculator Handbook for additional information.

17 $21^2 =$ $31^2 =$
$22^2 =$ $32^2 =$
$23^2 =$ $33^2 =$
$24^2 =$ $34^2 =$
$25^2 =$ $35^2 =$

18 $\sqrt{441} =$ $\sqrt{3,481} =$
$\sqrt{625} =$ $\sqrt{3,721} =$
$\sqrt{841} =$ $\sqrt{4,225} =$
$\sqrt{1,225} =$ $\sqrt{4,761} =$
$\sqrt{1,521} =$ $\sqrt{5,041} =$

MEASUREMENT AND GEOMETRY

Volume of Rectangular Solids and Cubes

EXAMPLE Celia works for a shipping company. She figures out how many cubic feet each shipment will be. In the diagram, the shipment is 12 feet by 8 feet by 11 feet. To find out the cubic feet of this shipment, Celia needs to find its volume.

○ **A cubic foot** is a box that measures 1 foot on each side. The box is 1 foot long, 1 foot wide, and 1 foot high. The volume of the box is 1 cubic foot.

○ **Volume** measures the space inside an object. The space is measured in cubic units such as cubic inches or cubic feet. The formula for the volume of a rectangular solid is written:

$V = l \times w \times h$, or $V = l\,w\,h$, which means
Volume = length × width × height.

Which is the correct expression to find the volume of the shipment in the example above?
(1) $12 + 8 + 11$
(2) $12 \times 8 \times 11$
(3) 12×8
(4) 12×11
(5) $12 + 8$

Answer **(2)** is correct. To find the volume of a rectangular solid, multiply the length by the width by the height: $V = 12 \times 8 \times 11 = 1{,}056$ **cubic feet.**

EXAMPLE Another shipment is in the shape of a cube. Each side measures 8 feet. Celia needs to find the volume of the cube before loading it.

You can multiply the length by the width by the height to find the **volume** of a **cube**, but there is another way to write the formula.

$V = s^3$, which means
Volume = side × side × side.

Which is the correct expression to find the volume for the cube?
(1) $8 + 8$
(2) $8 + 8 + 8$
(3) 8×8
(4) $8 \times 8 \times 8$
(5) $8 + 1$

Answer **(4)** is correct. Since the shipment is a cube, all sides have the same measurement. To find the volume of a cube, multiply the length of the side by itself three times: $V = 8 \times 8 \times 8 = 512$ **cubic feet.**

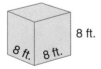

PRACTICE

Circle the best answer for each item.

1 The food compartment of a refrigerator is 3 feet long, 2 feet wide, and 5 feet high. Which is the correct expression to find the volume of the food compartment?

(1) 3 + 2 + 5
(2) 3 × 2 + 5
(3) 3 × 2 × 5
(4) 3 + 2 × 5
(5) 3 × 5

2 Tony is going to buy sand for his children's sandbox. The measurements for the sandbox are shown below.

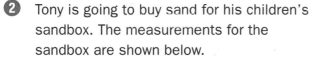

Which expression shows how to find the volume of the sandbox?

(1) 6 × 4
(2) 6 × 4 × 1
(3) 6 + 4 + 1
(4) 6 × 1
(5) 4 × 1

3 Joann works for a packing and mailing service. A customer brings in a gift-wrapped box that measures 7 inches on each side. Which is the correct expression to find the volume of the box?

(1) 7^1
(2) 7^2
(3) 7^3
(4) 7^4
(5) 7^5

4 Josefina is shipping a carton that is 12 inches long, 10 inches wide, and 8 inches tall. Which expression shows how to find the volume of the shipping carton?

(1) 10 × 8
(2) 12 + 10
(3) 12 × 10
(4) 12 × 10 × 8
(5) 12 + 10 + 8

5 Juan is shipping a carton that is 28 inches long and 40 inches wide. In cubic inches, what is the volume of the carton?

(1)   12
(2)   40
(3)   68
(4) 1,120
(5) Not enough information is given.

6 Gail has filled the delivery truck with 1,056 cubic feet of cartons. The inside of the truck measures 16 feet long, 8 feet wide, and 14 feet high. How many cubic feet of space is left on the truck?

(1)   320
(2)   736
(3) 1,376
(4) 1,792
(5) 2,528

7 **Reason** Explain why you can find the volume of a cube by measuring only one side. On a separate piece of paper, write your explanation in a sentence or two. Include a drawing with your explanation.

Using Formulas

EXAMPLE Linda works for a trucking company. She drives an average of 50 miles per hour. If Linda drives for 3 hours at that rate, how many miles could she drive?

To find how many miles Linda could drive, you can use the **distance formula.** A **formula** is an algebraic equation. To solve a formula, you need to know 2 of the 3 variables.

$$d = rt, \text{ or } d = r \times t, \text{ which means } \underline{\text{distance}} = \underline{\text{rate}} \times \underline{\text{time}}.$$

Follow these steps when using a formula:

Step 1: Substitute the numbers you know for variables, or letters, in the formula.

Step 2: Do the operations (add, subtract, multiply, or divide) shown in the formula.

Which expression shows how many miles Linda could drive in 3 hours?
(1) $50 + 3$
(2) $50 - 3$
(3) 50×3
(4) $50 \div 3$
(5) $3 \div 50$

Answer **(3)** is the correct choice. The distance formula tells you to multiply the rate (in this case, speed) by the time. Substitute the numbers you know into the formula. $d = rt = 50 \times 3 = \textbf{150 miles,}$ so at 50 miles per hour, Linda could drive **150 miles** in 3 hours.

EXAMPLE Mai-Ling wants to buy flooring for her kitchen. The kitchen is 14 feet long by 10 feet wide. You learned in Section 3 the formula for area of a rectangle is $A = l \times w$, or $A = l\,w$.

Which is the correct expression to find the area of Mai-Ling's kitchen?
(1) $14 + 10$
(2) $14 \div 10$
(3) $14 - 10$
(4) 14×10
(5) $10 - 14$

Answer **(4)** is correct. The area formula tells you to multiply the length by the width. Substitute the numbers you know into the formula.
$A = l\,w = 10 \times 14 = \textbf{140 square feet,}$ so the area of the kitchen is **140 square feet.**

Circle the best answer for each item.

1 Richard is putting wood trim along the walls of his living room as shown in the picture.

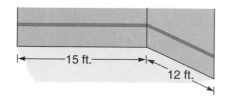

The formula for perimeter (*P*) can be written as $P = 2 \times l + 2 \times w$. Which is the correct expression to find how long the trim will be?

(1) $2 \times 15 \times 2 \times 12$
(2) $2 \times 15 - 2 \times 12$
(3) $2 \times 15 + 2 \times 12$
(4) $2 + 15 \times 2 + 12$
(5) $2 \times 15 \times 2 + 12$

2 Lilia is a salesperson for Advanced Technologies. On average she drives 5 hours a day. If she drives at a rate of 45 miles per hour, which expression finds how many miles Lilia drives in one day? (*d* = *rt*)

(1) 45×5
(2) $45 + 5$
(3) $45 - 5$
(4) $45 \div 5$
(5) $5 \div 45$

3 John's customer wants to put a 1-foot-wide strip of contrasting color carpet around the edges of her living room. The room measures 20 feet on each side. The formula for the perimeter (*P*) of a square is $P = 4s$, or $P = 4 \times s$. Which is the correct expression to find the perimeter of the room?

(1) 20^4
(2) 4×20
(3) $20 + 20$
(4) 2×20
(5) $20 + 20 + 20 + 20$

4 Kathleen is training for a two-day 150-mile bicycle ride to benefit the children's hospital. She wants to ride six hours each day. How many miles does she need to ride each hour to finish the ride in 12 hours? (Use $r = d \div t$.)

(1) 12×150
(2) $12 \div 150$
(3) $150 \div 12$
(4) 150×12
(5) Not enough information is given.

5 Kathy has been driving for 7 hours. She left at 9:00 and put 12 gallons of gas in her car. How many miles has she driven so far?

(1) 19
(2) 28
(3) 54
(4) 756
(5) Not enough information is given.

6 Emily sells storage sheds in many different styles and sizes. The most popular shed measures 10 feet wide by 12 feet long by 8 feet high. In cubic feet, what is the volume of this type of storage shed?

(1) 80
(2) 96
(3) 120
(4) 960
(5) Not enough information is given.

7 **Connect** Recall the distance formula is $d = rt$. Use what you know about solving equations to rewrite the distance formula to find *t* (time), given *d* (distance) and *r* (rate).

Section 4 Review

Write each expression as a multiplication problem and multiply.

1 8^2

2 1^3

3 3^3

4 9^2

Find the value of each expression.

5 17^2

6 15^2

7 $\sqrt{25}$

8 $\sqrt{49}$

 Use your calculator to find the value of each expression.

9 215^2

10 19^3

11 $\sqrt{324}$

12 $\sqrt{1,089}$

Solve each problem. Show your work.

13 The Mahers' living room measures 20 feet on each side. Use the formula $A = s^2$ to find the area of the living room.

14 A rectangular storage bin is 15 feet long, 12 feet wide, and 5 feet deep. Use the volume formula, $V = l \times w \times h$, to find the volume of the bin.

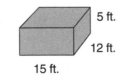

5 ft.

12 ft.

15 ft.

15 Willis installs pools. Today he is beginning to install an in-ground pool that is 30 feet long, 6 feet wide, and 4 feet deep. What is the volume, in cubic feet, of the dirt that he must remove before he installs the pool?

16 Marsha bought 9 square yards of carpet for her bedroom. The carpet cost $14 per square yard. What was the total cost of the carpet?

17 Viola is a carpet salesperson. She measures a customer's family room and finds its area is 576 square feet. What is the area in square yards?

18 A trash hauler places large trash bins at apartment complexes. Each bin is 12 feet long, 8 feet wide, and 6 feet high. What is the volume of each bin in cubic feet?

Check your answers on page 271.

Circle the best answer for each item.

19 Which is the correct expression to find the volume of the cube shown in cubic inches?

6 in.
6 in. 6 in.

(1) 6 + 6 + 6
(2) 6 × 6 × 6
(3) 6 × 6 + 20
(4) 6 × 6 × 6 ÷ 6
(5) 6 + 6 + 6 ÷ 6

20 Dave orders vinyl flooring for his kitchen, which measures 15 feet on each side. Which is the correct expression to find the area of the kitchen in square feet?

(1) 15 + 15
(2) $15^2 \div 9$
(3) 15 × 15 × 15
(4) 15 + 15 + 15 + 15
(5) 15^2

21 Dennis drives a delivery van. The cargo area of the van is 8 feet long, 5 feet wide, and 4 feet high. What is the volume of the cargo area in cubic feet?

(1) 17
(2) 40
(3) 52
(4) 105
(5) 160

8 ft. 5 ft. 4 ft.

22 Dwayne measures the conference room of his office building and finds it is 42 feet on each side. Which is the correct expression to find the area in square yards?

(1) 42^2
(2) $42^2 \times 9$
(3) $42^2 \div 9$
(4) 42 ÷ 9
(5) 42 × 4

23 Marcus needs to buy bricks for a patio he is building. The patio will measure 10 feet on each side. What will be the area of the patio in square feet?

(1) 20
(2) 40
(3) 100
(4) 200
(5) 1,000

24 Joan needs to fill a rectangular carton with foam pieces. The carton is 18 inches long, 12 inches wide, and 3 inches high. Which is the correct expression to find the volume of the carton in cubic inches?

(1) 18 × 12 × 3
(2) 18 + 12 + 3
(3) 18 × 12 + 3
(4) 18 + 12 × 3
(5) 18 × 12 ÷ 3

25 What is the volume in cubic inches of a shipping carton that is 18 inches wide, 18 inches long, and 20 inches tall?

(1) 324
(2) 360
(3) 720
(4) 5,832
(5) 6,480

26 Pat measures the hallways and bathroom in her house so she can buy new vinyl flooring. She finds a total area of 108 square feet. What is the area in square yards?

(1) 9
(2) 12
(3) 36
(4) 108
(5) 972

Math at Work

Transportation: Truck Driver

A tractor-trailer driver inspects his big rig.

Do you enjoy driving and seeing new places? If so, a job in the transportation industry may interest you. Because these types of jobs involve moving goods or people from place to place, truck drivers must have good map reading and math skills.

Every day truck drivers and other transportation workers use many of the math skills you've studied in this section. Drivers must **add, subtract, multiply, and divide whole numbers.** They often round numbers to make them easier to work with. Planning their routes and judging the time it will take to drive the distances requires drivers to rely heavily on good **map reading skills.**

Look at the Some Careers in Transportation chart.

- Do any of the careers interest you? If so, which ones?

- What information would you need to find out more about those careers? **On a separate piece of paper, write some questions that you would like answered.** You can find out more information about those careers in the *Occupational Outlook Handbook* at the library.

Transportation workers must have good map reading skills. Use the map below to answer the questions that follow.

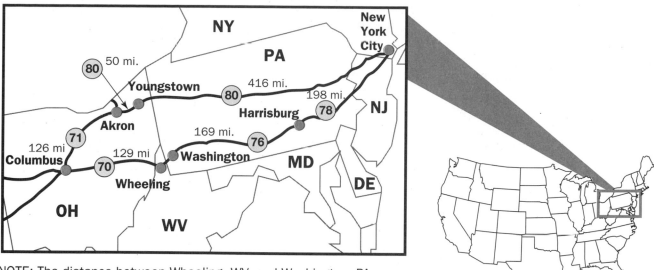

NOTE: The distance between Wheeling, WV, and Washington, PA, is 50 miles.

Mike drives a tractor-trailer truck. His job requires him to haul fruits and vegetables from Ohio to New York City. Because his cargo is perishable, he must be very careful about how long it takes him to make his runs.

EXAMPLE According to the mileage written on the map, how far would Mike have to drive from Youngstown to Akron, Ohio?

Read the mileage on the map to see that it is **50 mi.** (miles) from Youngstown to Akron.

PRACTICE

1 According to the interstate map above, how many miles is it between Harrisburg, Pennsylvania, and New York City?
(1) 50
(2) 129
(3) 169
(4) 198
(5) 416

Use a separate piece of paper and a calculator to answer the questions below.

2 There are two routes between Columbus, Ohio, and New York City.

a. Find the miles round trip from Columbus to New York City on each route.

b. Compare the routes. Which is shorter?

c. Mike needs to get this cargo to New York City as quickly as possible. If there is more than one hour of delays on Route 76, should he take the northern or southern route? Explain.

Check your answers on page 272.

Write the place value of each underlined digit.

1 2<u>3</u>4,957 _____

2 <u>6</u>,030,495 _____

Compare each pair of numbers. Write >, <, or =.

3 417 _____ 1,740

4 54,972 _____ 54,927

Follow the directions for each problem.

5 Round 4,375,429 to the nearest ten thousand. _____

6 Write 23,012 in words. _____

Solve. Show your work.

7 $x + 8 = 19$

8 $4x = 20$

9
```
  426
   84
+253
```

10
```
 $248
×   4
```

11
```
 $954
- 378
```

12 $27\overline{)785}$

13
```
  726
×308
```

14 $4^3 =$

15
```
 500
-243
```

16 $\sqrt{64} =$

17 $\$2,504 \div 8 =$

18 $3,976 \div 13 =$

19 The heights of the starting players on a basketball team are 80 inches, 76 inches, 75 inches, 80 inches, and 79 inches. What is the mean height of the players?

20 The temperatures for one week in Chicago were 35, 42, 37, 31, 24, 21, and 18. What was the median temperature for that week?

Use your calculator to find the value of each expression.

21 4,203 ⊗ 3,056

22 160,208 ⊘ 2,356

23 83^2

24 √ 55,225

Circle the best answer for each item.

25 Martin is building a fence around his rectangular vegetable garden. The garden is 48 feet long and 20 feet wide. Which is the correct expression to find the perimeter of the garden?

(1) 48×20
(2) 48^2
(3) $48^2 + 20^2$
(4) $48 + 20$
(5) $48 + 48 + 20 + 20$

26 On Friday Lorene spent $57 for groceries, $13 for dry cleaning, and $21 for an oil change. Estimate the total amount of money Lorene spent on Friday.

(1) $70
(2) $80
(3) $90
(4) $100
(5) $110

27 Roberto earns $14 an hour as a cook. He worked 38 hours last week and 43 hours this week. Which is the correct expression to find how much Roberto earned this week?

(1) $14 × 38
(2) $14 × 43
(3) 38 + 43
(4) $14 × 38 + 43
(5) $14 + 38 + 43

Item 28 refers to the following graph.

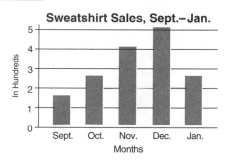

28 Find the approximate difference between the number of sweatshirts sold during the month with the greatest number of sales and the number of sweatshirts sold during the month with the least number of sales.

(1) 100
(2) 250
(3) 300
(4) 350
(5) 400

29 Carol is a carpet salesperson. She measures a bedroom and finds each side is 14 feet long. What is the area of this room in square feet?

(1) 18
(2) 28
(3) 56
(4) 196
(5) 38,416

Math Extension

Look through today's newspaper. Find ten different features of the paper that use whole numbers. Make a list of the features and give an example of how the whole number is used. For example, Classified Ads—Apartment for Rent, 3 bedrooms, 2 baths.

Math Connection: Whole Numbers and Economics

The Y2K Scare

Y2K. Do you know what it means? Y2K stands for the Year 2000. More specifically, it stands for January 1, 2000. "K" is a symbol borrowed from the metric system. It means 1,000.

Before January 1, 2000, Y2K made many people very nervous. People worried because computers only used the last two digits of the year to record a date and assumed that the first two digits were 19. Since the last two digits of the year 2000 are "00," many computers would see "00" only as a placeholder, not as the actual date. Computers would assume that "00" meant the year 1900, not the year 2000.

Governments and businesses that depended on computers would shut down or make serious mistakes if the Y2K problem was not fixed in time. Without computers operating properly, companies would not be able to bill their customers or pay their employees. Governments could not issue food stamps or social security checks. Power plants could not produce electricity. Banks would find errors in their records.

Governments and businesses worked hard and spent lots of money to try to solve this problem. The U.S. government alone spent over $5.4 billion in an effort to correct the problem before it occurred. State governments and businesses also spent billions of dollars to get their computer systems ready for Y2K.

Economics: Scarcity

Computers used to run governments, businesses, schools, transportation systems, and other important services had to be ready for the year 2000. Otherwise, people around the globe could not have gotten all of the goods and services they needed. When there are not enough goods or services to meet people's needs, it is called **scarcity.**

Scarcity can sometimes cause major problems. If people can't get essentials they need, panic may occur or crime may increase. If certain medicines become scarce, disease may spread. Keeping people healthy may become a problem.

Even if scarcity does not cause a major crisis, it can cause great inconveniences or increased prices for goods and services. For these reasons, businesses and governments worked to avoid the Y2K crisis.

Estimated Cost to States to Avoid Y2K Crisis

Top 12 States in Alphabetical Order

State	Estimated Cost
Alabama	$92 million
Arizona	$100 million
California	$200 million
Florida	$75–90 million
Indiana	$87 million
Maryland	$100 million
Missouri	$75 million
New York	$250 million
North Carolina	$82 million
Ohio	$60 million
Oregon	$75 million
Texas	$207 million

Source: *USA Today* 8/03/98

Use the information and the chart on the previous page to answer the questions below.

1 Why was it important for businesses and government to fix the Y2K problem?
   (1) Businesses would use too many computers on January 1, 2000.
   (2) Businesses and government would need to buy new computers.
   (3) Governments would have to adopt the metric system.
   (4) Governments and businesses would be scarce.
   (5) Computers would make mistakes interpreting the "00" in 2000.

2 Business and government invested billions of dollars to avoid the Y2K crisis because
   (1) people feel good in times of scarcity.
   (2) they didn't want goods and services to be scarce.
   (3) people were becoming very sick.
   (4) goods and services were becoming more expensive.
   (5) around the globe, crime was increasing.

3 The table on page 66 lists the states in alphabetical order. Rank the estimated costs for each state from <u>highest to lowest</u> and complete the table below. The first entry has been done for you.

Estimated Cost to States to Avoid the Y2K Crisis
Top 12 States – Highest to Lowest

State	Estimated Cost	State	Estimated Cost
New York	$250 million		

4 Why was it important for power companies to fix their computer systems before Y2K? Use a separate piece of paper to write your answer.

5 Computers are everywhere. They let you make phone calls, watch the news, and get electricity. Think about three ways in which you use or find computers in your life. Pick <u>one</u> situation and write a paragraph on a separate sheet of paper to describe what could have happened if computers were shut down by Y2K problems.

UNIT 2

Fractions

Fractions are used to show part of a whole amount. Fractions are all around us, often seen on highway signs and in recipes. Many tasks for making things, from sewing to construction, call for fractions. You already know a lot about fractions. You round and compare fractions to make decisions every day. Fractions are most often used when you are measuring something.

After you have made measurements that include fractions, you may need to use those measurements to solve problems. Frequently you will need to add, subtract, multiply or divide these fractions.

⊙ What did you do today where you saw fractions? How did you use them? Did you add hours on your timecard to find the total time worked? Working with fractions is all around us at home and at work.

⊙ Think of situations such as cooking, finding distances using road signs, figuring out your hours using a timecard, laying carpet, and so on. Did you have to multiply fractions for a recipe or to find the total weight of several items?

SECTIONS

⑤ **Fraction Basics**

⑥ **Adding and Subtracting Fractions**

⑦ **Multiplying and Dividing Fractions**

Fraction Basics

Reading Measurement Tools

EXAMPLE Rosa's chicken soup recipe calls for 10 ounces of chicken broth. Below is a measuring cup used for liquid measures.

Which of the following is the number of cups equal to the 10 ounces Rosa needs?

(1) 1 cup

(2) $1\frac{1}{4}$ cups

(3) $1\frac{1}{3}$ cups

(4) $1\frac{1}{2}$ cups

(5) $1\frac{3}{4}$ cups

Answer **(2)** is correct. First, find 10 ounces on the measuring cup. Find the same corresponding line for cups. **10 ounces = $1\frac{1}{4}$ cups,** so Rosa needs $1\frac{1}{4}$ cups of chicken broth.

EXAMPLE Mr. Asano is a framer. A customer wants $2\frac{1}{2}$ inch matting around a photograph. Mr. Asano used his ruler to find the exact width of the matte.

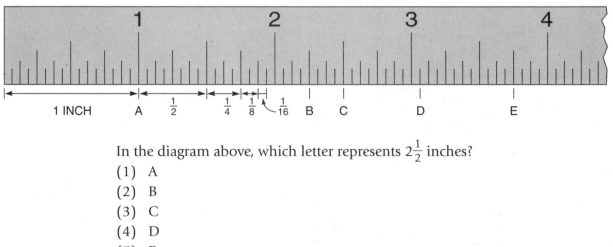

In the diagram above, which letter represents $2\frac{1}{2}$ inches?

(1) A

(2) B

(3) C

(4) D

(5) E

Answer **(3)** is correct. The $\frac{1}{2}$-inch mark is midway between two one-inch marks. The mark for $2\frac{1}{2}$ is in the middle of the 2-inch and 3-inch marks. The $2\frac{1}{2}$**-inch** mark is at letter **C.**

PRACTICE

Use the measuring cup or ruler from page 70 to answer Items 1–5.
Circle the best answer for each item.

1 Mr. Asano measured a postcard that had a width of $3\frac{1}{16}$ inches. Which letter is at the $3\frac{1}{16}$-inch mark on Mr. Asano's ruler?

(1)　A
(2)　B
(3)　C
(4)　D
(5)　E

2 Andrea is making chicken-rice soup for her dinner party. The recipe calls for $\frac{3}{4}$ cup chicken broth. Which expression is correct about the fraction $\frac{3}{4}$?

(1)　$\frac{3}{4}$ is less than $\frac{1}{3}$

(2)　$\frac{3}{4}$ is greater than $\frac{1}{2}$

(3)　$\frac{3}{4}$ is less than $\frac{2}{3}$

(4)　$\frac{3}{4}$ is greater than 1

(5)　$\frac{3}{4}$ is greater than $1\frac{1}{4}$

3 Jaime wants to buy new tiling for his bathroom. The new tile has sides of $2\frac{1}{4}$ inches. Which expression is correct about the mixed number $2\frac{1}{4}$?

(1)　$2\frac{1}{4}$ is less than $2\frac{1}{8}$

(2)　$2\frac{1}{4}$ is equal to $2\frac{2}{8}$

(3)　$2\frac{1}{4}$ is less than $1\frac{3}{4}$

(4)　$2\frac{1}{4}$ is greater than $2\frac{3}{8}$

(5)　$2\frac{1}{4}$ is greater than $2\frac{1}{2}$

4 Rosa's measuring cup does not show the $\frac{1}{8}$ mark. Which statement best describes where a $\frac{1}{8}$ mark would go?

(1)　below the $\frac{1}{4}$ mark

(2)　between the $\frac{1}{4}$ and $\frac{1}{3}$ marks

(3)　between the $\frac{1}{3}$ and $\frac{1}{2}$ marks

(4)　between the $\frac{1}{2}$ and $\frac{2}{3}$ marks

(5)　between $\frac{3}{4}$ and 1 marks

5 When measuring length in inches, which of the following is true of $3\frac{1}{4}$?

(1)　$3\frac{1}{4}$ is less than 3

(2)　$3\frac{1}{4}$ is greater than $3\frac{1}{2}$

(3)　$3\frac{1}{4}$ is less than $3\frac{1}{16}$

(4)　$3\frac{1}{4}$ is greater than $3\frac{1}{8}$

(5)　$3\frac{1}{4}$ is greater than $3\frac{3}{4}$

6 Dean bought bananas, pears, and grapes at the local grocery store. The grapes weighed $1\frac{1}{2}$ pounds and the pears weighed $2\frac{1}{4}$ pounds. What is the total weight of the fruit Dean bought?

(1)　$1\frac{1}{2}$ pounds

(2)　$2\frac{1}{4}$ pounds

(3)　$3\frac{3}{4}$ pounds

(4)　4 pounds

(5)　Not enough information is given.

7 **Reason**   In a sentence or two, explain how you know $1\frac{1}{2} = 1\frac{2}{4}$. Include a drawing with your explanation.

Check your answers on page 274.

Facts About Fractions

A **fraction** is part of a whole. When a whole is broken into parts, the parts are fractions of the whole.

A fraction is written with one number over another number. The top number is called the **numerator.** The bottom number is called the **denominator.**

$$\frac{1}{8} \longleftarrow \text{numerator}$$
$$\phantom{\frac{1}{8}} \longleftarrow \text{denominator}$$

Proper Fractions

EXAMPLE   A pizza is cut into 8 equal pieces. After dinner, only 1 piece is left. So $\frac{1}{8}$ of the pizza is left, and $\frac{7}{8}$ of the pizza is gone.

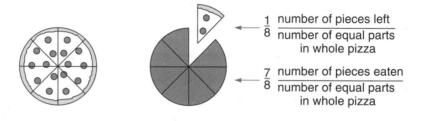

$\frac{1}{8}$ $\dfrac{\text{number of pieces left}}{\text{number of equal parts in whole pizza}}$

$\frac{7}{8}$ $\dfrac{\text{number of pieces eaten}}{\text{number of equal parts in whole pizza}}$

In the example above, the fractions $\frac{1}{8}$ and $\frac{7}{8}$ show part of one whole thing—the pizza. A fraction can also describe part of a group.

EXAMPLE   Margo bought 4 apples. She ate 3 apples. She ate $\frac{3}{4}$ of the apples. She saved $\frac{1}{4}$ of the apples.

$\frac{3}{4}$ $\dfrac{\text{number of apples eaten}}{\text{number of apples in group}}$

Fractions like $\frac{1}{8}$, $\frac{7}{8}$, $\frac{1}{4}$, and $\frac{3}{4}$ are **proper fractions.** In a proper fraction, the numerator (top number) is less than the denominator (bottom number). A proper fraction has a value less than 1.

EXAMPLE   The rectangle above has been divided into 8 parts. Three parts are shaded. So $\frac{3}{8}$ of the rectangle is shaded. In the proper fraction, $\frac{3}{8}$, the numerator, 3, is less than the denominator, 8. The value of $\frac{3}{8}$ is less than 1 because the shaded part is less than the whole rectangle.

A. Write the fraction that names the shaded part. The first one is done for you.

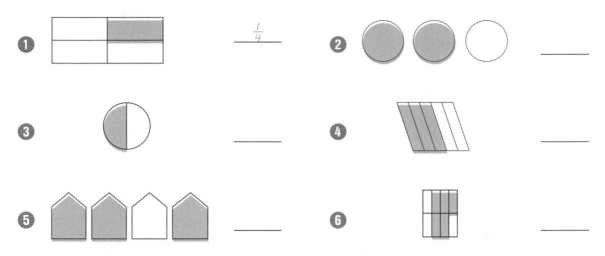

1 _____ $\frac{1}{4}$

2 _____

3 _____

4 _____

5 _____

6 _____

B. Divide and shade the figure to represent the given fraction. The first one is started for you.

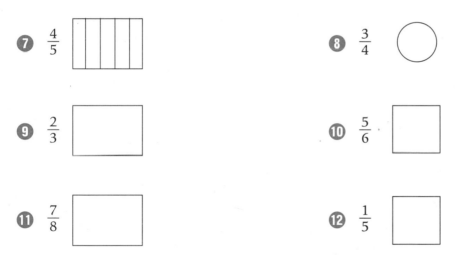

7 $\frac{4}{5}$

8 $\frac{3}{4}$

9 $\frac{2}{3}$

10 $\frac{5}{6}$

11 $\frac{7}{8}$

12 $\frac{1}{5}$

C. Solve each problem.

13 Mr. Martinez makes $1,200 a month. He pays $400 in rent. What fraction of his salary does Mr. Martinez pay in rent?

14 Harry's Restaurant employs 25 people. Eight employees are bus boys. What fraction of the employees are bus boys?

15 Cliff needs $24 for a new movie video. So far he has saved $18. What fraction of the price has Cliff saved so far?

16 Donna's basketball team won 8 out of the last 15 games. What fraction of the games did her team win?

Improper Fractions and Mixed Numbers

There are two other kinds of fractions in addition to **proper fractions.** They are improper fractions and mixed numbers.

The value of an **improper fraction** is equal to or greater than 1. The numerator is the same as or greater than the denominator. Fractions such as $\frac{4}{4}$, $\frac{4}{3}$, and $\frac{3}{2}$ are examples of improper fractions. Note that when the whole number 1 is written as an improper fraction, the numerator equals the denominator.

A **mixed number** is another way to show a value greater than 1. A mixed number is the sum of a whole number and a proper fraction. Fractions such as $1\frac{1}{2}$, $2\frac{1}{6}$, and $3\frac{3}{4}$ are examples of mixed numbers.

Improper fractions can be written as mixed numbers. Each of the 4 squares below has been divided into 4 parts. The shaded part can be called $\frac{13}{4}$ (an improper fraction) or $3\frac{1}{4}$ (a mixed number).

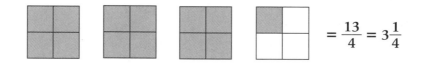

Changing an Improper Fraction to a Mixed Number

EXAMPLE Write $\frac{9}{5}$ as a mixed number.

Step 1. Divide the numerator by the denominator.

Step 2. If there is a remainder, write the remainder over the original denominator. When there is no remainder, the improper fraction equals a whole number.

$$\frac{9}{5} = 1\frac{4}{5}$$

$$5\overline{)9}$$
$$\underline{-5}$$
$$4$$

Changing a Mixed Number to an Improper Fraction

A mixed number can always be written as an improper fraction because both name a value greater than 1.

EXAMPLE Write $2\frac{3}{8}$ as an improper fraction.

Step 1. Multiply the whole number part of the mixed number by the denominator of the fraction. Put this product over the original denominator.

Step 2. Add the numerator of the fraction to the result.

Step 3. Write the total over the original denominator.

$$2\frac{3}{8} = \frac{2 \times 8}{8} + \frac{3}{8}$$

$$= \frac{16}{8} + \frac{3}{8}$$

$$= \frac{19}{8}$$

A. Write the improper fraction and mixed number that name the shaded part. The first one is done for you.

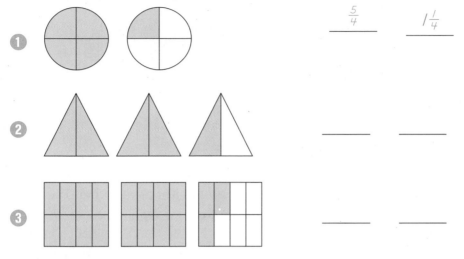

① $\dfrac{5}{4}$ $1\dfrac{1}{4}$

② _____ _____

③ _____ _____

B. Change each improper fraction to a whole or mixed number. Show your work. The first one is done for you.

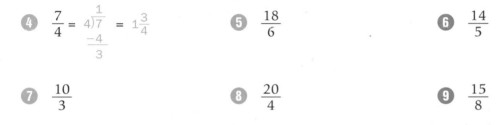

④ $\dfrac{7}{4} = 4\overline{)7}^{\;1} = 1\dfrac{3}{4}$
$\quad\quad \dfrac{-4}{3}$

⑤ $\dfrac{18}{6}$

⑥ $\dfrac{14}{5}$

⑦ $\dfrac{10}{3}$

⑧ $\dfrac{20}{4}$

⑨ $\dfrac{15}{8}$

C. Change each whole or mixed number to an improper fraction. Show your work. The first one is started for you.

⑩ $1\dfrac{5}{6}$

$\dfrac{1 \times 6}{6} + \dfrac{5}{6} = \dfrac{6 + 5}{6} =$

⑪ $2\dfrac{1}{2}$

⑫ 7

⑬ $3\dfrac{4}{5}$

⑭ 12

⑮ $10\dfrac{3}{4}$

D. Solve each problem.

⑯ A recipe calls for $3\dfrac{1}{2}$ cups of flour. Write an improper fraction to show the total amount in $\dfrac{1}{2}$ cups.

⑰ Stephen worked 6 hours on Monday and $\dfrac{1}{3}$ hour Monday night at home. Write a mixed number that describes the number of hours Stephen worked on Monday.

Equivalent Fractions

Different fractions can represent the same value. The shaded fraction bars shown below are equivalent.

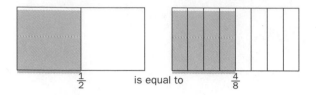

$\frac{1}{2}$ is equal to $\frac{4}{8}$

$\frac{1}{2}$ and $\frac{4}{8}$ are different names for the same value. Fractions that name the same value are called **equivalent fractions.**

You can **cross-multiply** to find if two fractions are equivalent. Multiply the numerator of each fraction by the denominator of the other. The results are called **cross-products.** If the cross-products are equal, then the fractions are equivalent.

TIP

The fraction $\frac{1}{2}$ names the same value as $\frac{2}{4}$ and $\frac{4}{8}$. Therefore, $\frac{1}{2}$ is equivalent to $\frac{2}{4}$ and $\frac{4}{8}$.

EXAMPLE Are $\frac{6}{8}$ and $\frac{3}{4}$ equivalent fractions?

Step 1.   Cross-multiply. $\frac{6}{8} \diagdown \diagup \frac{3}{4}$ $6 \times 4 = 24$
                                                           $8 \times 3 = 24$

Step 2.   Look at the results. If the results (24 and 24) are the same, then the fractions are equivalent.

$\frac{6}{8}$ and $\frac{3}{4}$ **are equivalent fractions.**

PRACTICE

Determine if the fractions are equivalent. Show your work. The first one is done for you.

1 $\frac{2}{4}$ $\frac{6}{12}$ **2** $\frac{2}{3}$ $\frac{12}{18}$ **3** $\frac{1}{2}$ $\frac{5}{10}$ **4** $\frac{3}{6}$ $\frac{5}{12}$

     $\frac{2}{4} \diagdown \diagup \frac{6}{12}$ $2 \times 12 = 24$
                     $4 \times 6 = 24$
     $\frac{2}{4}$ and $\frac{6}{12}$ are equivalent.

5 $\frac{4}{6}$ $\frac{8}{12}$ **6** $\frac{4}{7}$ $\frac{8}{14}$ **7** $\frac{5}{6}$ $\frac{25}{30}$ **8** $\frac{3}{5}$ $\frac{6}{15}$

9 $\frac{6}{10}$ $\frac{24}{40}$ **10** $\frac{3}{5}$ $\frac{20}{25}$ **11** $\frac{1}{2}$ $\frac{4}{8}$ **12** $\frac{2}{3}$ $\frac{6}{9}$

13 $\frac{7}{8}$ $\frac{21}{24}$ **14** $\frac{3}{8}$ $\frac{6}{16}$ **15** $\frac{8}{9}$ $\frac{16}{18}$ **16** $\frac{1}{4}$ $\frac{8}{16}$

Reducing a Fraction to Lower Terms

TIP

You may want to look for the largest number to divide into the numerator and denominator. In the example, you can divide both the top and bottom numbers by 12. (This is the same as dividing by 6 in step 1 and dividing by 2 in step 2.)

$$\frac{12 \div 12}{24 \div 12} = \frac{1}{2}$$

The numerator and denominator of a fraction are sometimes called **terms**. Thus, **reducing** a fraction to **lower terms** means finding an equivalent fraction with a smaller numerator and denominator.

To reduce a fraction, divide the numerator and denominator by the same number. The fraction has been reduced to its **lowest terms** when no number except 1 will divide evenly into both the numerator and denominator.

EXAMPLE   Reduce $\frac{12}{24}$ to lowest terms.

Step 1.   Choose a number that will divide into both. Try 6. $\frac{12 \div 6}{24 \div 6} = \frac{2}{4}$

Step 2.   You may need to divide more than once to reduce a fraction to lowest terms. Use 2. $\frac{2 \div 2}{4 \div 2} = \frac{1}{2}$

✔ Check your answer by cross-multiplying.

$$\frac{12}{24} \times \frac{1}{2} \quad \begin{array}{l} 12 \times 2 = 24 \\ 24 \times 1 = 24 \end{array}$$

Since the cross-products are the same, the fractions are equivalent.

PRACTICE

Reduce each fraction to lowest terms. Check each answer by cross-multiplying. The first one is done for you.

1 $\frac{3}{6} = \frac{3 \div 3}{6 \div 3} = \frac{1}{2}$ **2** $\frac{8}{12}$ **3** $\frac{4}{10}$ **4** $\frac{6}{8}$

 $\frac{3}{6} \times \frac{1}{2} \quad \begin{array}{l} 3 \times 2 = 6 \\ 6 \times 1 = 6 \end{array}$

5 $\frac{20}{25}$ **6** $\frac{12}{36}$ **7** $\frac{6}{9}$ **8** $\frac{8}{16}$

9 $\frac{16}{20}$ **10** $\frac{10}{14}$ **11** $\frac{5}{15}$ **12** $\frac{20}{24}$

13 $\frac{18}{27}$ **14** $\frac{18}{24}$ **15** $\frac{22}{33}$ **16** $\frac{21}{28}$

Raising a Fraction to Higher Terms

Raising a fraction to higher terms means finding an equivalent fraction with a greater numerator and denominator.

To raise a fraction to higher terms, multiply the numerator and the denominator by the same number.

TIP

To find the number that should be multiplied by 4 to get 20, you can divide 4 into 20.

$20 \div 4 = 5$

EXAMPLE   Raise $\frac{1}{4}$ to an equivalent fraction with a denominator of 20.

Step 1.   Think, "What number should be multiplied by 4 to get 20?" The number 5 is multiplied by 4 to get 20.

$4 \times ? = 20$

Step 2.   Multiply both the numerator and the denominator by 5.

$\frac{1}{4} = \frac{1 \times 5}{4 \times 5} = \frac{5}{20}$

✔ Check your work by cross-multiplying.

$$\frac{1}{4} \quad \frac{5}{20} \qquad \begin{array}{l} 1 \times 20 = 20 \\ 4 \times 5 = 20 \end{array}$$

Since the cross-products are the same, the fractions $\frac{1}{4}$ and $\frac{5}{20}$ are equivalent.

PRACTICE

Find an equivalent fraction with the given denominator. Show your work.
Check each answer by cross-multiplying. The first one is done for you.

① $\frac{3}{4} = \frac{?}{8}$

$\frac{3 \times 2}{4 \times 2} = \frac{6}{8}$

$\frac{3}{4} \quad \frac{6}{8} \quad \begin{array}{l} 3 \times 8 = 24 \\ 4 \times 6 = 24 \end{array}$

② $\frac{2}{3} = \frac{?}{12}$

③ $\frac{1}{6} = \frac{?}{18}$

④ $\frac{7}{10} = \frac{?}{20}$

⑤ $\frac{5}{8} = \frac{?}{24}$

⑥ $\frac{2}{5} = \frac{?}{30}$

⑦ $\frac{4}{9} = \frac{?}{18}$

⑧ $\frac{1}{4} = \frac{?}{12}$

⑨ $\frac{7}{8} = \frac{?}{32}$

⑩ $\frac{3}{5} = \frac{?}{25}$

⑪ $\frac{7}{12} = \frac{?}{36}$

⑫ $\frac{3}{10} = \frac{?}{40}$

⑬ $\frac{4}{9} = \frac{?}{45}$

⑭ $\frac{9}{16} = \frac{?}{32}$

⑮ $\frac{3}{7} = \frac{?}{56}$

⑯ $\frac{18}{25} = \frac{?}{100}$

Comparing Fractions

TIP

Recall the following:

> means *greater than*

< means *less than*

= means *equals*

Recall the symbols >, <, and = that were used to compare whole numbers. You wrote these symbols between two numbers to show how the value of the numbers compare. You can also use these symbols for comparing fractions.

When comparing two fractions with the same denominator, the fraction with the greater numerator is the greater fraction.

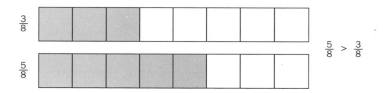

$\frac{5}{8} > \frac{3}{8}$

To compare two fractions with different denominators, cross-multiply and compare the results:

EXAMPLE Compare $\frac{3}{4}$ and $\frac{2}{3}$.

Step 1. Cross-multiply.

$$\frac{3}{4} \quad \frac{2}{3} \qquad 3 \times 3 = 9 \quad \text{numerator of first fraction}$$
$$2 \times 4 = 8 \quad \text{numerator of second fraction}$$

Step 2. Look at the results. Since 9 is greater than 8, then the first fraction is greater than the second fraction.

So $\frac{3}{4} > \frac{2}{3}$.

PRACTICE

Compare each pair of fractions. Write >, <, or =. The first one is done for you.

① $\frac{4}{5}$ __>__ $\frac{5}{7}$

$4 \times 7 = 28; 5 \times 5 = 25$

Since $28 > 25$, $\frac{4}{5} > \frac{5}{7}$

② $\frac{3}{4}$ _____ $\frac{3}{8}$

③ $\frac{9}{16}$ _____ $\frac{12}{16}$

④ $\frac{2}{4}$ _____ $\frac{2}{3}$

⑤ $\frac{2}{3}$ _____ $\frac{3}{5}$

⑥ $\frac{8}{10}$ _____ $\frac{4}{5}$

⑦ $\frac{7}{8}$ _____ $\frac{3}{4}$

⑧ $\frac{5}{9}$ _____ $\frac{8}{9}$

⑨ $\frac{11}{12}$ _____ $\frac{8}{12}$

⑩ $\frac{5}{6}$ _____ $\frac{7}{8}$

⑪ $\frac{3}{8}$ _____ $\frac{4}{5}$

⑫ $\frac{3}{4}$ _____ $\frac{2}{3}$

⑬ $\frac{1}{3}$ _____ $\frac{3}{9}$

⑭ $\frac{15}{21}$ _____ $\frac{13}{21}$

⑮ $\frac{15}{24}$ _____ $\frac{18}{24}$

⑯ $\frac{3}{4}$ _____ $\frac{18}{24}$

Section 5 Review

Write the proper or improper fraction that names the shaded part.

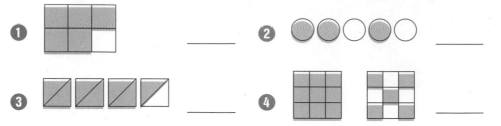

1 _____   **2** _____

3 _____   **4** _____

Write the mixed number that names the shaded part.

5 _____   **6** _____

Change each fraction to a whole or mixed number.

7 $\dfrac{11}{6}$ **8** $\dfrac{25}{5}$ **9** $\dfrac{19}{8}$ **10** $\dfrac{32}{7}$

Change each whole or mixed number to an improper fraction.

11 7   **12** $1\dfrac{7}{8}$ **13** $4\dfrac{3}{5}$ **14** $6\dfrac{2}{3}$

Reduce each fraction to lowest terms.

15 $\dfrac{5}{10}$ **16** $\dfrac{9}{12}$ **17** $\dfrac{16}{24}$ **18** $\dfrac{24}{32}$

Find an equivalent fraction with the given denominator.

19 $\dfrac{1}{2} = \dfrac{?}{18}$ **20** $\dfrac{5}{6} = \dfrac{?}{24}$ **21** $\dfrac{3}{4} = \dfrac{?}{32}$ **22** $\dfrac{4}{7} = \dfrac{?}{28}$

Compare each pair of fractions. Write >, <, or =.

23 $\dfrac{4}{5}$ _____ $\dfrac{3}{4}$ **24** $\dfrac{7}{5}$ _____ $\dfrac{5}{6}$

25 $\dfrac{4}{8}$ _____ $\dfrac{12}{24}$ **26** $\dfrac{8}{9}$ _____ $\dfrac{9}{12}$

Check your answers on pages 276–277.

Solve each problem.

27 There are 40 children who come to the day-care center where Ramona works. Of these 13 come only for the after-school program. What fraction of the children come only after school?

28 Elmer baked 36 muffins for a bake sale. His children ate 5 muffins. What fraction of the muffins did his children eat?

Circle the best answer for each item.

29 Juan completed $\frac{6}{8}$ of his workday before taking a break. In lowest terms, how much of his workday did Juan complete?

(1) $\frac{2}{2}$

(2) $\frac{3}{2}$

(3) $\frac{3}{4}$

(4) $\frac{4}{6}$

(5) $\frac{12}{16}$

30 Val is making a tile collage to sell at a craft show. There are four big squares. Each big square of tile has four small squares. Fifteen of the small squares have a design. Write a mixed number that describes the number of squares that have a design.

(1) $1\frac{5}{4}$

(2) 3

(3) $3\frac{3}{4}$

(4) 4

(5) $4\frac{1}{4}$

31 Betty lives 7 miles from work. Which improper fraction describes this distance?

(1) $\frac{7}{1}$

(2) $\frac{1}{7}$

(3) $\frac{7}{7}$

(4) $\frac{10}{7}$

(5) $\frac{7}{10}$

32 Lupe needs 6 ounces of water for a cake recipe. Use the measuring cup on page 70 to find how many cups of water Lupe needs.

(1) $\frac{1}{3}$

(2) $\frac{1}{2}$

(3) $\frac{2}{3}$

(4) $\frac{3}{4}$

(5) 1

33 On Friday Stock 1 dropped $\frac{3}{4}$ point and Stock 2 dropped $\frac{5}{8}$ point. Based on this information, which statement is true?

(1) Stock 2 dropped more.
(2) Stock 1 dropped more.
(3) The stocks dropped the same amount.
(4) The stocks dropped at the same time.
(5) Both stocks sold at the same price.

34 There are 32 members in an investment club. Eight of the members are senior citizens. In lowest terms, what fraction of the club are senior citizens?

(1) $\frac{1}{2}$

(2) $\frac{1}{4}$

(3) $\frac{1}{8}$

(4) $\frac{1}{16}$

(5) $\frac{1}{32}$

Check your answers on page 277.

Adding and Subtracting Fractions

DATA ANALYSIS ## Reading a Table

Number information or **data** can be organized in tables. The table is built with **rows** and **columns**. The rows go across and the columns go up and down. The rows and columns have **labels.** You can find the information you need where the row you want meets the column you need.

EXAMPLE David, a tailor, is sewing bridesmaid dresses for Jan and Sue. He will use the table below to find how much fabric to buy. David decides to make dress style C in size 10 for Jan and in size 14 for Sue. Which is the correct expression to find the total number of yards of fabric he will need?

(1) $10 - 3\frac{1}{4}$

(2) $3\frac{1}{4} + 3\frac{5}{8}$

(3) $3\frac{1}{4} - 3\frac{5}{8}$

(4) $10 + 3\frac{1}{4}$

(5) $3\frac{5}{8} - 3\frac{1}{4}$

Yardage Requirements

Sizes	10	12	14
Dress A	$2\frac{7}{8}$	3	$3\frac{1}{8}$
Dress B	3	$3\frac{1}{8}$	$3\frac{1}{4}$
Dress C	$3\frac{1}{4}$	$3\frac{3}{8}$	$3\frac{5}{8}$

Answer **(2)** is correct. David uses the table to find the yardage he needs: the row for Style C meets the column for size 10 at $3\frac{1}{4}$ and for size 14 at $3\frac{5}{8}$. David adds $3\frac{1}{4}$ (the yardage for Jan's size 10 dress) and $3\frac{5}{8}$ (the yardage for Sue's size 14 dress) to find the total yardage needed: $3\frac{1}{4} + 3\frac{5}{8} = 3\frac{2}{8} + 3\frac{5}{8} = 6\frac{7}{8}$ **yards,** so David needs to buy $6\frac{7}{8}$ yards.

EXAMPLE David may use dress style B instead. David knows he needs $6\frac{7}{8}$ yards for the dresses in style C and $6\frac{1}{4}$ yards for the dresses in style B. Which is the correct expression to find how many more yards of fabric he will need for dresses in style C?

(1) $6\frac{7}{8} \div 6\frac{1}{4}$

(2) $6\frac{7}{8} \times 6\frac{1}{4}$

(3) $6\frac{7}{8} + 6\frac{1}{4}$

(4) $6\frac{7}{8} - 6\frac{1}{4}$

(5) $6 - 6 - \frac{7}{8} - \frac{1}{4}$

Answer **(4)** is correct. To find how many more yards of fabric he will need for dresses in style C David needs to subtract. $6\frac{7}{8} - 6\frac{1}{4}$.

Circle the best answer for each item.

Items 1–6 refer to the table below.

**Yardage Requirements
Child's Jumper and Blouse**

Sizes	4	5	6
Jumper	$1\frac{3}{4}$	$1\frac{7}{8}$	2
Blouse A	$1\frac{1}{8}$	$1\frac{1}{4}$	$1\frac{3}{8}$
Blouse B	$1\frac{1}{8}$	$1\frac{1}{8}$	$1\frac{1}{4}$
Facing, Blouse B	$\frac{3}{4}$	$\frac{3}{4}$	$\frac{7}{8}$

1 Charnelle decides to make matching jumpers for her two daughters. One daughter wears size 4 and the other wears size 6. Which is the correct expression to find how many yards of fabric she needs?

(1) $1\frac{3}{4} + 1\frac{7}{8}$

(2) $1\frac{3}{4} - 2$

(3) $1\frac{3}{4} + 2$

(4) $2 - 1\frac{1}{3}$

(5) Not enough information is given.

2 Misha decides to make blouse B and a jumper for her daughter in size 5. Which is the correct expression to find how many yards of fabric and facing she needs?

(1) $1\frac{3}{4} + 1\frac{1}{8} + \frac{3}{4}$

(2) $1\frac{7}{8} + 1\frac{1}{8} + \frac{3}{4}$

(3) $1\frac{7}{8} - 1\frac{1}{8} + \frac{3}{4}$

(4) $1\frac{7}{8} + 1\frac{1}{4} - \frac{3}{4}$

(5) $1\frac{7}{8} + 1\frac{1}{4} + \frac{7}{8}$

3 Russ has a fabric remnant that is 5 yards long. The remnant will be used to make a jumper in size 4. Which is the correct expression to find how many yards of fabric will be left?

(1) $5 - 1\frac{1}{8}$

(2) $5 + 1\frac{3}{4}$

(3) $5 - 1\frac{7}{8}$

(4) $5 + \frac{3}{4}$

(5) $5 - 1\frac{3}{4}$

4 A tailor is making a jumper in size 4 and blouse A in size 6. Which is the correct expression to find the total number of yards of fabric he needs?

(1) $1\frac{3}{4} + 1\frac{3}{8}$

(2) $1\frac{3}{4} - 1\frac{3}{8}$

(3) $1\frac{3}{4} + 1\frac{1}{4}$

(4) $1\frac{7}{8} + 1\frac{3}{8}$

(5) $1\frac{3}{4} + 1\frac{1}{8}$

5 Yvonne has a piece of fabric that is $1\frac{3}{4}$ yards long. Choose the pattern for which Yvonne would not have enough fabric.

(1) blouse A, size 4
(2) blouse A, size 6
(3) blouse B, size 6
(4) jumper, size 4
(5) jumper, size 5

6 **Connect** Look for tables in the sports section of your local newspaper. Choose one table and explain what three different row and/or column labels mean.

Finding Common Denominators

Like fractions are fractions that have the same, or a common, denominator. The fractions $\frac{1}{4}$ and $\frac{3}{4}$ have a **common denominator** 4. The fractions $\frac{1}{6}$ and $\frac{2}{3}$ have different denominators, so they are **unlike fractions**.

The Least (or lowest) Common Denominator

To add or subtract unlike fractions, you need to find a common denominator. The smallest number that each denominator will divide into evenly is called the **least common denominator.**

EXAMPLE   Find the common denominator for $\frac{1}{2}$ and $\frac{3}{4}$.

Look at the denominators for both fractions. Think, "What number will 2 and 4 divide into evenly?" Try 4.

$$4 \div 2 = 2$$
$$4 \div 4 = 1$$

The lowest number 2 and 4 will divide into evenly is 4. So 4 is the least common denominator for $\frac{1}{2}$ and $\frac{3}{4}$. With a common denominator of 4, $\frac{1}{2}$ becomes $\frac{2}{4}$, while $\frac{3}{4}$ remains the same.

$$\frac{1}{2} = \frac{1 \times 2}{2 \times 2} = \frac{2}{4} \qquad\qquad \frac{3}{4} = \frac{3}{4}$$

EXAMPLE   Find the least common denominator for $\frac{3}{4}$ and $\frac{5}{6}$.

Step 1.   Think of multiples of 4 (numbers that 4 divides into evenly).

4, 8, 12, 16

Step 2.   Think of multiples of 6.

6, 12, 18

Step 3.   Choose the first multiple that both 4 and 6 have in common as the lowest common denominator.

4, 8, ⑫, 16

6, ⑫, 18

The lowest multiple that both 4 and 6 will divide into evenly is 12. So the least common denominator for $\frac{3}{4}$ and $\frac{5}{6}$ is 12; $\frac{3}{4}$ becomes $\frac{9}{12}$ and $\frac{5}{6}$ becomes $\frac{10}{12}$.

$$\frac{3}{4} = \frac{3 \times 3}{4 \times 3} = \frac{9}{12} \qquad\qquad \frac{5}{6} = \frac{5 \times 2}{6 \times 2} = \frac{10}{12}$$

EXAMPLE   Find the least common denominator for $\frac{2}{3}$, $\frac{5}{8}$, and $\frac{15}{24}$.

Step 1.   Think of multiples of 3, 8, and 24.

3: 3, 6, 9, 12, 15, 18, 21, ㉔, 27

Step 2.   Choose the first multiple 3, 8, and 24 have in common. Use 24.

8: 8, 16, ㉔ 32

24: ㉔ 48

Step 3.   Convert each fraction to an equivalent fraction with the common denominator.

$$\frac{2}{3} = \frac{2 \times 8}{3 \times 8} = \frac{16}{24} \qquad\qquad \frac{5}{8} = \frac{5 \times 3}{8 \times 3} = \frac{15}{24} \qquad\qquad \frac{15}{24}$$

TIP

The lowest common multiple of two prime numbers is the product of the two numbers. Given $\frac{3}{5}$ and $\frac{2}{7}$, 5 and 7 are prime numbers. Multiply 5 by 7. 35 is the lowest common multiple.

$\frac{3}{5}$ becomes $\frac{21}{35}$

$\frac{2}{7}$ becomes $\frac{10}{35}$

A. Find the lowest common multiple for each set of fractions. Do not add or subtract the fractions. The first one is done for you.

① $\frac{1}{2}$ and $\frac{4}{6}$ 6

multiples of 2: 2, 4, ⑥, 8
multiples of 6: ⑥, 12, 18

② $\frac{3}{4}$ and $\frac{4}{9}$

③ $\frac{1}{2}$ and $\frac{4}{9}$

④ $\frac{1}{3}$ and $\frac{2}{7}$

⑤ $\frac{1}{4}$ and $\frac{7}{12}$

⑥ $\frac{2}{3}$ and $\frac{4}{5}$

⑦ $\frac{1}{2}, \frac{2}{5}$ and $\frac{9}{10}$

⑧ $\frac{2}{3}, \frac{3}{4}$, and $\frac{5}{8}$

⑨ $\frac{1}{4}, \frac{2}{5}$, and $\frac{1}{10}$

B. Find the least common denominator for each pair of fractions. Then change the fractions to have common denominators. The first one is started for you.

⑩ $\frac{3}{5}$ and $\frac{5}{6}$

multiples of 5: 5, 10, 15, 20, 25, ㉚
multiples of 6: 6, 12, 18, 24, ㉚

$\frac{3}{5} = \frac{3 \times 6}{5 \times 6} = \frac{18}{30}$

$\frac{5}{6} =$

⑪ $\frac{1}{5}$ and $\frac{6}{7}$

⑫ $\frac{1}{2}$ and $\frac{1}{5}$

⑬ $\frac{4}{5}$ and $\frac{7}{8}$

⑭ $\frac{3}{4}$ and $\frac{4}{16}$

⑮ $\frac{2}{3}$ and $\frac{6}{8}$

⑯ $\frac{4}{6}$ and $\frac{6}{9}$

⑰ $\frac{2}{5}$ and $\frac{7}{8}$

⑱ $\frac{2}{3}$ and $\frac{9}{15}$

⑲ $\frac{3}{4}$ and $\frac{11}{14}$

⑳ $\frac{10}{12}$ and $\frac{3}{16}$

㉑ $\frac{5}{6}$ and $\frac{7}{8}$

C. Find an equivalent fraction with the given denominator.

㉒ $\frac{5}{8} = \frac{?}{24}$

㉓ $\frac{3}{4} = \frac{?}{28}$

㉔ $\frac{4}{9} = \frac{?}{36}$

㉕ $\frac{7}{10} = \frac{?}{50}$

㉖ $\frac{2}{5} = \frac{?}{30}$

㉗ $\frac{2}{3} = \frac{?}{18}$

㉘ $\frac{5}{6} = \frac{?}{48}$

㉙ $\frac{4}{7} = \frac{?}{35}$

㉚ $\frac{5}{11} = \frac{?}{44}$

Adding Fractions and Mixed Numbers

To add like fractions, add only the numerators. Write the sum over the common denominator. Reduce your answer to lowest terms. If the answer is an improper fraction, change it to a mixed number.

EXAMPLE Add: $\frac{5}{8} + \frac{7}{8}$

Step 1. Add the numerators. $\qquad\qquad\qquad\qquad\qquad\qquad 5 + 7 = 12$

Step 2. Write 12 over the common denominator 8. $\qquad\qquad \frac{12}{8}$

Step 3. Change $\frac{12}{8}$ to a mixed number; reduce.

$\qquad\qquad\qquad\qquad\qquad\qquad\qquad\qquad\qquad\qquad\qquad \frac{12}{8} = 1\frac{4}{8} = 1\frac{1}{2}$

So $\frac{5}{8} + \frac{7}{8} = \frac{12}{8} = 1\frac{4}{8} = 1\frac{1}{2}$.

To add unlike fractions, change the unlike fractions to like fractions.

EXAMPLE Add: $\frac{1}{3} + \frac{3}{4}$

Step 1. Use multiples of 3 and 4 to find a common denominator. Think, "What number will both 3 and 4 divide into evenly?"

$\qquad\qquad\qquad\qquad\qquad\qquad\qquad\qquad\qquad\qquad$ 3: 3, 6, 9, ⑫, 15

The lowest common denominator is 12. $\qquad\qquad$ 4: 4, 8, ⑫, 16

Step 2. Change $\frac{1}{3}$ to a like fraction with a denominator of 12. $\qquad\qquad\qquad\qquad\qquad \frac{1 \times 4}{3 \times 4} = \frac{4}{12}$

Step 3. Change $\frac{3}{4}$ to a like fraction with a denominator of 12. $\qquad\qquad\qquad\qquad\qquad \frac{3 \times 3}{4 \times 3} = \frac{9}{12}$

Step 4. Add. Change $\frac{13}{12}$ to a mixed number. $\qquad \frac{4}{12} + \frac{9}{12} = \frac{13}{12} = 1\frac{1}{12}$

Adding Mixed Numbers

When you add mixed numbers, add the fractions first, changing unlike fractions to like fractions. Add the whole numbers. If the sum of the fractions is an improper fraction, change it to a whole or mixed number. Add any whole numbers from the fraction total to the whole number total. Reduce your answer to lowest terms.

> **TIP**
>
> To add a mixed number and a whole number, add the whole numbers. Write the fraction next to the whole number total.
>
> $2\frac{5}{6} + 3 = 5\frac{5}{6}$

EXAMPLE Add: $2\frac{5}{6} + 3\frac{1}{2}$

Step 1. Find a common denominator and change to like fractions. $\qquad\qquad\qquad\qquad\qquad\qquad\qquad 2\frac{5}{6} \qquad 2\frac{5}{6}$

Step 2. Add the fractions and then add the whole numbers. $\qquad\qquad\qquad\qquad\qquad \dfrac{+3\frac{1}{2}}{} = \dfrac{+3\frac{3}{6}}{5\frac{8}{6}}$

Step 3. Change the improper fraction $\left(\frac{8}{6}\right)$ to a mixed number $\left(1\frac{2}{6}\right)$. Add the whole number from the fraction total to the whole number total $\left(5 + 1\frac{2}{6}\right)$. Reduce your answer to lowest terms.

$\qquad\qquad\qquad\qquad\qquad\qquad\qquad\qquad\qquad 5\frac{8}{6} = 5 + 1\frac{2}{6}$

$\qquad\qquad\qquad\qquad\qquad\qquad\qquad\qquad\qquad\quad = 6\frac{2}{6} = 6\frac{1}{3}$

A. **Add the fractions below. Reduce your answers to lowest terms. The first one is done for you.**

1 $\dfrac{2}{9} + \dfrac{5}{9} = \dfrac{7}{9}$

2 $\dfrac{7}{10} + \dfrac{9}{10} =$

3 $\dfrac{1}{4} + \dfrac{3}{5} =$

4 $\dfrac{5}{6} + \dfrac{2}{3} =$

5 $\dfrac{3}{8} + \dfrac{2}{3} =$

6 $\dfrac{2}{7} + \dfrac{4}{5} =$

B. **Add the mixed fractions below. Reduce your answers to lowest terms. The first one is started for you.**

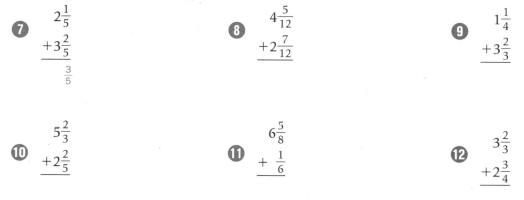

7
$2\dfrac{1}{5}$
$+3\dfrac{2}{5}$
___
$\dfrac{3}{5}$

8
$4\dfrac{5}{12}$
$+2\dfrac{7}{12}$

9
$1\dfrac{1}{4}$
$+3\dfrac{2}{3}$

10
$5\dfrac{2}{3}$
$+2\dfrac{2}{5}$

11
$6\dfrac{5}{8}$
$+\dfrac{1}{6}$

12
$3\dfrac{2}{3}$
$+2\dfrac{3}{4}$

C. **Solve each problem. Show your work. Reduce your answers to lowest terms.**

13 Angela hiked $4\dfrac{3}{8}$ miles one day and $2\dfrac{5}{16}$ miles the next day. How many total miles did Angela hike?

14 Keith works at Whigam Woods. He needs to saw two pieces of wood for a customer. One piece of wood is $12\dfrac{1}{4}$ feet long, the other is $9\dfrac{3}{16}$ feet long. How long are the two pieces of wood?

15 One weekend Raul worked $5\dfrac{3}{4}$ hours on Saturday and $3\dfrac{2}{3}$ hours on Sunday. How many hours did Raul work that weekend?

16 To make banana nut bread, Jeanne needs $1\dfrac{3}{4}$ cups flour, $1\dfrac{1}{4}$ cups bran, $\dfrac{1}{2}$ cup nuts, and $\dfrac{2}{3}$ cup sugar. How many cups do these dry ingredients total?

Subtracting Fractions and Mixed Numbers

To subtract like fractions, subtract only the numerators. Write the difference over the common denominator.

EXAMPLE Subtract: $\frac{7}{8} - \frac{1}{8}$

Step 1.	Subtract the numerators.	$7 - 1 = 6$
Step 2.	Write 6 over the common denominator 8.	$\frac{6}{8}$
Step 3.	Reduce $\frac{6}{8}$ to lowest terms.	$\frac{6 \div 2}{8 \div 2} = \frac{3}{4}$

To subtract unlike fractions, change the unlike fractions to like fractions.

EXAMPLE Subtract: $\frac{4}{5} - \frac{3}{10}$

Step 1.	Use multiples to find a common denominator. Think, "What number will both 5 and 10 divide into evenly?" 10 is the lowest common denominator.	5: 5, ⑩, 15 10: ⑩, 20, 30
Step 2.	Change $\frac{4}{5}$ to a like fraction with a denominator of 10.	$\frac{4 \times 2}{5 \times 2} = \frac{8}{10}$
Step 3.	Subtract and reduce the answer to lowest terms.	$\frac{8}{10} - \frac{3}{10} = \frac{5}{10} = \frac{1}{2}$

To subtract a fraction or mixed number from a whole number, regroup 1 from the whole number and rewrite it as a fraction.

EXAMPLE Subtract: $5 - 2\frac{3}{8}$

Step 1.	Regroup 1 from the whole number 5 and rewrite it as a fraction with a common denominator of 8.	
Step 2.	Subtract. The answer, $2\frac{5}{8}$, is in lowest terms.	

$$\begin{array}{r} 5 \\ -2\frac{3}{8} \\ \hline \end{array} \quad \begin{array}{r} 4\frac{8}{8} \\ -2\frac{3}{8} \\ \hline 2\frac{5}{8} \end{array}$$

Subtracting Mixed Numbers

When you subtract mixed numbers, sometimes the fraction being subtracted is larger than the fraction it is being subtracted from.

EXAMPLE Subtract: $3\frac{1}{4} - 1\frac{5}{6}$

Step 1.	Find a common denominator and write equal fractions.	
Step 2.	Regroup 1 $(\frac{12}{12})$ from the whole number column. Add it to $\frac{3}{12}$ in the fraction column. $(\frac{12}{12} + \frac{3}{12} = \frac{15}{12})$	
Step 3.	Subtract. The answer, $1\frac{5}{12}$, is in lowest terms.	

$$\begin{array}{r} 3\frac{1}{4} \\ -1\frac{5}{6} \\ \hline \end{array} = \begin{array}{r} 3\frac{3}{12} \\ -1\frac{10}{12} \\ \hline \end{array} = \begin{array}{r} 2\frac{15}{12} \\ -1\frac{10}{12} \\ \hline 1\frac{5}{12} \end{array}$$

TIP

Any fraction with the same number in both the numerator and denominator has a value of 1. So, $\frac{3}{3}$, $\frac{5}{5}$, and $\frac{100}{100}$ each has a value of 1.

A. Subtract the fractions below. Reduce your answers to lowest terms.
The first one is done for you.

1 $\dfrac{11}{12} - \dfrac{5}{12} = \dfrac{6}{12} = \dfrac{1}{2}$

2 $\dfrac{5}{6} - \dfrac{3}{8} =$

3 $\dfrac{9}{16} - \dfrac{1}{4} =$

4 $6 - 2\dfrac{2}{3} =$

5 $\dfrac{7}{8} - \dfrac{2}{3} =$

6 $\dfrac{4}{5} - \dfrac{3}{4}$

B. Subtract the mixed fractions below. Reduce your answers to lowest terms.
The first one is started for you.

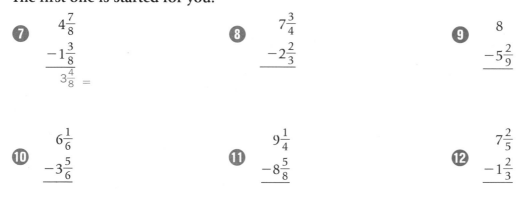

7 $\begin{array}{r} 4\frac{7}{8} \\ -1\frac{3}{8} \\ \hline 3\frac{4}{8} \end{array} =$

8 $\begin{array}{r} 7\frac{3}{4} \\ -2\frac{2}{3} \\ \hline \end{array}$

9 $\begin{array}{r} 8 \\ -5\frac{2}{9} \\ \hline \end{array}$

10 $\begin{array}{r} 6\frac{1}{6} \\ -3\frac{5}{6} \\ \hline \end{array}$

11 $\begin{array}{r} 9\frac{1}{4} \\ -8\frac{5}{8} \\ \hline \end{array}$

12 $\begin{array}{r} 7\frac{2}{5} \\ -1\frac{2}{3} \\ \hline \end{array}$

C. Solve each problem. Show your work. Reduce your answers to lowest terms.

13 Chin swims $9\dfrac{3}{4}$ miles a week. So far this week he has swum $3\dfrac{1}{4}$ miles. How many more miles does Chin need to swim this week?

14 At the beginning of the week, the cook at Pasta Unlimited had 35 pounds of pasta. On Monday he used $8\dfrac{9}{16}$ pounds. How many pounds of pasta are left?

15 Elena mailed two packages at the post office. One package weighed $4\dfrac{3}{4}$ pounds. The other package weighed $2\dfrac{1}{2}$ pounds. How many more pounds did the heavier package weigh?

16 Yuki needs a shelf. She cut a board $4\dfrac{2}{3}$ feet long from a board that is $8\dfrac{1}{2}$ feet long. How much board is left?

Check your answers on page 279.

Using a Map

EXAMPLE Kim drives a delivery truck for a laundry service. The map shows her daily route. Kim begins at point A, drives east to point B, then north to point C. Next she goes west and south through points D, E, and F. Finally, Kim drives south to point A.

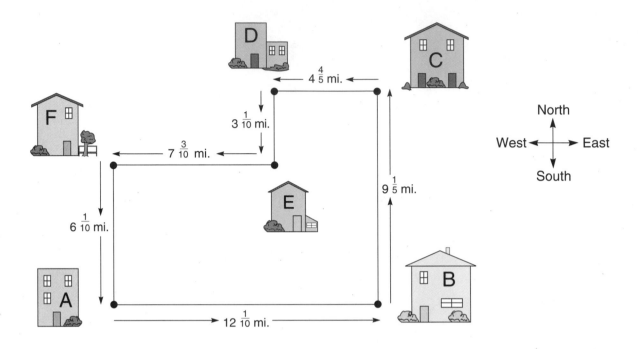

Kim's boss, Miguel, wants her to be at point D on the map by noon. How many miles will Kim drive by noon?

(1) $16\frac{1}{10}$

(2) $24\frac{11}{10}$

(3) $26\frac{1}{10}$

(4) $26\frac{6}{10}$

(5) $26\frac{6}{20}$

Answer **(3)** is correct. Kim covers three distances to get from point A to point D by noon: $12\frac{1}{10}$ miles from A to B, $9\frac{1}{5}$ miles from B to C, $4\frac{4}{5}$ miles from C to D.

$$12\frac{1}{10} + 9\frac{1}{5} + 4\frac{4}{5} = 12\frac{1}{10} + 9\frac{2}{10} + 4\frac{8}{10} = 25\frac{11}{10} = 25 + 1\frac{1}{10} = 26\frac{1}{10} \text{ miles}$$

PRACTICE

Circle the best answer for each item.

<u>Items 1–6</u> refer to the following information.

José restocks soda machines for a vending company. The map below shows the route José follows each day.

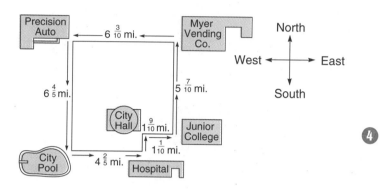

1 During the morning, José restocks machines for all the customers between Myer Vending Co. and the hospital. How many miles does he drive during the morning?

(1) $13\frac{1}{10}$

(2) $16\frac{1}{2}$

(3) $16\frac{4}{5}$

(4) $16\frac{9}{10}$

(5) $17\frac{1}{2}$

2 One-Stop Gas and Food is $3\frac{3}{10}$ miles past the city pool. How many miles does José drive to get from One-Stop Gas and Food to the hospital?

(1) $7\frac{7}{10}$

(2) $7\frac{1}{5}$

(3) $1\frac{1}{5}$

(4) $1\frac{1}{10}$

(5) $1\frac{1}{15}$

3 During the afternoon, José drives from the hospital back to Myer Vending Co. How many miles does he drive during the afternoon?

(1) $8\frac{7}{10}$

(2) $8\frac{1}{10}$

(3) $7\frac{7}{10}$

(4) $7\frac{17}{30}$

(5) 3

4 How many miles closer to Myer Vending Co. is Junior College than Precision Auto?

(1) $1\frac{3}{5}$

(2) $1\frac{2}{5}$

(3) $1\frac{1}{10}$

(4) $\frac{3}{5}$

(5) $\frac{2}{5}$

5 José lives near Myer Vending Co. How many miles does José drive from home to work at Myer Vending Co.?

(1) $1\frac{1}{10}$

(2) $4\frac{2}{5}$

(3) $5\frac{7}{10}$

(4) $6\frac{3}{10}$

(5) Not enough information is given.

6 **Connect** Explain how you would estimate the total number of miles of José's route. Show your work on a separate piece of paper.

Filling in Timecards

Kareem is an office clerk in a grocery store. His job is to complete the timecards for the store's employees. At the end of the day, Kareem records the time employees worked to the nearest quarter of an hour.

NAME:	Jan Lopez
WEEK OF:	3/4 – 3/11

IN	MON.	3/4	8:15 A.M.
OUT	MON.	3/4	12:36 P.M.
IN	MON.	3/4	1:28 P.M.
OUT	MON.	3/4	5:17 P.M.
IN	TUES.	3/5	8:05 A.M.

- 15 minutes $= \frac{1}{4}$ hour, because $\frac{15}{60} = \frac{1}{4}$
- 30 minutes $= \frac{1}{2}$ hour, because $\frac{30}{60} = \frac{1}{2}$
- 45 minutes $= \frac{3}{4}$ hour, because $\frac{45}{60} = \frac{3}{4}$

EXAMPLE First, Kareem finds the number of hours Jan worked in the morning. To the nearest quarter of an hour, how many hours did Jan work in the morning?

(1) $3\frac{1}{4}$

(2) 4

(3) $4\frac{1}{6}$

(4) $4\frac{1}{4}$

(5) $5\frac{1}{4}$

Answer **(4)** is correct. Jan started at 8:15 A.M. Kareem thinks of 8:15 as $8\frac{1}{4}$. Jan went to lunch at 12:36 P.M. Since 36 is close to 30 minutes, Kareem thinks of 12:36 as $12\frac{1}{2}$. Then he subtracts the smaller mixed number from the larger one.

$12\frac{1}{2} - 8\frac{1}{4} = 12\frac{2}{4} - 8\frac{1}{4} = 4\frac{1}{4}$, so Jan worked about $4\frac{1}{4}$ **hours** in the morning.

EXAMPLE Next, Kareem finds about how many hours Jan worked after lunch. To the nearest quarter of an hour, how many hours did Jan work after lunch?

(1) $4\frac{3}{4}$

(2) $4\frac{1}{2}$

(3) 4

(4) $3\frac{3}{4}$

(5) $3\frac{1}{2}$

Answer **(4)** is correct. Jan came back from lunch at 1:28 P.M. Kareem changes 1:28 to $1\frac{1}{2}$. Jan left work at 5:17 P.M. Since 17 is close to 15, Kareem changes 5:17 to $5\frac{1}{4}$. Then he subtracts the smaller mixed number from the larger one.

$5\frac{1}{4} - 1\frac{1}{2} = 5\frac{1}{4} - 1\frac{2}{4} = 4\frac{5}{4} - 1\frac{2}{4} = 3\frac{3}{4}$ **hours,** so Jan worked $3\frac{3}{4}$ **hours** after lunch.

Circle the best answer for each item.

1 Kareem knows Jan worked $4\frac{1}{4}$ hours in the morning and $3\frac{3}{4}$ hours in the afternoon. What was the total number of hours Jan worked during the day?

(1) 7

(2) $7\frac{3}{4}$

(3) 8

(4) $8\frac{1}{4}$

(5) $8\frac{1}{2}$

Items 2–4 refer to the timecard at right.

NAME:	Paul Wagner	
WEEK OF:	3/12 – 3/19	
IN	TUES. 3/13	8:50 A.M.
OUT	TUES. 3/13	12:32 P.M.
IN	TUES. 3/13	1:24 P.M.
OUT	TUES. 3/13	5:10 P.M.

2 How many hours did Paul work in the morning?

(1) $3\frac{1}{4}$

(2) $3\frac{1}{2}$

(3) $3\frac{3}{4}$

(4) $4\frac{1}{4}$

(5) $5\frac{1}{4}$

3 How many hours did Paul work in the afternoon?

(1) $3\frac{3}{4}$

(2) 4

(3) $4\frac{1}{2}$

(4) $4\frac{3}{4}$

(5) $6\frac{3}{4}$

4 How many total hours did Paul work on Tuesday?

(1) $7\frac{1}{2}$

(2) 8

(3) $8\frac{1}{4}$

(4) 9

(5) $9\frac{1}{2}$

5 On Wednesday Paul arrived to work at 8:25 A.M. Later he took a 30-minute lunch and then worked the rest of the afternoon. How many hours did Paul work on Wednesday morning?

(1) $\frac{1}{2}$ hour

(2) $3\frac{1}{2}$ hours

(3) 4 hours

(4) 8 hours

(5) Not enough information is given.

6 Maria worked $38\frac{1}{4}$ hours one week and $40\frac{3}{4}$ hours the following week. How many more hours did Maria work the second week?

(1) 2

(2) $2\frac{1}{4}$

(3) $2\frac{1}{2}$

(4) $2\frac{3}{4}$

(5) Not enough information is given.

7 **Reason** On a separate piece of paper, explain how you would change the following minutes to fractional parts of an hour. What fraction did you get for each?
20 minutes
40 minutes
50 minutes

Section 6 Review

Add or subtract. Reduce your answers to lowest terms.

1 $\dfrac{15}{16} - \dfrac{9}{16} =$

2 $\dfrac{1}{5} + \dfrac{7}{10} =$

3 $8 - 2\dfrac{1}{6} =$

4 $\dfrac{5}{6} + \dfrac{3}{8} =$

5 $\begin{array}{r} 3\frac{7}{9} \\ +4\frac{2}{9} \\ \hline \end{array}$

6 $\begin{array}{r} 4\frac{9}{10} \\ -2\frac{3}{10} \\ \hline \end{array}$

7 $\begin{array}{r} 2\frac{5}{8} \\ -2\frac{3}{4} \\ \hline \end{array}$

8 $\begin{array}{r} 9\frac{1}{5} \\ -8\frac{3}{5} \\ \hline \end{array}$

9 $\begin{array}{r} 3 \\ -\ \frac{7}{8} \\ \hline \end{array}$

10 $\begin{array}{r} 6\frac{1}{3} \\ -2\frac{3}{4} \\ \hline \end{array}$

Items 11–13 refer to the table at the right.

11 Jenny would like to make dress B in size 12. How many yards of fabric will she need?

   (1) $3\frac{1}{4}$

   (2) $6\frac{5}{8}$

   (3) $7\frac{1}{8}$

   (4) $8\frac{1}{2}$

   (5) $9\frac{1}{8}$

Yardage Requirements
Prom Dress

Sizes	8	10	12
Dress A	$5\frac{3}{4}$	$5\frac{7}{8}$	$6\frac{3}{8}$
Dress B Main Color Contrasting Color	$5\frac{1}{8}$ $1\frac{3}{8}$	$5\frac{1}{4}$ $1\frac{3}{8}$	$5\frac{5}{8}$ $1\frac{1}{2}$
Lace A	$2\frac{1}{2}$	$2\frac{5}{8}$	$2\frac{3}{4}$
Lace B	$2\frac{1}{8}$	$2\frac{1}{8}$	$2\frac{1}{4}$

12 Caryn is trying to decide whether she wants to make dress A or dress B. How many more yards of lace are needed if she chooses dress A in size 8 rather than dress B in size 8?

   (1) $\frac{1}{8}$

   (2) $\frac{1}{6}$

   (3) $\frac{3}{8}$

   (4) $\frac{5}{8}$

   (5) $4\frac{5}{8}$

13 Amanda is making two style A prom dresses in sizes 8 and 10. What is the total amount of lace she will need for the two dresses?

   (1) $2\frac{1}{4}$

   (2) $2\frac{3}{4}$

   (3) $5\frac{1}{8}$

   (4) $10\frac{3}{8}$

   (5) $11\frac{5}{8}$

Check your answers on pages 280–281.

Solve each problem.

Items 14–17 refer to the following information.

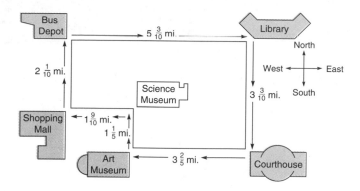

Sundra is a city bus driver. The map above shows the route she drives starting at the bus depot.

14 The busiest part of Sundra's route is between the courthouse and the shopping mall. About how many miles is this part of the route?

_____

15 Exactly how many miles is the part of Sundra's route between the courthouse and the shopping mall?

_____

16 Sundra then returns to the bus depot. What is the total distance from the courthouse to the bus depot?

_____

17 Sundra's entire route is $17\frac{1}{5}$ miles long. Estimate how many miles she has left to drive after she has reached the library.

_____

Items 18–20 refer to the timecard below.

NAME:	Joy Chan		
WEEK OF:	4/4 – 4/11		
IN	MON.	4/4	8:24 A.M.
OUT	MON.	4/4	12:13 P.M.
IN	MON.	4/4	12:58 P.M.
OUT	MON.	4/4	5:43 P.M.

18 How many hours did Joy work in the morning?

_____

19 How many hours did Joy work in the afternoon?

_____

20 How many total hours did Joy work on Monday?

_____

Multiplying and Dividing Fractions

MEASUREMENT AND GEOMETRY

Mixing Ingredients

EXAMPLE Mike is taking baked beans to a picnic. His recipe serves 5 people. He needs enough for 15 people. He knows to triple the recipe ($3 \times 5 = 15$).

The recipe calls for $1\frac{1}{3}$ cups of catsup. Which expression finds the number of cups of catsup Mike should use?

(1) $1 \times 1\frac{1}{3}$

(2) $1\frac{1}{3} \times 3$

(3) $1\frac{1}{3} \div 3$

(4) $1\frac{1}{3} \times \frac{1}{3}$

(5) $3 \div 1\frac{1}{3}$

Answer **(2)** is correct. Since Mike wants to make three times the amount of the recipe, he must multiply the amount of catsup in the recipe, $1\frac{1}{3}$ cups, by 3.

$1\frac{1}{3} \times 3 = 1 \times 3 + \frac{1}{3} \times 3 = 3 + \frac{3}{3} = 3 + 1 = 4$ **cups**

EXAMPLE Anna is making chili for the picnic. Her recipe serves 20 people. She wants to make enough for 10 people. Her recipe calls for $3\frac{1}{2}$ pounds of ground beef. Which expression finds how much ground beef she will need?

(1) $\frac{1}{2} \times 3 \times 2$

(2) $3\frac{1}{2} \times 2$

(3) $3\frac{1}{2} \div \frac{1}{2}$

(4) $3\frac{1}{2} \div 2$

(5) $\frac{1}{2} \times 3\frac{1}{2}$

Answer **(4)** is correct. Anna needs $\frac{1}{2}$ the amount of meat for 10 people. Dividing by 2 separates the whole amount into two equal parts: $3\frac{1}{2} \div 2$. Another way to find $\frac{1}{2}$ the amount is to multiply by $\frac{1}{2}$: $3\frac{1}{2} \times \frac{1}{2}$. However, this is not given as an answer option.

Circle the best answer for each item.

1 Vince has a recipe for scalloped potatoes. The recipe calls for $5\frac{1}{2}$ pounds of potatoes. He decides to make one third of this recipe for his family. Which is the correct expression to find the amount of potatoes he should use?

(1) $\frac{1}{3} \div 5\frac{1}{2}$

(2) $5\frac{1}{2} \div \frac{1}{3}$

(3) $5\frac{1}{2} \times 1$

(4) $5\frac{1}{2} \times 3$

(5) $5\frac{1}{2} \div 3$

2 Kasia wants to make one loaf of bread. Her recipe makes 2 loaves and calls for $7\frac{1}{4}$ cups of flour. Which is the correct expression to find the number of cups of flour Kasia should use?

(1) $\frac{1}{2} \div 7\frac{1}{4}$

(2) $2 \div 7\frac{1}{4}$

(3) $7\frac{1}{4} \times \frac{1}{2}$

(4) $7\frac{1}{4} \div \frac{1}{2}$

(5) $7\frac{1}{4} \times 2$

3 Ginny is making fudge to sell at a fair. To make 1 pound of fudge, she needs $\frac{1}{3}$ cup of cocoa. She has $4\frac{2}{3}$ cups of cocoa. Which is the correct expression to find the number of pounds of fudge Ginny can make with the cocoa she has?

(1) $\frac{1}{3} \div 4\frac{2}{3}$

(2) $3 \div 4\frac{2}{3}$

(3) $4\frac{2}{3} \times \frac{1}{3}$

(4) $4\frac{2}{3} \div \frac{1}{3}$

(5) $4\frac{2}{3} \div 1$

4 Luis is making macaroni and cheese for supper. His recipe calls for $1\frac{1}{4}$ cups of milk, but he plans to double the recipe. Which is the correct expression to find the number of cups of milk he should use?

(1) $\frac{1}{2} \div 1\frac{1}{4}$

(2) $2 \div 1\frac{1}{4}$

(3) $1\frac{1}{4} \times \frac{1}{2}$

(4) $1\frac{1}{4} \div 2$

(5) $1\frac{1}{4} \times 2$

5 Rosa is making chiles rellenos. Her recipe serves 6 people. She wants to serve 18 people. How many cups of chiles should Rose use?

(1) 3

(2) 6

(3) 12

(4) 18

(5) Not enough information is given.

6 Armand is mixing a cleaning solution. The directions say to add $1\frac{1}{2}$ cups of cleaning solution to 1 gallon of water. Armand plans to use 5 gallons of water. Which is the correct expression to find the number of cups of cleaning solution he needs?

(1) $1\frac{1}{2} + 5$

(2) $5 - 1\frac{1}{2}$

(3) $1\frac{1}{2} \times 5$

(4) $5 \div \frac{1}{2}$

(5) Not enough information is given.

7 **Reason** On a separate piece of paper, explain why multiplying the amount of each ingredient by $\frac{1}{2}$ is the same as dividing each amount by 2.

Multiplying Fractions and Mixed Numbers

Multiplying Fractions

You do not need to find a common denominator to multiply and divide fractions. To multiply one fraction by another, multiply the numerators. Then multiply the denominators.

EXAMPLE Multiply: $\frac{2}{3} \times \frac{3}{8}$

Step 1. Multiply the numerators.

Step 2. Multiply the denominators.

Step 3. Reduce the answer to lowest terms.

$$\frac{2}{3} \times \frac{3}{8} = \frac{6}{?}$$
$$\frac{2}{3} \times \frac{3}{8} = \frac{6}{24}$$
$$\frac{6}{24} = \frac{6 \div 6}{24 \div 6} = \frac{1}{4}$$

A shortcut called **canceling** will make your work easier. Let's work the same problem again using the shortcut.

EXAMPLE Multiply: $\frac{2}{3} \times \frac{3}{8}$

Step 1. Look at the numerator of one fraction and the denominator of the other fraction. Think of a number that will evenly divide into both. Both 2 and 8 can be divided by 2. Divide both by 2. Cross out the 2 and write 1 to show $2 \div 2 = 1$. Cross out the 8 and write 4 to show $8 \div 2 = 4$.

$$\overset{1}{\cancel{2}} \times \frac{3}{\underset{4}{\cancel{8}}}$$

Step 2. Look at the numerator of the second fraction and the denominator of the other fraction and repeat the process. Since both numbers are 3, you can divide each by 3.

$$\frac{\overset{1}{\cancel{2}}}{\underset{1}{\cancel{3}}} \times \frac{\overset{1}{\cancel{3}}}{\underset{4}{\cancel{8}}} = \frac{1}{4}$$

Step 3. Multiply the new numerators. Multiply the new denominators.

> **TIP**
>
> Canceling is a way to simplify a problem with fractions before you multiply.
>
> $$\frac{1}{\cancel{2}} \times \frac{\overset{2}{\cancel{4}}}{5} = \frac{2}{5}$$

Multiplying Mixed Numbers

You can also multiply fractions by mixed numbers and whole numbers.

EXAMPLE Multiply: $4\frac{1}{2} \times 2\frac{1}{3}$

Step 1. Change the mixed numbers to improper fractions.

Step 2. Use canceling to reduce fractions.

Step 3. Multiply. If your answer is an improper fraction, simplify it by changing the improper fraction to a mixed whole number.

$$4\frac{1}{2} \times 2\frac{1}{3} = \frac{9}{2} \times \frac{7}{3}$$

$$\frac{\overset{3}{\cancel{9}}}{2} \times \frac{7}{\underset{1}{\cancel{3}}} = \frac{21}{2} = 10\frac{1}{2}$$

$$\underset{\text{Change.}}{4\frac{1}{2} \times 2\frac{1}{3}} \quad = \quad \underset{\text{Cancel.}}{\frac{9}{2} \times \frac{7}{3}} \quad = \quad \underset{\text{Multiply.}}{\frac{\overset{3}{\cancel{9}}}{2} \times \frac{7}{\underset{1}{\cancel{3}}}} \quad = \quad \underset{\text{Simplify.}}{\frac{21}{2}} \quad = \quad 10\frac{1}{2}$$

A. Multiply the fractions. Use canceling before you multiply, or reduce your answers to lowest terms. The first one is done for you.

1 $\dfrac{\overset{1}{\cancel{2}}}{3} \times \dfrac{1}{\underset{2}{\cancel{4}}} = \dfrac{1}{6}$

2 $\dfrac{5}{6} \times \dfrac{5}{8} =$

3 $\dfrac{3}{8} \times \dfrac{4}{5} =$

4 $\dfrac{3}{7} \times \dfrac{2}{9} =$

B. Multiply the fractions by whole numbers or mixed numbers. Use canceling before you multiply, or reduce your answers to lowest terms. The first one is started for you.

5 $\dfrac{3}{8} \times 6 = \dfrac{3}{8} \times \dfrac{6}{1} =$

6 $5 \times 2\dfrac{3}{10} =$

7 $\dfrac{5}{8} \times 2\dfrac{2}{5} =$

8 $3\dfrac{1}{2} \times \dfrac{8}{9} =$

9 $2\dfrac{1}{3} \times 1\dfrac{1}{2} =$

10 $1\dfrac{4}{5} \times 6\dfrac{2}{3} =$

C. Solve each problem. Show your work. Use canceling before you multiply, or reduce your answers to lowest terms.

11 On Saturday 24 people bought cars at Best Car Sales. Of the 24 customers, $\dfrac{3}{4}$ bought new cars. How many people bought new cars?

13 The Sugar Bowl ordered $15\dfrac{3}{4}$ pounds of almonds. The nut supplier sent only $\dfrac{2}{3}$ of the order. How many pounds of almonds did the supplier send?

12 Donna can walk at the rate of $3\dfrac{1}{2}$ miles per hour. If she keeps up this pace, how far can she walk in $1\dfrac{1}{2}$ hours?

14 Tony can pick $4\dfrac{1}{4}$ pints of berries per hour. How many pints can he pick in 6 hours?

Check your answers on page 282.

Dividing Fractions and Mixed Numbers

Dividing Fractions

Dividing fractions is the same as multiplying fractions—with one important difference. You must **invert** the fraction you are dividing by. When you invert the fraction you are dividing by, also change the division sign to a multiplication sign. **Note:** Do not cancel before inverting.

TIP

Invert means to turn over or to switch the numerator and the denominator. For example:

$\frac{2}{3}$ becomes $\frac{3}{2}$

$\frac{1}{4}$ becomes $\frac{4}{1} = 4$

$2 = \frac{2}{1}$ becomes $\frac{1}{2}$

EXAMPLE Divide: $\frac{3}{8} \div \frac{1}{2}$

Step 1. Invert the fraction you are dividing by. Change the $\div$ sign to $\times$.

$$\frac{3}{8} \div \frac{1}{2} = \frac{3}{8} \times \frac{2}{1}$$

Step 2. Use canceling to reduce both fractions. Multiply the fractions.

$$\frac{3}{\cancel{8}_4} \times \frac{\cancel{2}^1}{1} = \frac{3}{4}$$

Dividing Mixed Numbers

To divide by a mixed number or a whole number, change it to an improper fraction before you invert.

EXAMPLE Divide: $6 \div 2\frac{2}{5}$

Step 1. Change the whole number and mixed number to improper fractions.

$$6 \div 2\frac{2}{5}$$
$$6 = \frac{6}{1}$$
$$2\frac{2}{5} = \frac{2 \times 5}{5} + \frac{2}{5} = \frac{12}{5}$$
$$6 \div 2\frac{2}{5} = \frac{6}{1} \div \frac{12}{5}$$

Step 2. Invert the fraction you are dividing by, and change the $\div$ sign to $\times$.

$$\frac{6}{1} \times \frac{5}{12}$$

Step 3. Use canceling to reduce. Multiply.

$$\frac{\cancel{6}^1}{1} \times \frac{5}{\cancel{12}_2} = \frac{5}{2}$$

Step 4. Reduce your answer to lowest terms.

$$\frac{5}{2} = 2\frac{1}{2}$$

EXAMPLE Divide: $5\frac{1}{3} \div 4\frac{1}{2}$

Step 1. Change the mixed numbers to improper fractions. **Note:** Do not cancel before inverting.

$$5\frac{1}{3} \div 4\frac{1}{2}$$
$$5\frac{1}{3} = \frac{5 \times 3}{3} + \frac{1}{3} = \frac{16}{3}$$
$$4\frac{1}{2} = \frac{4 \times 2}{2} + \frac{1}{2} = \frac{9}{2}$$
$$5\frac{1}{3} \div 4\frac{1}{2} = \frac{16}{3} \div \frac{9}{2}$$

Step 2. Invert the fraction you are dividing by, and change the $\div$ sign to $\times$.

$$\frac{16}{3} \times \frac{2}{9}$$

Step 3. Multiply. (Note that there is no canceling to be done.)

$$\frac{16}{3} \times \frac{2}{9} = \frac{32}{27}$$

Step 4. Reduce your answer to lowest terms.

$$\frac{32}{27} = 1\frac{5}{27}$$

A. Divide the fractions. Use canceling whenever possible. Reduce your answers to lowest terms. The first one is done for you.

1 $\dfrac{4}{5} \div \dfrac{1}{3} = \dfrac{4}{5} \times \dfrac{3}{1}$

$= \dfrac{12}{5} = 2\dfrac{2}{5}$

2 $\dfrac{5}{6} \div \dfrac{5}{8} =$

3 $\dfrac{4}{15} \div \dfrac{4}{5} =$

4 $\dfrac{2}{3} \div \dfrac{5}{6} =$

B. Divide by whole numbers or mixed numbers. Use canceling whenever possible. Reduce your answers to lowest terms. The first one is started for you.

5 $\dfrac{7}{12} \div 3 = \dfrac{7}{12} \times \dfrac{1}{3} =$

6 $2 \div 1\dfrac{1}{7} =$

7 $\dfrac{5}{8} \div 1\dfrac{5}{6} =$

8 $3\dfrac{3}{4} \div \dfrac{3}{10} =$

9 $1\dfrac{2}{5} \div 2\dfrac{1}{4} =$

10 $4\dfrac{1}{4} \div 1\dfrac{1}{2} =$

C. Solve each problem. Show your work. Use canceling whenever possible. Reduce your answers to lowest terms.

11 The 15 acres of vacant land across the street from Dominic's house are being developed as building lots. Each lot is to be $\dfrac{5}{8}$ acre. How many building lots will there be?

12 Sharla bought $5\dfrac{1}{2}$ pounds of ground beef. Before freezing the ground beef, she plans to make it into hamburger patties that weigh $\dfrac{1}{4}$ pound each. How many patties can Sharla make?

13 A hiking trail in a national park is $6\dfrac{1}{4}$ miles long. Roger averages $2\dfrac{1}{2}$ miles per hour. How many hours will it take him to hike the trail?

14 Gina wants to have music playing continuously in her store during business hours. The store is open $12\dfrac{1}{2}$ hours every day. Each music tape lasts $1\dfrac{1}{4}$ hours. How many tapes does she need for each day?

Check your answers on page 282.

Make a Diagram

Sometimes making a diagram may help you understand and solve a problem.

EXAMPLE Tien works for a company that makes and repairs furniture. She cuts boards into lengths needed to make sofas. Tien has an order form that tells her to cut four pieces of lumber measuring $16\frac{1}{8}$ inches each. Tien knows that each cut she makes will waste $\frac{1}{8}$ inch of the board. Tien makes a diagram of the situation.

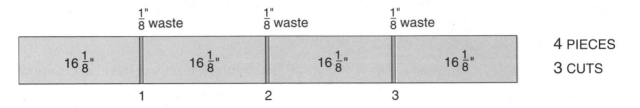

Look at the diagram above. Which expression finds the total amount of waste that will result from cutting the board?

(1) $3 \times \frac{1}{8}$

(2) $3 \div \frac{1}{4}$

(3) $3 \times 16\frac{1}{8}$

(4) $4 \times \frac{1}{8}$

(5) $4 \times 16\frac{1}{8}$

Answer (1) is correct. Tien will make three cuts; each cut will waste $\frac{1}{8}$ inch of wood. $3 \times \frac{1}{8} = \frac{3}{1} \times \frac{1}{8} = \frac{3}{8}$ **inch**, so $\frac{3}{8}$ **inch** will be wasted in the cutting.

EXAMPLE Tien wants to use 1 board for this order. Think of the number of pieces needed, the length of the pieces, and the amount of wood that will be wasted. In inches, what is the minimum length of board she could use?

(1) 48

(2) $64\frac{1}{2}$

(3) $64\frac{7}{8}$

(4) $64\frac{6}{8}$

(5) 65

Answer (3) is correct. Tien needs $64\frac{1}{2}$ inches for the four pieces of wood ($16\frac{1}{8} \times 4$) and $\frac{3}{8}$ inch for the waste. $64\frac{1}{2} + \frac{3}{8} = 64\frac{4}{8} + \frac{3}{8} = 64\frac{7}{8}$ **inches**, so Tien needs a board with a minimum length of $64\frac{7}{8}$ inches.

Circle the best answer for each item.

Items 1–3 refer to the following information.

Tien needs to cut five shelves for a cabinet. Each shelf is to be $1\frac{2}{3}$ feet long. She would like to cut the shelves from one piece of board and not have any wood left over. Each cut will waste $\frac{3}{16}$ inch of wood.

1 How many inches of wood are needed for five shelves?

   (1) 180
   (2) 100
   (3)   96
   (4)   60
   (5)   8

2 How many cuts are needed to make five shelves from one board? Make a diagram to support your answer.

   (1) 1
   (2) 2
   (3) 3
   (4) 4
   (5) 5

3 What is the total amount of waste (in inches) Tien will have from cutting the shelves?

   (1) $\frac{1}{4}$
   (2) $\frac{3}{8}$
   (3) $\frac{9}{16}$
   (4) $\frac{5}{8}$
   (5) $\frac{3}{4}$

Items 4–6 refer to the following information.

Miriam works at the same company Tien does. She needs to cut four boards, each $6\frac{7}{8}$-feet long. Miriam would like to cut the four boards from a single, longer board.

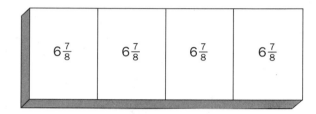

4 In feet, what is the total length of the four boards Miriam needs to cut?

   (1) $6\frac{7}{8}$
   (2) 24
   (3) $24\frac{7}{8}$
   (4) $27\frac{1}{2}$
   (5) 52

5 If a total of $2\frac{1}{4}$ inches will be wasted by the cuts made to the longer board, how many inches will be wasted with each cut?

   (1) $\frac{2}{3}$
   (2) $\frac{3}{4}$
   (3) $2\frac{1}{4}$
   (4) $2\frac{3}{4}$
   (5) $6\frac{3}{4}$

6 **Communicate** On a separate piece of paper, draw a diagram that shows how many cuts you would make to a board to get six smaller, but equal, boards.

Section 7 Review

Multiply or divide. Reduce your answers to lowest terms.

1 $\dfrac{4}{5} \times \dfrac{1}{3} = $ _____

2 $\dfrac{4}{9} \div \dfrac{1}{6} = $ _____

3 $\dfrac{5}{8} \div \dfrac{3}{5} = $ _____

4 $\dfrac{2}{7} \times \dfrac{5}{12} = $ _____

5 $7 \div \dfrac{7}{8} = $ _____

6 $6 \times \dfrac{3}{4} = $ _____

7 $\dfrac{4}{15} \times \dfrac{9}{10} = $ _____

8 $\dfrac{2}{5} \div \dfrac{4}{9} = $ _____

9 $\dfrac{11}{12} \times 8 = $ _____

10 $4 \times 2\dfrac{5}{6} = $ _____

11 $\dfrac{3}{8} \div 4 = $ _____

12 $5 \div 1\dfrac{1}{9} = $ _____

13 $\dfrac{7}{12} \times 3\dfrac{3}{5} = $ _____

14 $\dfrac{5}{6} \div 2\dfrac{1}{2} = $ _____

15 $2\dfrac{3}{4} \times \dfrac{2}{3} = $ _____

16 $4\dfrac{1}{2} \div 1\dfrac{1}{5} = $ _____

17 $3\dfrac{1}{5} \div 2\dfrac{2}{5} = $ _____

18 $2\dfrac{5}{12} \times 1\dfrac{5}{7} = $ _____

19 $2\dfrac{6}{7} \times 1\dfrac{5}{16} = $ _____

20 $3\dfrac{1}{3} \div 4\dfrac{1}{6} = $ _____

Check your answers on page 283.

Solve each problem. Show your work.

21 Delia is a secretary at Auto Insurance Agency. She works $7\frac{1}{2}$ hours a day. Delia answers the telephone $\frac{1}{3}$ of each day. How many hours does Delia spend answering the telephone each day?

22 Mahmud planted $4\frac{2}{3}$ dozen flowering plants in his garden. During the drought, he lost $\frac{1}{7}$ of his plants. How many dozen plants survived?

23 Miguel's recipe for jelly calls for $1\frac{3}{4}$ ounces of pectin. He has 14 ounces of pectin. How many batches of jelly could he make with this?

24 Maggie makes braided rugs to sell at craft shows. She needs to cut strips of fabric $1\frac{1}{8}$ inches wide for the rugs. How many strips can she cut from a piece of fabric 45 inches wide?

Circle the best answer for each item.

25 Richard cooks at a restaurant. He is making meatballs, and the recipe calls for $\frac{3}{4}$ cup of bread crumbs. He plans to double the recipe. Which is the correct expression to find the number of cups of bread crumbs he should use?

(1) $\frac{3}{4} \times \frac{1}{2}$

(2) $\frac{3}{4} \times 1$

(3) $\frac{3}{4} \times 2$

(4) $\frac{3}{4} \div 2$

(5) $2 \div \frac{3}{4}$

26 Carla's recipe for potato salad serves 18 people and calls for $1\frac{7}{8}$ cups of mayonnaise. Carla wants to make enough salad to serve 6 people. How many cups of mayonnaise should she use?

(1) $\frac{5}{16}$

(2) $\frac{5}{8}$

(3) $\frac{15}{16}$

(4) $3\frac{3}{4}$

(5) $5\frac{5}{8}$

27 John wants seven shelves for his bookcase. Each shelf is to be $2\frac{1}{2}$ feet long. From one long board, how many cuts are needed to make the seven shelves? Make a diagram to support your answer.

(1) 3

(2) 4

(3) 5

(4) 6

(5) 7

28 Brenda works at a fabric store. A customer is buying a fabric remnant that is $3\frac{1}{2}$ feet long. How long is the remnant in inches?

(1) 18

(2) 24

(3) 32

(4) 36

(5) 42

Math at Work

Construction: Carpenter

Some Careers in Construction

Carpet Installer removes any existing carpet and lays new padding and carpeting

Electrician installs, repairs, and rewires buildings for electricity

Painter prepares walls for painting; selects color and applies paint to surface

Tilesetter prepares floor for new tile; lays tiles in specific pattern

Carpenters use a variety of tools to construct a building.

Carpenters build many different types of structures. They measure, cut and fit together the materials and pieces used to make the building's framework—the walls, floors, roofs and spaces inside. Carpenters also work on the finishing touches inside buildings. They hang doors or cabinets, put on molding and baseboards, and build bookcases. Carpenters may also work on renovations and additions to existing buildings.

Carpenters must have excellent **math** and **measuring skills.** They apply these skills when they **read blueprints,** and measure and cut the materials used in the building's construction. Blueprints are line drawings that show the exact place and measurements of each part of the building. Blueprints frequently include **fractional measurements** that the carpenter must be able to interpret and use.

Look at the chart showing some careers in construction.

- Do any of the careers interest you? If so, which ones?

- What information would you need to find out more about those careers? On a separate piece of paper, write some questions that you would like answered. You can find out more information about those careers in the *Occupational Outlook Handbook* at your local library.

Carpenters often purchase the materials to be used on the job. Use the diagram below to answer the questions that follow.

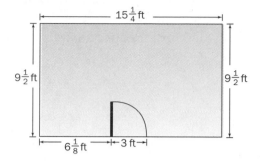

Jake has been hired to put baseboards and ceiling molding around the bedroom in the diagram above.

Definitions you will need:

Baseboard: decorative wood strips along the perimeter of a room. Strips do not go across doorways, but they do go under windows.

Ceiling molding: decorative wood strips placed along the ceiling of a room. Strips do run on ceiling area over doorways and windows.

1 What is the measurement of the wall to the right of the doorway?

(1) $15\frac{1}{4}$ feet

(2) $9\frac{1}{8}$ feet

(3) $6\frac{1}{8}$ feet

(4) 3 feet

(5) Not enough information is given.

2 How many feet of baseboard does he need for this room?

(1) $49\frac{1}{2}$

(2) $46\frac{1}{2}$

(3) $30\frac{1}{2}$

(4) 19

(5) Not enough information is given.

3 What is the total length of ceiling molding and baseboard that Jake needs for this room? Use a separate piece of paper. Explain how you got your answer.

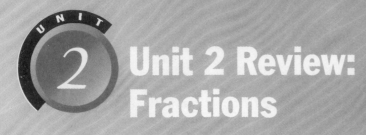

Unit 2 Review: Fractions

Solve. Show your work. Reduce answers to lowest terms.

1 $\dfrac{5}{6} + \dfrac{3}{4}$

2 $\dfrac{9}{10} - \dfrac{2}{3}$

3 $\begin{aligned} 4\tfrac{7}{8} \\ +\,2\tfrac{1}{3} \end{aligned}$

4 $\begin{aligned} 3\tfrac{2}{5} \\ -\,1\tfrac{3}{4} \end{aligned}$

5 $\dfrac{7}{12} \div \dfrac{5}{8}$

6 $\dfrac{15}{16} \times \dfrac{8}{9}$

7 $3\tfrac{1}{3} \times 2\tfrac{1}{2}$

8 $3\tfrac{3}{8} \div 3\tfrac{3}{5}$

Write your answers in the blanks. Show your work.

9 The Rockets lost 3 of the 20 games played this season. What fraction of the games did the Rockets lose?

_____

10 Raul works at a garden supply store. The store receives flower seeds in 30-ounce boxes. Raul repackages the seeds in $\tfrac{3}{4}$-ounce packets. How many packets can Raul fill from each box?

_____

11 Keiko needs a piece of wood that is $55\tfrac{1}{2}$ inches long to trim a window. How many feet of wood does she need?

_____

12 Bob bought a jumbo roll of wrapping paper. The paper on the roll is $8\tfrac{1}{3}$ yards long. How many feet of paper are on the roll?

_____

Check your answers on page 284.

Item 13 refers to the following map.

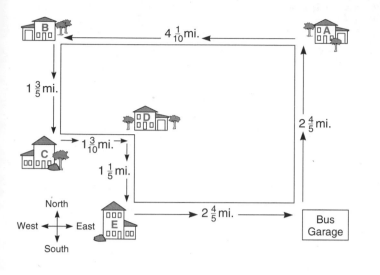

13 Henry drives a school bus. He completes his route at point C. Find how many miles he must drive to get from the end of his route to the bus garage.

Item 14 refers to the following timecard.

NAME: GERI D' ANGELO
WEEK OF: 6/21–6/27

IN	MON. 6/21	8:51 A.M.
OUT	MON. 6/21	12:24 P.M.
IN	MON. 6/21	1:08 P.M.
OUT	MON. 6/21	5:03 P.M.

14 How many hours did Geri work on Monday?

Circle the best answer for each item.

Items 15 and 16 refer to the following table.

Yardage Requirements: Skirt and Blouse

Sizes	10	12	14
Blouse A	$2\frac{3}{4}$	$2\frac{7}{8}$	3
Blouse B	$2\frac{7}{8}$	3	$3\frac{1}{4}$
Skirt A, B	$1\frac{1}{4}$	$1\frac{1}{4}$	$1\frac{5}{8}$

15 Gloria is making a skirt and blouse A in size 12. How many yards of fabric does she need for the skirt and blouse?

(1) $1\frac{3}{4}$

(2) $3\frac{1}{8}$

(3) $4\frac{1}{8}$

(4) $4\frac{1}{4}$

(5) $4\frac{5}{8}$

16 Pearl is trying to decide whether to make blouse B in size 14 or skirt B in size 14. Which is the correct expression to find how many more yards of fabric she will need for the blouse than for the skirt?

(1) $3\frac{1}{4} + 1\frac{5}{8}$

(2) $3\frac{1}{4} + 1\frac{1}{4}$

(3) $3\frac{1}{4} - 1\frac{1}{4}$

(4) $3\frac{1}{4} - 1\frac{5}{8}$

(5) $3 - 1\frac{5}{8}$

17 Clara's recipe for bread pudding serves 8 people. The recipe calls for $2\frac{1}{3}$ cups of milk. Clara wants to serve 16 people. Which is the correct expression to find the number of cups of milk she should use?

(1) $2\frac{1}{3} \times \frac{1}{2}$

(2) $2\frac{1}{3} \times 2$

(3) $2\frac{1}{3} \div 2$

(4) $2\frac{1}{3} \times 16$

(5) $16 \div 2\frac{1}{3}$

Math Extension Find 10 different examples showing fractions used in everyday life. Start with recipes and food labels. You might also look at instructions for home-improvement projects and sewing patterns. Find out how fractions are used in lumber yards and in sports.

Math Connection: Fractions and Drama

Do you like going to the movies? As the numbers below show, lots of us do. Movie studios count on us coming to see the films they produce. They plan that the production costs for a movie should only account for a <u>fraction of</u> the amount the film *grosses*, or earns, at the box office through the sale of tickets.

The most expensive movie made through 1998 was *Titanic*. Its titanic production costs are said to have exceeded $200 million.

The amount of money movies earn can vary a great deal. Compare the earnings of the #1 ranked film, *Titanic*, to the #2 ranked film, *Star Wars*. Produced and released twenty years earlier, *Star Wars* earned $\frac{1}{4}$ less than *Titanic*. Now compare the earnings of the #1 ranked film, *Titanic*, to the #10 ranked film, *Home Alone*. Although they were released in the same decade, *Home Alone* earned less than $\frac{1}{2}$ of the amount that *Titanic* earned.

Drama: The Art of Film

Motion pictures, or movies as we know them, first made their appearance in the late 19th century. These first films, made by individuals, usually lasted only a few minutes. Occasionally, a film would run as long as 10 minutes.

Nowadays, big movie studios produce most of the movies we see in the theaters. They run an average of 100 minutes or more. Thus, most of the films made during the first 20 years of motion picture history ran only $\frac{1}{50}$ to $\frac{1}{10}$ of the time that today's movies run.

Movie Facts as of January 1, 1999

Rank	Title	Box Office Earnings (in millions)	Running Time (in minutes)
1	*Titanic* (1997)	$600	201
2	*Star Wars** (1977)	461	125
3	*E. T. the Extra-Terrestrial* (1982)	400	115
4	*Jurassic Park* (1993)	357	127
5	*Forrest Gump* (1994)	329	120
6	*The Lion King* (1994)	313	88
7	*Return of the Jedi** (1983)	309	135
8	*Independence Day* (1996)	305	145
9	*The Empire Strikes Back** (1980)	290	127
10	*Home Alone* (1990)	285	102

* Part of the *Star Wars* Trilogy

Use the material on the previous page and a calculator to answer the questions that follow.

1 Which fraction expresses the number of the top ten movies released before 1990? Hint: Start this way: $\frac{\text{Number Before 1990}}{\text{Total Number of Movies}} = $ _____

(1) $\frac{1}{10}$

(2) $\frac{2}{5}$

(3) $\frac{1}{2}$

(4) $\frac{3}{4}$

(5) $\frac{7}{10}$

2 *Titanic* represents <u>approximately</u> what fraction of the total dollars earned by the top 5 movies? (Round the **Box Office Earnings** for each movie to the nearest hundred dollars to make the numbers easier to work with.)

(1) $\frac{1}{2}$

(2) $\frac{1}{3}$

(3) $\frac{1}{4}$

(4) $\frac{1}{8}$

(5) $\frac{1}{10}$

3 Three of the all time, top-grossing movies comprise the *Star Wars* trilogy. Use your calculator and rounding skills to find the fraction that best represents the <u>dollars earned</u> by the *Star Wars* trilogy compared to the <u>total earnings</u> of the top 10 movies?

(1) $\frac{1}{6}$

(2) $\frac{1}{3}$

(3) $\frac{1}{2}$

(4) $\frac{2}{3}$

(5) $\frac{3}{4}$

4 The fraction $\frac{1}{2}$ best represents the <u>running time relationship</u> between which pair of movies? Use estimation to find your answer.

(1) *Titanic/Star Wars*

(2) *Home Alone/Titanic*

(3) *The Lion King/Forrest Gump*

(4) *Independence Day/Titanic*

(5) *Return of the Jedi/The Empire Strikes Back*

5 Have you seen any of the top 10 movies? Is one of the movies a favorite of yours? Using a separate piece of paper, name the movie and explain why it you like it.

Cumulative Review

Compare each pair of numbers. Write >, <, or = to make each expression true.

1 32,774 _____ 3,274

2 948,526 _____ 948,256

Solve. Show your work.

3 $x + 19 = 53$

4 $\frac{x}{9} = 4$

5
$$\begin{array}{r} \$852 \\ -\ 398 \end{array}$$

6
$$\begin{array}{r} 594 \\ \times 607 \end{array}$$

7
$$\begin{array}{r} 900 \\ -417 \end{array}$$

8 $38\overline{)7{,}843}$

9 $684 + 35 + 257 =$

10 $\$3{,}807 \div 9 =$

11 $\frac{2}{3} + \frac{5}{8} =$

12 $\frac{5}{9} \div \frac{5}{6} =$

13
$$\begin{array}{r} 3\frac{4}{5} \\ +5\frac{1}{3} \end{array}$$

14
$$\begin{array}{r} 6\frac{2}{3} \\ -2\frac{3}{4} \end{array}$$

15 $6\frac{2}{3} \div 1\frac{7}{9} =$

16 $2\frac{5}{8} \times 1\frac{5}{7}$

Check your answers on pages 285–286.

Write your answers in the blanks. Show your work.

17 Use the chart on page 55 to find the value of $\sqrt{324}$.

18 Round 5,648,312 to the nearest hundred thousand.

Item 19 refers to the following graph.

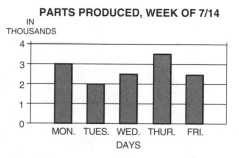

PARTS PRODUCED, WEEK OF 7/14

19 What is the approximate difference in the number of parts produced on the day with the highest production and the day with the lowest production?

20 Eric is putting a wallpaper border around his rectangular dining room. The room is 12 feet wide and 18 feet long. How many feet of border does he need?

21 What is the volume in cubic inches of a packing carton shown here?

Circle the best answer for each item.

22 On Friday, Lupe bought a pair of overalls for $36. He also put on layaway a jacket that costs $85. He paid $15 to hold the jacket. Which is the correct expression to find how much Lupe spent on Friday?
  (1) $36 + $85 + $15
  (2) $36 + $85 − $15
  (3) $36 + $85
  (4) $36 + $15
  (5) $85 − $15

23 Sally is a cook at The Corner Deli. On Monday, she made 63 sandwiches on wheat bread, 89 sandwiches on rye bread, and 45 sandwiches on white bread. Estimate the total number of sandwiches she made.
  (1) 185
  (2) 190
  (3) 200
  (4) 215
  (5) 225

24 Tawanna works at *I Was Framed*. She is making a picture frame and needs $54\frac{3}{4}$ inches of wood. Which is the correct expression to find the number of feet of wood she needs?
  (1) $54\frac{3}{4} \div 3$
  (2) $54\frac{3}{4} \div 12$
  (3) $54\frac{3}{4} \times 12$
  (4) $54\frac{3}{4} \div 36$
  (5) $54\frac{3}{4} \times 36$

25 During one week, the employees at Bronson Books worked the following hours: 38, 42, 35, 33, and 27. What is the mean number of hours worked?
  (1) 25
  (2) 34
  (3) 35
  (4) 37
  (5) 175

Check your answers on page 286.

3

Decimals

You already know a lot about **decimals.** In fact, you probably use decimals every day. You use decimals every time you use or make decisions about money. You even compare decimal amounts when you compare prices.

Think about how often you use decimals to count or measure something. Miles on road signs and on the odometer in your car are measured in decimals. Gasoline for your car is pumped in decimal amounts. Food is often measured in decimals, although many of our measurements, like ounces and pounds (and inches and feet), are not. However, the metric system of measurement, as we will learn later, is based completely on decimals.

◐ The next time you shop for groceries, take a closer look at how decimals are used. Besides the cost of an item, how are decimals used?

◐ Decimal use isn't limited to money and measurements. Scan a copy of your local newspaper. How are decimals used? What do those decimals describe?

SECTIONS

⑧ **Decimal Basics**

⑨ **Adding and Subtracting Decimals**

⑩ **Multiplying and Dividing Decimals**

Decimal Basics

Money and Decimals

EXAMPLE Suppose you stop to put gas in your car at the local gas station. You look at the pump and find you owe $9.75 for 8.2 gallons of gasoline.

 A decimal, like a fraction, shows part of a whole number. In fact, decimals and fractions can represent the same amount.

- Decimals are fractions that use the place value system.

Our money system is based on decimals. Dollars are represented by whole numbers. Cents represent 100 parts of a dollar and are written as decimals.

Which is the correct statement to describe the 7 in $9.75?

(1)   7 cents
(2)   7 dollars
(3)   70 cents
(4)   70 dollars
(5)   700 cents

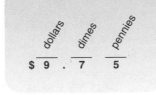

Answer **(3)** is correct. The 75 means 75 cents or 75 out of 100 parts of a dollar. The 7 has a value of **70/100** of a dollar, or **70 cents.** This also means 7 dimes since a dime is 10 out of 100 parts of a dollar.

EXAMPLE The chart below shows the decimal place values when the decimal is not a money amount.

Which is the value of the 5 in 9.75?

(1) five tenths
(2) five hundredths
(3) five thousandths
(4) five ten thousandths
(5) five hundred thousandths

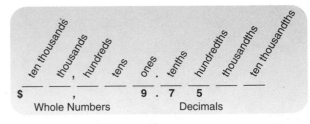

Answer **(2)** is correct. Look at the place value chart. The 5 is two places to the right of the decimal point. It has a value of **five hundredths.**

Circle the best answer for each item.

1 What is the value of 6 in $6.54?

(1)   6 cents
(2)   6 dimes
(3)   6 dollars
(4)   60 dollars
(5) 600 dollars

2 What is the value of 4 in $1.45?

(1)   4 cents
(2)   4 dollars
(3)   40 cents
(4)   45 cents
(5) 400 cents

3 What is the value of 2 in $28.31?

(1)   2 dollars
(2)   20 cents
(3)   20 dollars
(4) 200 cents
(5) 200 dollars

4 Ramon wrote an amount of money on a check. How should Ramon write out this amount?

1665

$49.55

DOLLARS

(1) 49 dollars and 5 cents
(2) 49 dollars and 50 cents
(3) 49 dollars and 55 cents
(4) 49 dollars and 550 cents
(5) Not enough information is given.

Items 5–7 refer to the place value chart on page 116.

5 What is the value of 6 in the decimal 14.1736?

(1) six tenths
(2) six hundredths
(3) six thousandths
(4) six ten thousandths
(5) six hundred thousandths

6 What is the value of 2 in the decimal 183.0625?

(1) two tenths
(2) two hundredths
(3) two thousandths
(4) two ten thousandths
(5) two hundred thousandths

7 What is the value of 1 in the decimal 364.1507?

(1) one tenth
(2) one hundredth
(3) one thousandth
(4) one ten thousandth
(5) one hundred thousandth

8 Communicate On a separate piece of paper, explain how the number of zeros in multiples of ten (10, 100, 1,000, etc.) can help you remember the decimal place names in the decimal system. For example:

Ten (10) has one zero, so one place after the decimal point (.1) is the *ten*ths place.

Explain the next three places to the right of the decimal point (.01, .001, .0001).

Reading and Writing Decimals

You can use what you know about place value to read and write decimals.

EXAMPLE   Follow these steps to read the decimal 3.82.

Step 1:	Say the whole number (3).	"Three"
Step 2:	Say the word *and* for the decimal point.	"Three and"
Step 3:	Say the number to the right of the decimal point (82) as though it were a whole number.	"Three and eighty-two"
Step 4:	Say the place name for the last digit.	"Three and eighty-two hundredths"

TIP

The place value of the last digit of a decimal names its value.

.003 means
3 thousandths

EXAMPLE   Write *sixteen and twelve thousandths* using digits.

Step 1:	Write the whole number (16).	16
Step 2:	Write a decimal point for the word *and*.	16.
Step 3:	Think of how many places after the decimal point you need. "Thousandths" is three places after the decimal point. The zero in the tenths place has no value, but is needed as a placeholder.	16.012

TIP

A zero before (to the left of) the decimal point means the decimal has no whole number part.

0.135 is less than one

When the decimal has no whole number part, you just say the decimal name. 0.35 is "thirty-five hundredths."

PRACTICE

Write each decimal in words. The first one is done for you.

1. 4.17 _____ *four and seventeen hundredths* _____

2. 0.256 _____

3. 2.09 _____

4. 6.805 _____

Write each decimal using digits. The first one is started for you.

5. Twenty-four and three hundred fifty-six thousandths _____ 24 _____

6. Three and seventy-eight hundredths _____

7. Four hundred ninety-one thousandths _____

8. Two hundred sixty-seven and three tenths _____

9. Fourteen and one thousand seven hundred thirty-six ten

   thousandths _____

10. Five and seven thousand eighty-four ten thousandths _____

Check your answers on page 287.

Relating Decimals and Fractions

You can change decimals to fractions.

EXAMPLE Change 0.25 to a fraction in lowest terms.

Step 1: For the numerator of the fraction, write the original number without its decimal point.

$$\frac{25}{}$$

Step 2: The denominator is the place value of the decimal. The place value of 0.25 is hundredths.

$$\frac{25}{100}$$

Step 3: Reduce the fraction to lowest terms.

$$\frac{25}{100} = \frac{1}{4}$$

EXAMPLE Change 0.0064 to a fraction in lowest terms.

TIP

Remember the zeros in the decimal 0.0064 have no value; they are only placeholders.

Step 1: Write the numerator of the fraction.

$$\frac{64}{}$$

Step 2: Write the place value of the decimal. The place value of 0.0064 is ten thousandths.

$$\frac{64}{10,000}$$

Step 3: Reduce the fraction to lowest terms.

$$\frac{64}{10,000} = \frac{64 \div 16}{10,000 \div 16} = \frac{4}{625}$$

If the decimal is a mixed decimal, that is, has a whole number part, the whole number is placed in front of the fraction. $1.25 = 1\frac{1}{4}$

PRACTICE

Change each decimal to a fraction in lowest terms. The first one is done for you.

1 0.5
$0.5 = \frac{5}{10} = \frac{1}{2}$

2 0.45

3 0.15

4 0.64

5 0.125

6 1.8

7 0.10

8 0.005

9 0.32

10 0.375

11 2.319

12 0.004

13 3.028

14 0.85

15 0.625

16 4.0024

Comparing and Rounding Decimals

Comparing Decimals

Would you rather pay $0.79 or $0.99 per pound for apples? You would pick $0.79 per pound because it is the smaller amount. To answer the question, you compared two decimals and found the smaller one.

EXAMPLE   Compare 0.325 and 0.5

Step 1:	Line up the decimal points.	0|325 0|5
Step 2:	If the two decimals have a different number of places, add zeros to the end of the decimal with the fewer digits so that the two decimals have the same number of digits.	0.325 0.5**00**
Step 3:	Compare as you would with whole numbers. Since 325 is less than 500, 0.325 is less than 0.500.	0.325 0.500 **0.325 < 0.5** or **0.5 > 0.325**

> **TIP**
>
> You can add zeros after the last digit in a decimal without changing the value of the decimal.
>
> .500 is the same as .5

Rounding Decimals

You can follow the same steps to round decimals as you did with whole numbers.

EXAMPLE   Round 5.1628 to the nearest thousandth.

Step 1:	Underline the digit in the place you want to round the number. (2 is in the thousandths place.)	5.16<u>2</u>8
Step 2:	Look at the digit to the right of the underlined digit.	5.16<u>2</u>8
Step 3:	Decide if the underlined digit needs to change. Since 8 is greater than 5, add 1 to the underlined digit.	5.16<u>3</u> _
Step 4:	Drop the digit to the right of the underlined digit.	5.163

> **TIP**
>
> If the number to the right of the underlined digit is less than 5, do not change the underlined digit; if it is 5 or more, add 1 to the underlined digit.

A. Compare each pair of numbers. Write >, <, or =. The first one is done for you.

1 0.27 __<__ 0.72   **2** 0.43 _____ 0.09   **3** 0.73 _____ 0.542

4 8.058 _____ 8.58   **5** 2.58 _____ 2.580   **6** 53.005 _____ 52.008

7 0.863 _____ 0.9   **8** 2.001 _____ 2.01   **9** 1.32 _____ 1.319

10 5 _____ 5.00   **11** 0.95 _____ 0.954   **12** 0.005 _____ 0.05

13 0.43 _____ 0.34   **14** 0.54 _____ 0.054   **15** 5.2 _____ 5.200

B. Round each decimal to the given place value. The first one is done for you.

16 Round 6.3782 to the nearest hundredth   _6.38_

17 Round 8.276 to the nearest whole number   _____

18 Round 46.3518 to the nearest tenth   _____

19 Round 71.0483 to the nearest thousandth   _____

20 Round 56.2819 to the nearest hundredth   _____

21 Round 0.5462 to the nearest tenth   _____

22 Round 101.667 to the nearest whole number   _____

23 Round 25.0235 to the nearest thousandth   _____

C. Solve each problem.

24 Last season Adrian had a batting average of 0.215 and Angel had a batting average of 0.213. Who had the better batting average?

25 The population of Sumner City is 95.7 million. What is the population of Sumner City to the nearest million?

26 A box of Brand A cereal costs $2.83. A box of Brand B cereal costs $2.79. Both boxes weigh the same. Which cereal costs less?

27 Saban stocks the shelves at a local health food store. The store buys spices in large quantities, then sells them in smaller quantities. Saban has written down the weights of the newly packaged garlic powder. They are

0.75 oz., 0.57 oz., 0.69 oz., and 0.5 oz.

If Saban wants to record the weights of the garlic powder from lightest to heaviest, in what order would these amounts be listed?

MEASUREMENT AND GEOMETRY

Using Metric Measurement

EXAMPLE Charlie is building fences around flower beds that are in front of an office building. The measurements on the landscaping plans are in meters. The plan for one of the flower beds is shown at the right.

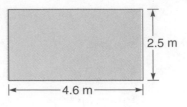

 A meter (m) is a measure of length in the metric system. A meter is a little longer than a yard.

 Parts of a meter are expressed as decimals.

Charlie sees that the flower bed is a rectangle that measures 4.6 meters long and 2.5 meters wide. Which is the correct expression to find the perimeter of the flower bed in meters? Remember, perimeter is the distance around a figure.

(1) 4.6 + 2.5
(2) 4.6 + 2.5 + 2.5 + 2.5
(3) 4.6 + 4.6 + 2.5 + 2.5
(4) 4.6 + 4.6 + 4.6 + 4.6
(5) 4.6 + 4.6 + 2.5 + 2.5 + 2.5

```
  4.6
  4.6
  2.5
 +2.5
 14.2
```

Answer **(3)** is correct. Charlie needs to find the total of all the sides, so he adds all four lengths. **4.6 + 4.6 + 2.5 + 2.5 = 14.2 meters,** so Charlie needs **14.2 meters** of fencing.

EXAMPLE Charlie decides to leave an 0.8-meter opening in the center of one of the 4.6-meter sides. He wants to find how much fencing he will need for the side with the opening. Which is the correct expression to find how much fencing Charlie will need for this side?

(1) 4.6 + 0.8
(2) 4.6 − 0.8
(3) 0.8 − 4.6
(4) 4.6 + 4.6 + 0.8 + 0.8
(5) 4.6 + 4.6 + 4.6 − 0.8

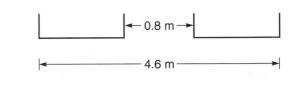

```
  3 10
  4̸.6
 −0.8
  3.8
```

Answer **(2)** is correct. Charlie needs to find the difference between the length of the side and the opening, so he subtracts. **4.6 − 0.8 = 3.8 meters,** so Charlie needs **3.8 meters** of fencing for the side.

Circle the best answer for each item.

Items 1–2 refer to the flower bed shown below.

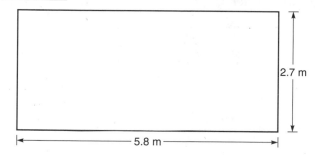

Items 4–5 refer to the playground shown below.

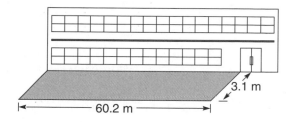

1 Shauna works for a landscaping service. She is putting edging around the flower bed shown here. Which is the correct expression to find the perimeter of the flower bed?

(1) 5.8 + 2.7
(2) 5.8 + 2.7 + 2.7
(3) 5.8 + 5.8 + 2.7
(4) 5.8 + 5.8 + 2.7 + 2.7
(5) 5.8 + 5.8 + 5.8 + 5.8

2 Shauna decides to leave a 0.7-meter opening in the center of one of the 2.7-meter sides. Which is the correct expression to find how many meters of edging Shauna will need for this side?

(1) 2.7 + 0.7
(2) 2.7 − 0.7
(3) 2.7 + 2.7 + 0.7 + 0.7
(4) 5.8 + 5.8 + 2.7 + 2.7
(5) 5.8 + 5.8 + 2.7 − 2.7

3 Luis wants to put a fence around a rectangular swimming pool that is 9.6 meters long. How many meters of fencing will he need?

(1)   9.6
(2) 19.2
(3) 28.8
(4) 38.4
(5) Not enough information is given.

4 Stan is a principal at Washington Elementary. He wants to build a playground area against the side of the school. If the playground area is roped off, how many meters of rope are needed?

(1) 60.2 + 3.1
(2) 60.2 − 3.1
(3) 60.2 + 3.1 + 3.1
(4) 60.2 + 60.2 + 3.1
(5) 60.2 + 60.2 + 3.1 + 3.1

5 Stan may want to paint a yellow strip (a place where teachers will stand) that is 15.8 meters long in the center of the 60.2-meter side. Which is the correct expression to find how many meters of the 60.2-meter side will not have a yellow strip?

(1) 15.8 + 60.2
(2) 15.8 − 60.2
(3) 60.2 − 15.8
(4) 60.2 + 3.1 + 3.1 − 15.8
(5) Not enough information is given.

6 **Extend** Measure the length and width of the top of your desk in meters. On a separate piece of paper, write these measurements to the nearest tenth and to the nearest hundredth of a meter.

Section 8 Review

Write the place value of each underlined digit.

1 $3.2<u>7</u> _____

2 14.078<u>2</u> _____

3 79.58<u>6</u> _____

4 $0.<u>5</u>1 _____

Write each decimal in words.

5 0.82 _____

6 3.513 _____

7 18.3406 _____

8 790.004 _____

Write each decimal using digits.

9 two and forty-three hundredths _____

10 two thousand three hundred forty-three ten thousandths _____

11 thirty-seven and six hundred twenty-one thousandths _____

12 four hundred twenty-nine and nine hundredths _____

Change each decimal to a fraction or mixed number in lowest terms.

13 0.05

14 0.28

15 0.75

16 0.025

17 0.96

18 3.205

19 5.0008

20 9.35

Compare each pair of numbers. Write >, <, or =.

21 0.759 _____ 0.795

22 0.326 _____ 0.54

23 0.6 _____ 0.60

24 32.574 _____ 32.547

Round each decimal to the given place value.

25 Round 3.196 to the nearest hundredth. _____

26 Round 6.453 to the nearest tenth. _____

27 Round 0.724 to the nearest whole number. _____

28 Round 42.9358 to the nearest thousandth. _____

Check your answers on page 288.

Circle the best answer for each item.

29 Julio's doctor prescribed sixteen thousandths of a gram of a medicine for him. How do you write *sixteen thousandths?*

(1) 0.0016
(2) 0.016
(3) 0.16
(4) 1,600
(5) 16,000

30 Lucy's vegetable garden is a rectangle that measures 5.2 meters long by 3.4 meters wide. Which is the correct expression to find the perimeter of the garden?

(1) 5.2 + 3.4
(2) 5.2 + 3.4 + 3.4
(3) 5.2 + 5.2 + 3.4
(4) 5.2 + 5.2 + 3.4 + 3.4
(5) 5.2 + 5.2 + 5.2 + 3.4

31 Calvin's Upholstery Service had 4.2 meters of fabric. He used 2.7 meters of fabric to recover a chair. Which expression shows how many meters of fabric Calvin has left?

(1) 4.2 − 2.7
(2) 4.2 + 2.7
(3) 4.2 ÷ 2.7
(4) 2.7 ÷ 4.2
(5) 4.2 × 2.7

32 Al ran a 100-meter race in 10.25 seconds; Ron finished in 10.06 seconds. Which statement is true about the race?

(1) Al finished before Ron.
(2) Both finished the race at the same time.
(3) Al's time is faster than Ron's time.
(4) Ron finished before Al.
(5) Ron's time is slower than Al's time.

33 Masami is a butcher at Fresh Cuts. He packaged a piece of meat that weighed 3.059 pounds. Which words best describe the number of pounds of meat in that package?

(1) three and fifty-nine hundredths
(2) three and fifty-nine thousandths
(3) three and five hundredths
(4) three and nine thousandths
(5) three and five hundred nine thousandths

34 Compare 24.6305 and 24.6350.

(1) 24.6305 > 24.6350
(2) 24.6350 = 24.6305
(3) 24.6350 < 24.6305
(4) 24.6350 > 24.6305
(5) 24.6305 = 24.6305

35 Last week Mary Lou paid $1.325 per gallon of gasoline. Rounded to the nearest cent, how much did she spend for a gallon of gasoline?

(1) $1.00
(2) $1.30
(3) $1.32
(4) $1.33
(5) $1.40

36 Betty finished reading 0.4 of a new mystery. What fraction of the book has Betty read so far?

(1) $\frac{4}{5}$

(2) $\frac{2}{5}$

(3) $\frac{1}{25}$

(4) $\frac{1}{250}$

(5) Not enough information is given.

Adding and Subtracting Decimals

Multi-Step Problems

The key to solving a problem with more than one step is to think through the problem and decide what to do in each step *before* you solve the problem.

EXAMPLE Eva works at a flower shop. She records each sale on a sales slip.

Ms. Morris is buying two flower arrangements: one costs $19.95 and the other costs $52.75. Eva needs to find three numbers: *SUBTOTAL*, *TAX*, and *TOTAL*.

● A subtotal is a total of part of a group of numbers; here, it is the sum of the cost of the two flower arrangements before tax.

● Tax is additional money people must pay based on the amount of the subtotal.

● The total is the complete cost of the order, including tax.

Compton Floral 444 W. Orange Grove Margate, FL 33063		
Sm. Arrangement	$19	95
Large Arrangement	52	75
SUBTOTAL		
TAX		
TOTAL	$	

First, Eva finds the subtotal. Which is the correct expression to find the subtotal?
(1) $19.95 − $52.75
(2) $19.95 + $52.75
(3) $52.75 − $19.95
(4) $52.75 × $19.95
(5) $52.75 ÷ $19.95

Answer (2) is correct. Eva adds the cost of the two arrangements to get the subtotal. $19.95 + $52.75 = $72.70, so the subtotal is $72.70.

EXAMPLE Next Eva needs to find the total. She knows the sales tax for this subtotal is $4.91. Which is the correct expression to find the total cost of this sale?
(1) subtotal − tax
(2) subtotal + tax
(3) tax − subtotal
(4) subtotal × tax
(5) subtotal ÷ tax

Answer (2) is correct. Eva adds the tax to the subtotal to find the total cost of this sale. $72.70 + $4.91 = $77.61, so the total cost of the sale is $77.61.

Circle the best answer for each item.

Items 1–3 refer to the following sales slip.

Frank's Coats 526 E. Pine Chicago, IL 60611		
Sm. Jacket	$39	99
Large Jacket	58	00
SUBTOTAL		
TAX		
TOTAL	$	

1 Mrs. Wilson wants to buy two winter jackets for her children. Which is the correct expression to find the subtotal of this sale before tax?

(1) $39.99 + $58.00
(2) $39.99 − $58.00
(3) $58.00 ÷ $39.99
(4) $58.00 − $39.99
(5) $58.00 × $39.99

2 The tax for the items is $6.86. Which is the correct expression to find the total including tax of this sale?

(1) subtotal − $6.86
(2) $6.86 − subtotal
(3) subtotal + $6.86
(4) subtotal × $6.86
(5) subtotal ÷ $6.86

3 Mrs. Wilson decides to put the coats on layaway and pays $45.00 to hold them. Which is the correct expression to find how much Mrs. Wilson still owes on the coats?

(1) total − $45.00
(2) total + $45.00
(3) $45.00 − total
(4) total × $45.00
(5) total ÷ $45.00

4 Ervin works at a clothing store. His customer is buying a pair of jeans that costs $30.00, a sweatshirt that costs $24.00, and a T-shirt for $12.00. Which is the correct expression to find the subtotal of this sale?

(1) $30.00 − $24.00 + $12.00
(2) $30.00 + $24.00 + $12.00
(3) $24.00 − $30.00 − $12.00
(4) $30.00 × $24.00 × $12.00
(5) $30.00 ÷ $24.00 + $12.00

5 Lucia is a sales clerk at a camping store. A customer bought 3 shirts for $18 each and 2 plastic water containers for $5.99 each. Which is the correct expression to find the total of this sale before tax?

(1) (3 + $18) × (2 + $5.99)
(2) (3 × $18) × (2 × $5.99)
(3) (3 × $18) − (2 + $5.99)
(4) (3 + $18) + (2 × $5.99)
(5) (3 × $18) + (2 × $5.99)

6 Lucia needs to find the total including tax of the sale in Item 5. What is the total cost of the sale, including tax?

(1) $54.00 + $11.98
(2) $54.00 × $11.98
(3) $54.00 − $11.98
(4) $54.00 ÷ $11.98
(5) Not enough information is given.

7 **Communicate** A customer buys three items at the following prices: $16.00, $21.50, and $32.25. The sales tax is $5.23, and the customer gives $80.00 to a cashier to pay for the purchase. Explain in three steps without calculation how to find the correct change to give the customer.

Adding and Subtracting Decimals

Adding and subtracting decimals, including money, is like adding and subtracting whole numbers. The important step is to line up the decimal points.

EXAMPLE   Add: 4.6 + 14 + 3.38 + 0.125

Step 1:   Write the problem. Line up the decimal points.

$$\begin{array}{r} 4.6 \\ 14. \\ 3.38 \\ +\ 0.125 \end{array}$$

Step 2:   Put zeros to the right of the last decimal digit so the numbers all have the same number of digits.

$$\begin{array}{r} 4.600 \\ 14.000 \\ 3.380 \\ +\ 0.125 \end{array}$$

Step 3:   Add, regrouping as needed. Put the decimal point in the answer directly under the decimal points in the problem.

$$\begin{array}{r} {\scriptstyle 1\ 1\ 1} \\ 4.600 \\ 14.000 \\ 3.380 \\ +\ 0.125 \\ \hline 22.105 \end{array}$$

> **TIP**
>
> If necessary, add zeros so that all the numbers have the same number of digits. Remember, you can add zeros after the last digit in a decimal without changing the value.
>
> 4.600 = 4.6

EXAMPLE   Subtract: 15.2 − 9.75

Step 1:   Write the problem. Line up the decimal points.

$$\begin{array}{r} 15.2 \\ -\ 9.75 \end{array}$$

Step 2:   Put zeros to the right of the last decimal digit so both numbers have the same number of digits.

$$\begin{array}{r} 15.20 \\ -\ 9.75 \end{array}$$

Step 3:   Subtract, regrouping as needed. Put the decimal point in the answer.

$$\begin{array}{r} {\scriptstyle 4\ 11\ 10} \\ 15.20 \\ -\ 9.75 \\ \hline 5.45 \end{array}$$

Adding and Subtracting Decimals on a Calculator

To add or subtract decimals on a calculator, use the decimal point key at the appropriate place between the digits of the decimal.

EXAMPLE   50.32 + 4.236 − 24.005 + 3.06

CE/C 50 **.** 32 **+** 4 **.** 236 **−** 24 **.** 005 **+** 3 **.** 06

= 33 **.** 611

> For more practice adding and subtracting decimals with the calculator, see Calculator Handbook pages 248–249.

A. Add or subtract. The first one is done for you.

1
```
  3.726
+4.915
  8.641
```

2
```
 17.368
−  8.415
```

3
```
 24.28
−13.7
```

4
```
 28.467
+37.29
```

5
```
 12.05
−  6.8
```

6
```
  3.7
 14.24
+ 2.3
```

7
```
 24.81
 35.7
+28.274
```

8
```
  8.4
−0.31
```

9
```
 14.8
−  6.753
```

10
```
 12
+ 5.88
```

B. Add or subtract. Rewrite the problems below. Line up the decimal points. The first one is started for you.

11 $2.43 + 0.57 + 3.18 =$
```
 2.43
 0.57
+3.18
```

12 $67.31 − 49.826 =$

13 $57.43 − 29.5 =$

14 $0.3 + 21.508 + 3.4 =$

15 $4.26 + 5.1 + 8.39 =$

16 $12.6 − 8.897 =$

C. Solve the problems below with a calculator. Refer to pages 248–249 in the Calculator Handbook for additional information.

17 Phalla had a balance of $256.81 in her checking account. She wrote checks for $2.80, $34.33, and $62.18. She deposited $122.77 and $83.93. Find Phalla's new balance.

18 One weekend Roberto worked a few hours of overtime. He earned $43.50 on Saturday and $72.50 on Sunday. How much did Roberto earn that weekend?

Order of Operations

There are four **operations** to use with numbers. You can add, subtract, multiply, or divide. Numbers and operation symbols together make a **mathematical expression.** When an expression has more than one operation, the *order* in which you do the operations can change the answer. To solve an expression, always follow the order of operations:

Step 1: Do any operations in parentheses first.
Step 2: Multiply and divide working from left to right.
Step 3: Add and subtract working from left to right.

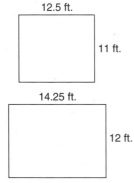

12.5 ft.

11 ft.

14.25 ft.

12 ft.

EXAMPLE Lee needs carpeting for two rooms. One room measures 12.5 feet by 11 feet. The other is 14.25 feet by 12 feet. Which is the correct expression to find the area of both rooms?
(1) $(12.5 + 11) + (14.25 + 12)$
(2) $(12.5 + 11) \times (14.25 + 12)$
(3) $(12.5 \times 11) - (14.25 \times 12)$
(4) $(12.5 \times 11) + (14.25 \times 12)$
(5) $(12.5 \times 11) \times (14.25 \times 12)$

Answer **(4)** is correct. Find the area of each room. (length $\times$ width) Then find the sum of the areas. (Add)

$(12.5 \times 11) + (14.25 \times 12) = 137.5 + 171 = 308.5$ square feet.

TIP

It may help to put the problem into words: "the total of two times the cost of one tire plus the oil change [in parentheses] is subtracted from the amount given to the clerk."

EXAMPLE Alicia bought 2 tires for $45.89 each. She also got an oil change for $22.25. Alicia gave the clerk $120. Which is the correct expression to find the amount of her change?
(1) $\$120 - 2 \times (\$45.89 + \$22.25)$
(2) $(\$120 - 2) \times (\$45.89 + \$22.25)$
(3) $\$120 - (2 \times \$45.89 + \$22.25)$
(4) $\$120 - 2 \times \$45.89 + \$22.25$
(5) $\$120 - (2 \times \$45.89) + \$22.25$

Answer **(3)** is correct. To find the amount of change, subtract the amount of the purchase from $120.

$$\$120 - (2 \times \$45.89 + \$22.25) = \$120 - (\$91.78 + \$22.25)$$
$$= \$120 - (\$114.03) = \$5.97.$$

Use the correct order of operations to show that the other four answer options give different incorrect answers.

Circle the best answer for each item.

1 Lorene earns $14.50 an hour. She worked 38 hours last week and 43 hours this week. Which is the correct expression to find Lorene's earnings in the 2-week period?

(1) $14.50 + (38 × 43)
(2) ($14.50 × 43) + 38
(3) ($14.50 × 38) + 43
(4) $14.50 × (38 + 43)
(5) $14.50 × 38 + 43

2 Miguel and Gina have $60 to spend on entertainment. They bought two movie tickets at $7.50 each and two drinks at $3.25 each. Which is the correct expression to find how much money they have left?

(1) $60 − 2 × $7.50 − $3.25
(2) $60 − (2 × $7.50) − (2 × $3.25)
(3) $60 − (2 × $7.50 − 2 × $3.25)
(4) $60 − (2 × $7.50 + $3.25)
(5) $60 + 2 × ($7.50 − $3.25)

Item 3 refers to the following table.

Bowling Scores	
Player 1	205
Player 2	165
Player 3	104
Player 4	190

3 Aida scored 40 points above the mean (average) score of the other four bowlers. Which is the correct expression to find Aida's score?

(1) 205 ÷ 4 + 40
(2) (205 + 40) ÷ 4 + 40
(3) (205 + 165 + 104 + 190) ÷ 4
(4) (205 + 165 + 104 + 190 + 40) ÷ 4
(5) (205 + 165 + 104 + 190) ÷ 4 + 40

4 Which expression describes the following: the sum of $6.50 and $9.25, multiplied by their difference?

(1) ($6.50 + $9.25) + ($9.25 − $6.50)
(2) ($6.50 + $9.25) × ($9.25 + $6.50)
(3) ($6.50 − $9.25) × ($9.25 − $6.50)
(4) ($6.50 + $9.25) + ($9.25 + $6.50)
(5) ($6.50 + $9.25) × ($9.25 − $6.50)

5 A customer at Fabric Factory bought 3 yards of fabric for $11.00 a yard and 5.5 yards of fabric for $8.40 a yard. Which is the correct expression to find the total cost of the fabric?

(1) (3 × $11.00) + (5.5 × $8.40)
(2) (3 + $11.00) + (5.5 + $8.40)
(3) (3 × $11.00) × (5.5 × $8.40)
(4) (3 × $11.00) − (5.5 × $8.40)
(5) Not enough information is given.

6 Glen has two part-time jobs. At one job he works 24 hours and earns $9.95 an hour. At the other job he works 18 hours and earns $11.75 an hour. Which expression finds how much Glen earns a week?

(1) 24 × ($9.95 + 18) × $11.75
(2) 24 × ($9.95 + $11.75) × 18
(3) $9.95 × (24 + 18) × $11.75
(4) (18 × $9.95) + (24 × $11.75)
(5) (24 × $9.95) + (18 × $11.75)

7 **Reason** On a separate piece of paper, use parentheses to rewrite and solve the expression 6 × 2 + 4 − 3 to give three different solutions.

DATA ANALYSIS   Reading a Line Graph

EXAMPLE   Dennis Stone works for the Department of Water and Power for the city of Los Angeles. His boss, Lupe Navarro, asks him to find the levels of rainfall for Los Angeles this year.

Dennis used the **line graph** at the right to answer questions about rainfall for Los Angeles. The graph shows the rainfall for the first 6 months of the year.

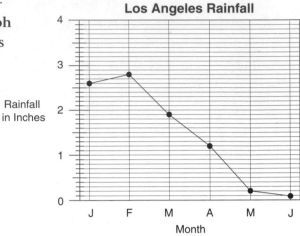

A line graph contains lines called **axes.** The **vertical axis** (up and down) is marked with a **scale.** The scale represents the number of inches of rainfall in Los Angeles. The longer marks on the scale represent whole numbers. The smaller marks divide each inch into tenths. The **horizontal axis** (across) is marked with letters representing the first 6 months of the year.

Ms. Navarro wants Dennis to find how many inches of rainfall the city received in March of this year. Which is the correct answer?

(1)   0.2
(2)   1.2
(3)   1.9
(4)   2.1
(5)   2.9

Answer **(3)** is correct. Dennis finds the point for March above the first M on the horizontal axis. He reads across to the scale on the vertical axis. The point represents **1.9 inches** of rainfall.

EXAMPLE   Next, Dennis needs to find the difference in rainfall between February and March. Which is the correct expression to find how many more inches of rainfall the city received in February than in March?

(1)   2.8 + 1.9
(2)   2.8 ÷ 1.9
(3)   2.8 × 1.9
(4)   2.8 − 1.9
(5)   1.9 ÷ 2.8

Answer **(4)** is correct. The rainfall for February is 2.8 inches. Subtract to find the difference between 2.8 inches and 1.9 inches (rainfall for March).
2.8 − 1.9 = 0.9 inch, so there was **0.9 inch** more rainfall in February than in March.

Circle the best answer for each item.

Items 1–3 refer to the following line graph.

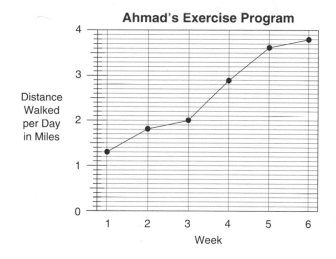

Ahmad's Exercise Program

① How many miles did Ahmad walk each day during week 4 of his exercise program?

(1) 2.4
(2) 2.9
(3) 3.1
(4) 3.9
(5) 29

② During which time period did Ahmad have the greatest increase in the distance he walked per day?

(1) from week 1 to week 2
(2) from week 2 to week 3
(3) from week 3 to week 4
(4) from week 4 to week 5
(5) from week 5 to week 6

③ How many miles farther did Ahmad walk each day during week 6 of his program than during week 1?

(1) 0.5
(2) 2.5
(3) 2.9
(4) 3.0
(5) 5.1

Items 4–6 refer to the following line graph.

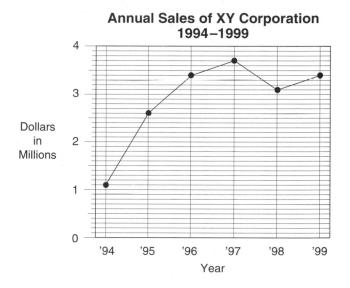

Annual Sales of XY Corporation 1994–1999

④ What were the annual sales of XY Corporation in 1995?

(1) $1.1 million
(2) $2.5 million
(3) $2.6 million
(4) $3.1 million
(5) $3.4 million

⑤ Diane wants to find the total sales from 1994 to 1999. What were the total sales from 1994 to 1999?

(1) $10.8 million
(2) $14.2 million
(3) $17.3 million
(4) $22.5 million
(5) Not enough information is given.

⑥ **Reason** Look at the graph above. On a separate piece of paper, explain what a steeper line means and what a flatter line means. (Hint: Think about increases and decreases.)

Section 9 Review

Add or subtract.

1 35.2
+41.84

2 74.3
−51.42

3 93.26
−14.336

4 15.93
+ 4.895

5 8.45
+15.316

6 4.26
+3.947

7 7.389
−0.49

8 5.9
−4.78

9 3.81
12.463
+ 5.4

10 42.18
53.7
+82.427

11 $12.372 - 9.473 =$

12 $7.37 - 5.893 =$

13 $8.25 + 9.47 + 3.06 =$

14 $15.43 + 2.8 + 3.816 =$

15 $4.245 + 1.97 + 2.8 =$

16 $3.8 - 1.924 =$

17 $124.897 + 23.8764 =$

18 $4,234.256 - 2,897.9825 =$

Circle the best answer for each item.

19 Robyn had a gift certificate for $100. She bought a pair of jeans for $30.00, a shirt for $24.00, and a scarf for $15.00. Sales tax was $5.18. Which is the correct expression to find how much money Robyn had left?

(1) $100 − $30.00 − $24.00

(2) $100 − ($30.00 + $5.18)

(3) $100 − ($30.00 + $24.00 − $5.18)

(4) $100 − ($30.00 + $24.00 + $15.00 + $5.18)

(5) $100 + ($30.00 − $24.00 − $5.18)

20 What is the value for the expression $4 + (2 \times 6) - (12 \div 2)$?

(1) 2

(2) 6

(3) 10

(4) 18

(5) 30

Check your answers on pages 290–291.

Items 21–24 refer to the following information.

George works in the electronics department of a large store. A customer is buying a portable tape player for $24.99 and a cassette tape for $4.99. Sales tax on this purchase is $2.10. The customer pays $40.00 in cash.

21 Estimate the subtotal before tax on this sale.

23 What is the total including tax on this sale?

22 What is the exact subtotal before tax on this sale?

24 How much change should George give his customer?

Circle the best answer for each item.

Items 25–27 refer to the following graph.

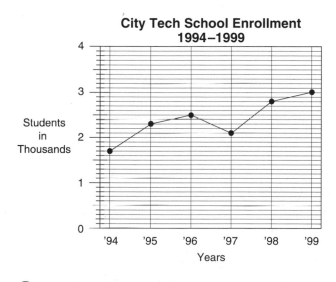

City Tech School Enrollment 1994–1999

26 Which best describes the greatest change in enrollment?
  (1) a decrease from 1994 to 1995
  (2) an increase from 1994 to 1995
  (3) a decrease from 1996 to 1997
  (4) a decrease from 1997 to 1998
  (5) an increase from 1997 to 1998

25 How many students were enrolled in 1995?
  (1) 1,600
  (2) 2,300
  (3) 2,500
  (4) 3,700
  (5) 23,000

27 How many more students were enrolled in 1999 than in 1994?
  (1) 1,300
  (2) 2,300
  (3) 2,700
  (4) 3,000
  (5) 4,700

Multiplying and Dividing Decimals

WORKPLACE MATH

Ordering Supplies

EXAMPLE Victor orders the supplies for the county clerk's office. He needs to order 50 computer disks and 3 cases of computer paper. Each case contains 10 packets of paper.

The price of one item is called the **unit cost. Unit** means one item. **Total cost** is found by multiplying the unit cost by the number of units you want.

Total cost = Unit cost × Number of units

Victor looks up the price of one case of computer paper in a catalog. The catalog states that each case costs $24.99. Which is the correct expression to find the total cost of the computer paper in Victor's order?
(1) $24.99 × 10
(2) $24.99 × 3
(3) $24.99 × 3 × 10
(4) $24.99 ÷ 3 × 10
(5) 3 × $24.99 ÷ 10

Answer **(2)** is correct. To find the total cost of the computer paper, Victor should multiply the unit cost ($24.99) by the number of units (3). **$24.99 × 3 = $74.97**, so the total cost of the 3 packages of computer paper is **$74.97**. (The fact that there are 10 packets of paper in each case is extraneous information.)

You can also find the unit cost of an item.

Unit cost = Total cost ÷ Number of units

EXAMPLE Victor looks up the price of computer disks. He can buy a box of 50 disks for $21.99. Which is the correct expression to find the cost of one disk?
(1) $21.99 × 50
(2) $21.99 × 1
(3) 50 ÷ $21.99
(4) $21.99 ÷ 50
(5) Not enough information is given.

Answer **(4)** is correct. To find the cost of each disk, Victor divides the total cost ($21.99) by the number of disks (50). Then he rounds his answer to the nearest cent. **$21.99 ÷ 50 = 0.4398**, so the cost of one computer disk is **$0.44**.

Circle the best answer for each item.

$0.99

1 Erica orders supplies for her department. She needs to order six daily calendar refills ($0.99 each). Which is the correct expression to find the total cost of her order?

(1) $0.99 ÷ 6
(2) $0.99 × 0.6
(3) 6 ÷ $0.99
(4) $0.99 × 6
(5) 1 ÷ $0.99

2 The Music Mart is having a sale. You can order five cassette tapes for $19.95. Fernando wants to buy one tape. Which is the correct expression to find the unit cost of one cassette tape?

(1) $19.95 ÷ 1
(2) $19.95 × 5
(3) $19.95 ÷ 5
(4) 5 ÷ $19.95
(5) $19.95 × 0.5

3 Rita is ordering supplies for her office. She can order six reams of color paper for $47.94. She only wants one ream of paper. Which is the correct expression to find the unit cost of one ream of paper?

(1) $47.94 ÷ 6
(2) 6 × $47.94
(3) 6 ÷ $47.94
(4) $47.94 × 0.6
(5) $47.94 ÷ 0.6

4 Kelvin is ordering school supplies for his students. Kelvin orders 25 colored pencils. The pencils cost $0.19 each. Which is the correct expression to find the total cost of the pencils?

(1) $0.19 × 0.25
(2) $0.19 ÷ 25
(3) $0.19 × 2.5
(4) 25 ÷ $0.19
(5) $0.19 × 25

5 Esther decides to stock up on paper towels for her day-care business. She wants to order 25 rolls of paper towels. Which is the correct expression to find the total cost of the paper towels?

(1) $0.59 × 25
(2) $0.59 ÷ 25
(3) 25 ÷ $0.59
(4) $0.59 × 2.5
(5) Not enough information is given.

6 Fantastic Lights is having a sale. You can buy a four-pack of frosted light bulbs for $1.69. Which is the correct expression to find the unit cost of one light bulb?

(1) $1.69 × 0.4
(2) 4 ÷ $1.69
(3) 4 × $1.69
(4) $1.69 ÷ 4
(5) Not enough information is given.

7 **Reason** Think about the equation Total cost = Unit cost × Number of units. On a separate piece of paper, use this information to explain how to get the following equation:

Unit cost = Total cost ÷ Number of units

Multiplying Decimals

Multiplying decimals is almost the same as multiplying whole numbers. There is only one difference. You have to place the decimal point in the answer.

EXAMPLE Multiply: 6.03×0.5

Step 1: Write the problem. In multiplication, you don't have to line up the decimal points. Line up the decimals on the right instead.

$$\begin{array}{r} 6.03 \\ \times\ 0.5 \\ \hline \end{array}$$

Step 2: Multiply as you would with whole numbers.

$$\begin{array}{r} 6.03 \\ \times\ 0.5 \\ \hline 3015 \end{array}$$

Step 3: Count the total number of decimal places in the problem. The number of decimal places is the number of digits to the right of the decimal point. There are two decimal places in 6.03 and one decimal place in 0.5. The answer will have 3 decimal places.

$$\begin{array}{rl} 6.03 & \leftarrow\quad 2 \text{ places} \\ \times\ 0.5 & \leftarrow +\ 1 \text{ place} \\ \hline 3015 & \leftarrow\quad 3 \text{ places} \end{array}$$

Step 4: Finally, place the decimal point in the answer. Start at the right and count the same number of decimal places from Step 3 to the left. Then write the decimal point.

$$\begin{array}{r} 6.03 \\ \times\ 0.5 \\ \hline 3.015 \end{array}$$
Count 3 places to the left.

TIP

Sometimes there are not enough digits in the answer. Then you need to use zeros as placeholders. Place the zeros to the *left* of the answer.

$$\begin{array}{rl} .014 & 3 \text{ places} \\ \times\ 0.6 & +1 \text{ place} \\ \hline 0.0084 & 4 \text{ places} \end{array}$$

PRACTICE

Multiply. The first one is done for you.

1.
$$\begin{array}{r} 13 \\ \times 0.4 \\ \hline 5.2 \end{array}$$

2.
$$\begin{array}{r} 9.1 \\ \times\ \ 8 \\ \hline \end{array}$$

3.
$$\begin{array}{r} 12.3 \\ \times\ 0.5 \\ \hline \end{array}$$

4.
$$\begin{array}{r} 1.04 \\ \times 0.07 \\ \hline \end{array}$$

5.
$$\begin{array}{r} 0.75 \\ \times\ 0.5 \\ \hline \end{array}$$

6.
$$\begin{array}{r} 136 \\ \times 0.006 \\ \hline \end{array}$$

7.
$$\begin{array}{r} 128 \\ \times\ 0.2 \\ \hline \end{array}$$

8.
$$\begin{array}{r} 17.3 \\ \times\ 1.6 \\ \hline \end{array}$$

9.
$$\begin{array}{r} 0.42 \\ \times 0.03 \\ \hline \end{array}$$

10.
$$\begin{array}{r} 2.05 \\ \times\ 0.9 \\ \hline \end{array}$$

11. 5.27×3.6

12. 6.2×0.08

13. 28.32×5.4

14. 16.5×0.063

15. 21.2×1.5

16. 8.3×0.025

Check your answers on page 292.

Reading Scales

You may have noticed that when you shop for vegetables, fruits, cheese, or items from a deli counter, the clerk weighs the item on a **digital scale.** Below is a picture of one such scale.

TIP

A digital scale uses decimals to show the weight (in pounds), unit price (in dollars), and total price (in dollars).

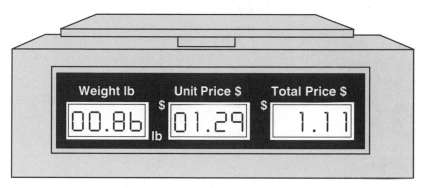

EXAMPLE Pedro purchased two red peppers. Red peppers cost $1.29 per pound. How much did the red peppers weigh?

Step 1: Find the column on the scale that says, "Weight lb."

Step 2: Read the display. *The display reads 00.86.*

The red peppers weighed 0.86 pounds.

PRACTICE

Solve each problem. The first one is done for you.

1 What is the weight of $\frac{1}{5}$ pound of reggianno parmesean cheese on a digital scale?

(Hint: To find the decimal value of a fraction, divide the numerator by the denominator.)

00.20 pounds

2 Write an expression that would give you the total price of temple oranges.

3 Fill in the display at the right to show the total cost of $2\frac{1}{4}$ pounds of ground round at $2.69 per pound.

4 Fill in the display at the right to show the total cost of $\frac{3}{4}$ pound of romaine lettuce at 69¢ per pound.

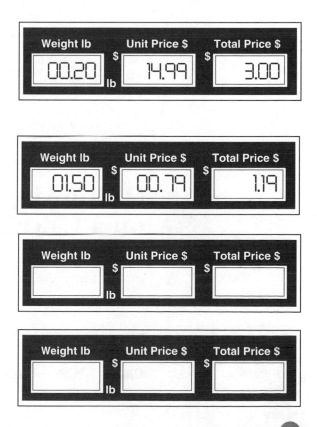

Dividing Decimals

Dividing a decimal by a whole number is the same as dividing whole numbers. You just need to place the decimal point in the answer.

EXAMPLE   Divide: 8.6 ÷ 2

Step 1: Write the problem. Put the decimal point in the answer space. It goes directly above the decimal point in the problem.

Step 2: Solve the problem. Divide as you would with whole numbers.

Step 3: Check by multiplying.

$$2)\overline{8.6}$$

$$\begin{array}{r} 4.3 \\ 2)\overline{8.6} \\ -8 \\ \hline 06 \\ -6 \\ \hline 0 \end{array}$$

$$\begin{array}{r} 4.3 \\ \times\ \ 2 \\ \hline 8.6 \end{array}$$

> **TIP**
>
> Sometimes you will need to put zeros at the end of a decimal to solve the problem.
>
> For example, 1.2 ÷ 5.
>
> $$\begin{array}{r} 0.24 \\ 5)\overline{1.2\mathbf{0}} \\ -10 \\ \hline 20 \\ -20 \\ \hline 0 \end{array}$$

To divide by a decimal, you need to change the decimal to a whole number by moving the decimal point as far to the right as you can. Then move the decimal point in the number you are dividing the same number of places to the right.

EXAMPLE   Divide: 2.032 ÷ 0.8

Step 1: Change 0.8 to a whole number. Move the decimal point one place to the right.

Step 2: Move the decimal point in 2.032 one place to the right.

Step 3: Place the decimal point in the answer space and divide.

$$0.8)\overline{2.0\ 32}$$

$$\begin{array}{r} 2.54 \\ 8)\overline{20.32} \\ -16 \\ \hline 4\ 3 \\ -4\ 0 \\ \hline 32 \\ -32 \\ \hline 0 \end{array}$$

> **TIP**
>
> Sometimes you will need to use zeros as placeholders in the answer.
>
> For example, 0.028 ÷ 4. Since 4 won't divide into 0 or 2, place zeros in the answer as needed.
>
> $$\begin{array}{r} 0.\mathbf{00}7 \\ 4)\overline{0.028} \\ -28 \\ \hline 0 \end{array}$$

Multiplying and Dividing Decimals on a Calculator

To multiply or divide decimals on a calculator, enter each decimal number including the decimal point, use the × or ÷ key between each number, then use the equals key to get the answer.

For more practice multiplying and dividing decimals with the calculator, see Calculator Handbook pages 250–251.

EXAMPLE   6.8 × 5.3 ÷ 0.02

CE/C   6.8   **×**   5.3   **÷**   0.02   **=**   1802

A. Divide. The first one is done for you.

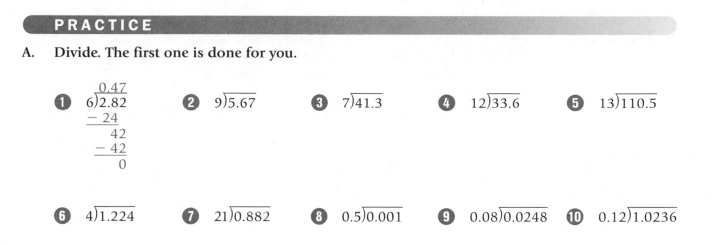

1
$$\begin{array}{r} 0.47 \\ 6\overline{)2.82} \\ -24 \\ \hline 42 \\ -42 \\ \hline 0 \end{array}$$

2 $9\overline{)5.67}$

3 $7\overline{)41.3}$

4 $12\overline{)33.6}$

5 $13\overline{)110.5}$

6 $4\overline{)1.224}$

7 $21\overline{)0.882}$

8 $0.5\overline{)0.001}$

9 $0.08\overline{)0.0248}$

10 $0.12\overline{)1.0236}$

B. Rewrite the problems below. Divide. The first one is started for you.

11 $0.105 \div 6 =$
$$\begin{array}{r} 0.01 \\ 6\overline{)0.1050} \\ -6 \\ \hline 45 \end{array}$$

12 $0.51 \div 0.012 =$

13 $12.6 \div 3.6 =$

14 $0.015 \div 0.25 =$

15 $64 \div 0.004 =$

16 $26.2701 \div 5.1 =$

C. Solve the problems below with a calculator. Refer to pages 250–251 in the Calculator Handbook for additional information.

17 The electric company charges $0.0892 per kilowatt-hour. For the month of December the Chou family used 194 kilowatt-hours. What amount will the Chou family be charged before taxes?

18 Last month Gabriel earned $2,541. He worked 38.5 hours each week. What was Gabriel's hourly pay? (Hint: Assume 4 weeks = 1 month)

Metric Conversion

EXAMPLE Jane works in a factory that makes hair-care products. One of the products contains a fruit extract. The extract is weighed in **kilograms**. Jane fills plastic tubes with small amounts of the extract. These amounts are so small they are weighed in **grams**.

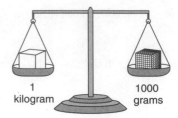

1 kilogram   1000 grams

⬤   **1 kilogram (kg) = 1,000 grams (g)**

The numbers 10, 100, and 1,000 are called **powers of ten.** A simple way to multiply by a power of ten is to move the decimal point one place to the right for each zero in the power of ten.

This morning Jane received a shipment weighing 25.6 kilograms. Which is the correct expression to find how many grams the shipment weighs?

(1) 25.6×10
(2) 25.6×100
(3) $25.6 \div 100$
(4) $25.6 \times 1,000$
(5) $25.6 \times 10,000$

Answer **(4)** is correct. To convert 25.6 kilograms to grams, Jane needs to multiply by 1,000. Since there are three zeros in 1,000, she moves the decimal point three places to the right. **$25.6 \times 1,000 = 25,600$**, so the shipment weighed **25,600 grams**. (The computation on the left shows that standard multiplication gives the same answer.)

EXAMPLE Jane's co-worker, Leon, prepares the liquid base for a hair-care product. The ingredients are measured in **milliliters.** After he finishes the job, he needs to record the amount in **liters** on a report form.

1 milliliter

1 liter

⬤   **1 liter (L) = 1,000 milliliters (ml)**

Today Leon prepares 45,000 milliliters of the liquid base. Which is the correct expression to find the amount in liters?

(1) $45,000 \div 10$
(2) $45,000 \div 100$
(3) $45,000 \div 1,000$
(4) $45,000 \times 1,000$
(5) $45,000 \times 10,000$

Answer **(3)** is correct. To convert 45,000 milliliters to liters, Leon needs to divide by 1,000. To divide by a power of ten, count the zeros. Move the decimal point the same number of places to the left. **$45,000 \div 1,000 = 45$**, so Leon has prepared **45 liters** of the liquid base. (See standard division at left.)

PRACTICE

Circle the best answer for each item. Refer to the chart below if necessary.

Length	10 millimeters (mm) 100 centimeters (cm) 1,000 meters (m)	= 1 centimeter (cm) = 1 meter (m) = 1 kilometer (km)
Mass (Weight)	1,000 milligrams (mg) 1,000 grams (g)	= 1 gram (g) = 1 kilogram (kg)
Volume	1,000 milliliters (ml)	= 1 liter (L)

1. Ana's personal best in the long jump is 17.5 meters. How many centimeters did she jump?

   (1) 0.0175 cm
   (2) 0.175 cm
   (3) 1.75 cm
   (4) 175 cm
   (5) 1,750 cm

2. Rick ate a bowl of chili that had 5.5 grams of fat. How many milligrams of fat were in the chili?

   (1) 0.0055 mg
   (2) 0.55 mg
   (3) 550 mg
   (4) 5,500 mg
   (5) 55,000 mg

3. A square measures 0.4 meter on each side. What is the perimeter of the square in centimeters?

   (1) 1.6 cm
   (2) 160 cm
   (3) 1,600 cm
   (4) 16,000 cm
   (5) 160,000 cm

4. A tube contains 100 milliliters of lotion. How many liters of lotion does the tube contain?

   (1) 0.001 L
   (2) 0.01 L
   (3) 0.1 L
   (4) 1 L
   (5) 10 L

5. Monica ran a race in 4.5 seconds. How many meters long is the race?

   (1) 0.1 m
   (2) 1 m
   (3) 10 m
   (4) 100 m
   (5) Not enough information is given.

6. A can of turkey gravy weighs 298 grams. What is the weight of the can in milligrams?

   (1) 298,000 mg
   (2) 29,800 mg
   (3) 2,900 mg
   (4) 290 mg
   (5) Not enough information is given.

7. **Connect** Look at three labels on foods that you commonly eat. Change any amounts listed in grams to kilograms or amounts listed as liters to milliliters.

Check your answers on page 294.

Filling Out Forms

EXAMPLE Rafael is a stock supervisor at a large department store. He orders items as shelf supplies run low. He started to fill out the following order form.

Item Number	Description	Quantity	Unit Price	Amount
23F	Cotton washcloths green mist	12	$4.49	
24A	Cotton hand towels ivory	8	$7.99	
26A	Medium-support polyester-fill pillow	15		$375
28B	Cappuccino/espresso maker, Model 781	3	$49.99	
31G	Ceramic hand-painted mug from Italy	6	$9.00	
45C	Black tote bag	4	$135.00	
			Total	

Rafael needs to complete the order form. What is the total amount for the cotton washcloths?

(1) $4.61
(2) $44.90
(3) $53.88
(4) $63.92
(5) $150.00

Answer **(3)** is correct. Rafael multiplies the quantity (12) by the unit price ($4.49). **$4.49 × 12 = $53.88**, so the total amount for the washcloths is **$53.88**.

EXAMPLE   Rafael is ordering 15 medium-support polyester-fill pillows that cost a total of $375. What is the unit price for each pillow?

(1) $15.00
(2) $25.00
(3) $37.50
(4) $360.00
(5) Not enough information is given.

Answer **(2)** is correct. Rafael divides the total amount of the pillows ($375) by the quantity (15). **$375 ÷ 15 = $25,** so the unit price for each pillow is **$25.**

PRACTICE

Circle the best answer for each item. Items 1–6 refer to the order form on page 144.

1. What is the total amount for the cappuccino/espresso makers?
   (1) $49.99
   (2) $52.99
   (3) $99.98
   (4) $149.97
   (5) $199.96

2. The total amount for the black tote bags is $540. How did Rafael arrive at that amount?
   (1) 6 × $9
   (2) 4 × $135
   (3) 5 + $135
   (4) 5 × $135
   (5) 5 ÷ $135

3. How many items did Rafael order all together?
   (1) 48
   (2) 38
   (3) 30
   (4) 45
   (5) 6

4. What is the total amount for the 12 cotton washcloths and the 8 cotton hand towels together?
   (1) $12.48
   (2) $53.88
   (3) $63.92
   (4) $117.80
   (5) $142.80

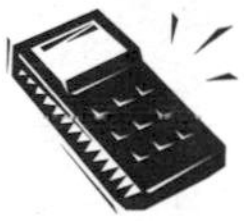

5. Use your calculator to complete the amount column. What is the TOTAL for this order?
   (1) $231.47
   (2) $642.77
   (3) $686.79
   (4) $1,182.77
   (5) $1,236.77

6. The following day Rafael's boss asked him to order four blenders. What is the total amount for the four blenders?
   (1) $279.96
   (2) $500.00
   (3) $636.00
   (4) $716.00
   (5) Not enough information is given.

7. The following week Rafael needed to order 15 more hand-painted mugs. Including the last order and this order, what is the total amount for the mugs?
   (1) $54
   (2) $135
   (3) $189
   (4) $324
   (5) Not enough information is given.

8. **Reason** Suppose the black tote bags were on back order. This means they were not shipped with the rest of the order, and the store was not charged for them. What is the new total, and how did you arrive at it?

Check your answers on page 294.

Section 10 Review

Multiply or divide.

1   3.2
    × 6

2   2.06
   ×0.04

3   4.17
   × 0.3

4   12.4
   × 2.7

5   0.752
   × 0.18

6   6)4.56   **7**   4)3.3   **8**   0.5)1.935   **9**   0.08)5.42   **10**   1.5)13.95

11 $0.56 \times 0.009 \div 0.02 =$

12 $37.6 \div 0.16 \times 0.554 =$

Solve each problem. Show your work.

13   A can of tuna contains 12.5 grams of protein. How many milligrams of protein does the tuna contain?

15   Risa bought $1\frac{3}{4}$ pounds of ground beef for $2.39 per pound. Fill in the display below to show the total cost of the ground beef.

Weight lb	Unit Price $	Total Price $
	$	$
lb		

14   Miguel works 5 days each week. He takes the same route to and from work each day. If he drives a total of 74.5 miles to and from work each week, how far does he drive to and from work each day?

16   A breakfast cereal costs $4.82 and contains 12 servings. How much does one serving cost? (Round to the nearest cent.)

Check your answers on pages 294–295.

Circle the best answer for each item.

17 Anna bought a set of 4 tires for $197.00. Which is the correct expression to find the cost of one tire?

(1) 4 × $197.00
(2) $197.00 ÷ 4
(3) $197.00 × 0.4
(4) 4 ÷ $197.00
(5) Not enough information is given.

18 Savemore Foods has bananas for $0.60 per pound. Kai bought 4 pounds of bananas. How much will the bananas cost?

(1) $0.15
(2) $1.50
(3) $2.40
(4) $8.32
(5) $11.20

19 Carolyn is a stock supervisor at a computer store. She filled out an order form with the following information: Item #1400S Scanner for PC, 4 at $149.00 each; Item #74051 Anti-Static Cleaning Wipes, 7 at $6.99 each. What is the subtotal amount of the order?

(1) $48.93
(2) $596.00
(3) $644.93
(4) $699.00
(5) $800.99

20 George drove 291.6 miles on 12 gallons of gas. How many miles per gallon does his car get?

(1) 1.2 miles per gallon
(2) 2.43 miles per gallon
(3) 12 miles per gallon
(4) 24.3 miles per gallon
(5) 29.16 miles per gallon

21 Mrs. Quan is ordering notebooks for her students. Each notebook costs $1.29. Which is the correct expression to find the cost of 31 notebooks?

(1) $1.29 ÷ 31
(2) $1.29 ÷ 31
(3) $1.29 × 0.31
(4) $1.29 × 31
(5) 31 ÷ $1.29

22 Raphael is building shelves. He needs to cut pieces of wood 1.2 meters long. Which is the correct expression to find how many pieces with that length can he cut from a board 5 meters long?

(1) 5 ÷ 1.2
(2) 5 × 1.2
(3) 1.2 ÷ 5
(4) 5 ÷ 1.02
(5) Not enough information is given.

23 Each dose of cold medicine contains 0.013 gram of decongestant. Which is the correct expression to find how many grams of decongestant are in each bottle of cold medicine?

(1) 100 × 0.013
(2) 0.013 ÷ 100
(3) 0.013 × 12
(4) 0.013 ÷ 12
(5) Not enough information is given.

24 A kitchen is 398 centimeters wide. How many meters wide is the kitchen?

(1) 0.398 meters
(2) 3.98 meters
(3) 39.8 meters
(4) 3,980 meters
(5) 39,800 meters

Math at Work

Service: Bank Teller

Some Careers in Service

Cashier
operates cash register, handles money, stocks items and maintains appearance of store

Fee Collector
welcomes visitors to museums and parks, collects fees, informs visitors of policies

Park Visitor Use Assistant
collects and tallies visitor fees, answers questions and provides assistance to visitors

Property Appraiser
calculates value of properties by examining their condition and comparing them to similar properties in the area

There are many service industries to help the public. One important service industry is banking. Do you have an excellent eye for detail? Do you enjoy working with numbers and money? If so, you may be interested in becoming a bank teller.

Bank tellers handle money and checks presented to them by the bank's customers. Tellers must be able to count money quickly and accurately. Because they are handling money, they must be able to **add and subtract decimals**. They may be required to do these calculations with actual coins and bills or on special pieces of paper called deposit and withdrawal slips. Tellers use **calculators** and **mental math** to make and check these calculations.

Bank tellers must be able to record the money brought into the bank (deposits) and the money taken out of the bank (withdrawals). All transactions performed by the teller must balance at the end of each day. Because they work constantly with customers, tellers should have good communication skills and a pleasant appearance.

Look at the *Some Careers in Service* chart.

- Do any of the careers interest you? If so, which ones?
- What information would you need to find out more about those careers?

On a separate piece of paper, write some questions that you would like answered. You can find out more information about those careers in the *Occupational Outlook Handbook* at your local library.

Josepha is a bank teller. She often helps her customers correctly fill out deposit tickets like the one below. **Use the deposit slip and material below to answer the questions that follow.**

Mr. Tong presents Josepha with his deposit ticket and four checks for deposit. He has three checks in the amount of $345.23 each. His fourth check is for $476.55. He tells her he also wishes to receive $125.50 in cash from his deposit. Josepha tells Mr. Tong to write the amount of each check in the spaces beside the word CHECKS and the money he wants back under Less Cash Received.

1 The total amount of Mr. Tong's checks is

    (1) $125.50
    (2) $345.23
    (3) $1,035.69
    (4) $1,386.74
    (5) $1,512.24

2 Mr. Tong should enter the total amount of the four checks he wants to deposit in the space marked

    (1) Cash
    (2) Checks
    (3) Subtotal
    (4) Less Cash Received
    (5) Total Deposit

3 In the space marked "Total Deposit," Mr. Tong should enter the amount

    (1) $125.50
    (2) $345.23
    (3) $1,035.69
    (4) $1,386.74
    (5) $1,512.24

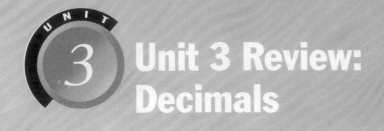
Write the value of the underlined digit in words.

1 5.394<u>6</u> _____   **2** 7.2<u>8</u>13 _____

Compare each pair of numbers. Write >, <, or =.

3 0.25 _____ 0.025   **4** 0.97 _____ 0.970

Round each decimal to the given place value. Show your work.

5 5.362 to the nearest hundredth   **6** 7.351 to the nearest tenth

Solve. Show your work.

7 Write 0.42 as a fraction.   **8** Write $\frac{19}{25}$ as a decimal.

Solve each problem. Show your work.

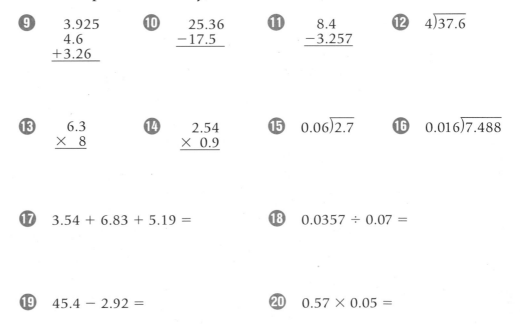

9
$$\begin{array}{r} 3.925 \\ 4.6 \\ +3.26 \\ \hline \end{array}$$

10
$$\begin{array}{r} 25.36 \\ -17.5 \\ \hline \end{array}$$

11
$$\begin{array}{r} 8.4 \\ -3.257 \\ \hline \end{array}$$

12 $4\overline{)37.6}$

13
$$\begin{array}{r} 6.3 \\ \times\ 8 \\ \hline \end{array}$$

14
$$\begin{array}{r} 2.54 \\ \times\ 0.9 \\ \hline \end{array}$$

15 $0.06\overline{)2.7}$

16 $0.016\overline{)7.488}$

17 3.54 + 6.83 + 5.19 =

18 0.0357 ÷ 0.07 =

19 45.4 − 2.92 =

20 0.57 × 0.05 =

Solve each problem. Show your work.

21 Bryant bought a pair of athletic shoes for $64.95 and a package of socks for $5.99. Sales tax on his purchases was $3.55. Bryant gave the clerk $80.00 in cash. How much change should the clerk give Bryant?

22 What is the value of $9 \div (4 - 1) + 8$?

23 Freddie can order a dozen pens for $0.96. He needs only five pens. How much would five pens cost at this rate?

24 What is the total cost for $\frac{2}{5}$ pound of cauliflower?

Weight lb	Unit Price $	Total Price $
$ 00.40 lb	$ 02.69	

25 What is the total amount for the two items shown below?

Item Number	Description	Quantity	Unit Price	Amount
35D	Twin sheets	4	$12.99	
26F	Down pillow	5	$20.00	

Circle the best answer for each item.

26 Auto Buys has tires on sale this week for $42.50 each. Lin needs to buy four tires. Lin also gets the tires balanced for $15.95. Tax amounted to $14.87. Which is the correct expression to find the total cost to Lin?

(1) $(4 \times \$42.50) + \$15.95 + \$14.87$
(2) $4 \times (\$42.50 + \$15.95) + \$14.87$
(3) $4 \times (\$42.50 + \$15.95 + \$14.87)$
(4) $4 \times (\$42.50 + \$15.95 - \$14.87)$
(5) $4 \times (\$42.50 - \$15.95) + \$15.95 + \$14.87$

Item 27 refers to the following graph.

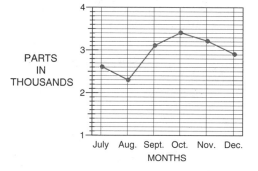

PARTS PRODUCED, JULY–DEC. 1999

27 How many parts were produced during the last two months of 1999?

(1) 0.3 thousand
(2) 4.9 thousand
(3) 5.1 thousand
(4) 5.11 thousand
(5) 6.1 thousand

Math Extension

Measure the dimensions of two rooms in your home. Convert to metric amounts.

Math Connection: Decimals and World History

Olympic Scores

Every four years, teams of athletes from countries around the world gather to compete in the Olympics. Representing their country in the Olympics is among the greatest honors athletes can achieve.

Many of the athletes train for years, even decades, to compete in events that may last only minutes or seconds. Running events can be won by hundredths of a second. Swimming events can be won by thousandths of a second.

Figure skating competitions are often won by tenths of a point. The nine judges in figure skating award each skater two marks. A 6.0 is a perfect mark. The two marks are added together and used to rank the skaters. The skater receiving the highest total mark from the judges "wins" that part of the competition.

World History: The Olympic Games

Olympic competitions were begun in Olympia, Greece, in 776 B.C. and ran for about twelve hundred years—until they were banned in 393 A.D. Fifteen hundred years later, Baron Pierre de Coubertin began the effort to renew the international athletic competitions. What we now know as the modern Olympic Games started in 1896 in Athens, Greece.

The Olympics were based on the idea of joining the world in "sport and fair competition." The vast majority of Olympics have been free from political influence. However, there have been some instances where world events have affected the Olympic Games. The 1916 games were canceled due to World War I. The Olympics scheduled in 1940 (in Tokyo, Japan) and 1944 were canceled due to World War II.

In recent decades, countries have boycotted the Olympic Games. The 1976 games in Montreal, Canada, were boycotted by 33 African nations protesting the South African policy of apartheid. A boycott by about 40 nations in 1980 (led by the United States) and the 1984 USSR-led boycott were two more important events in Olympic history.

Judges' scores for an Olympic event may differ by only a tenth of a point.

1 Which Olympic Games were canceled due to war?

(1) 776. B.C., 393 A.D., and 1896
(2) 393 A.D., 1896, and 1924
(3) 393 A.D., 1916, and 1940
(4) 1916, 1940, and 1944
(5) 1940, 1944, and 1976

2 In 1992 Bonnie Blair won the gold medal in women's 500-meter speed skating with a time of 40.33. In 1994 her winning time was 39.25. By how much time did she improve in 1994 over 1992?

(1) 8 hundredths of a second
(2) 92 hundredths of a second
(3) 1 and 8 tenths seconds
(4) 1 and 8 hundredths seconds
(5) 1 and 92 hundredths seconds

Look at the marks received by three figure skaters in the 1998 Olympics in Nagano, Japan. Use this information to answer the following questions.

Skaters	Marks								
	Judge 1	Judge 2	Judge 3	Judge 4	Judge 5	Judge 6	Judge 7	Judge 8	Judge 9
Chen Lu	5.5	5.5	5.6	5.5	5.5	5.5	5.7	5.5	5.5
	5.8	5.8	5.8	5.8	5.8	5.8	5.8	5.7	5.8
Total	11.3								
Michelle Kwan	5.7	5.7	5.8	5.7	5.8	5.8	5.7	5.7	5.8
	5.9	5.9	5.9	5.9	5.9	5.9	5.9	5.9	5.9
Total									
Tara Lipinski	5.9	5.9	5.9	5.8	5.8	5.9	5.9	5.8	5.9
	5.8	5.8	5.9	5.8	5.8	5.9	5.9	5.8	5.9
Total									

3 Complete the chart by finding the totals awarded to each skater by each judge. The first total has been done for you.

4 Under each *Judge's* column, circle the highest total points. Circle both scores if there is a tie.

5 The three top athletes in each Olympic event receive a medal. First-place winners receive a gold medal. Second-place winners receive a silver medal. Third-place winners receive a bronze medal. Which skater received the gold medal? Which skater received the silver, and which received the bronze medal?

Check your answers on page 297.

Cumulative Review

Write the value of the underlined digit in words.

1 5<u>2</u>6,438 _____

2 <u>7</u>,028,651 _____

3 8.3<u>2</u>96 _____

4 9.01<u>4</u>8 _____

Solve. Show your work.

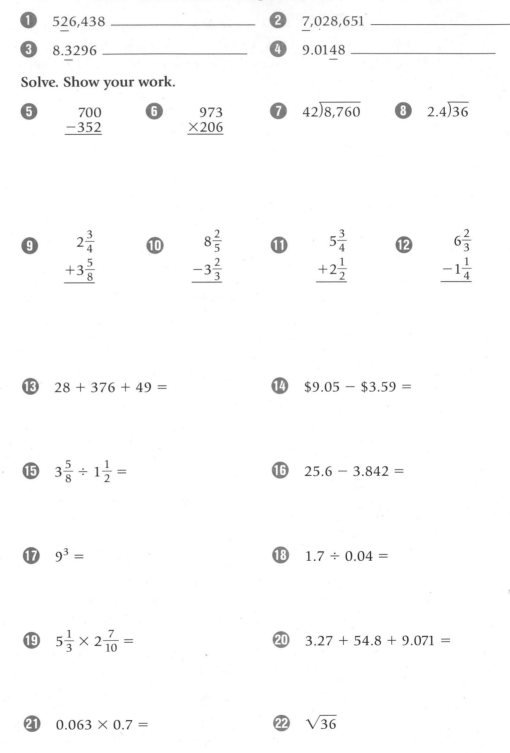

5
$$\begin{array}{r} 700 \\ -352 \\ \hline \end{array}$$

6
$$\begin{array}{r} 973 \\ \times 206 \\ \hline \end{array}$$

7 $42\overline{)8{,}760}$

8 $2.4\overline{)36}$

9
$$\begin{array}{r} 2\frac{3}{4} \\ +3\frac{5}{8} \\ \hline \end{array}$$

10
$$\begin{array}{r} 8\frac{2}{5} \\ -3\frac{2}{3} \\ \hline \end{array}$$

11
$$\begin{array}{r} 5\frac{3}{4} \\ +2\frac{1}{2} \\ \hline \end{array}$$

12
$$\begin{array}{r} 6\frac{2}{3} \\ -1\frac{1}{4} \\ \hline \end{array}$$

13 $28 + 376 + 49 =$

14 $\$9.05 - \$3.59 =$

15 $3\frac{5}{8} \div 1\frac{1}{2} =$

16 $25.6 - 3.842 =$

17 $9^3 =$

18 $1.7 \div 0.04 =$

19 $5\frac{1}{3} \times 2\frac{7}{10} =$

20 $3.27 + 54.8 + 9.071 =$

21 $0.063 \times 0.7 =$

22 $\sqrt{36}$

Check your answers on pages 297–298.

Solve each problem. Show your work.

<u>Items 23 and 24</u> refer to the following map.

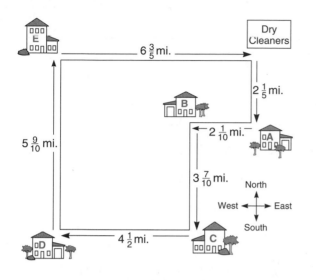

6 3/5 mi.

Dry Cleaners

2 1/5 mi.

B

2 1/10 mi.

A

5 9/10 mi.

3 7/10 mi.

North

West ◄──►East

South

4 1/2 mi.

D

C

23 Monica makes pickups and deliveries for a dry cleaner. She takes a lunch break after stopping at point C. How many miles does she drive before lunch?

24 Monica's entire route is 25 miles long. How far has she driven when she stops at point E?

Circle the best answer for each item.

25 Penny is buying new carpet for her living room. The living room is 24 feet long and 20 feet wide. What is the area of Penny's living room in square feet?

24 ft.

20 ft.

(1) 44

(2) 88

(3) 160

(4) 480

(5) 960

26 Rosa is making lasagna for a block party. Her recipe calls for $\frac{2}{3}$ pound of ricotta cheese. Rosa plans to triple the recipe. Which is the correct expression to find the number of pounds of ricotta cheese she should use?

(1) $\frac{1}{3} \div \frac{2}{3}$

(2) $\frac{2}{3} \times \frac{1}{3}$

(3) $\frac{2}{3} \times 3$

(4) $\frac{2}{3} \div 3$

(5) $3 + \frac{2}{3}$

27 Mike worked from 8:28 A.M. to 12:21 P.M. on Monday. He returned from lunch at 1:06 P.M. and worked until 5:36 P.M. How many hours did Mike work on Monday?

(1) $3\frac{3}{4}$

(2) $4\frac{1}{2}$

(3) $7\frac{1}{4}$

(4) 8

(5) $8\frac{1}{4}$

28 Of the 21 days Mario worked last month, he worked overtime 12 days. What fraction of the days did he work overtime?

(1) $\frac{1}{12}$

(2) $\frac{1}{21}$

(3) $\frac{4}{7}$

(4) $\frac{2}{3}$

(5) $1\frac{3}{4}$

UNIT 4

Ratios, Proportions, Percents and Special Topics

Ratios, proportions, and percents are used often in business. A **ratio** is a way to compare two numbers. You can use a ratio to compare the number of employees working in your department to the number of employees working in the entire company. A special ratio which compares unlike units, such as miles per gallon, or $/hr., is a **rate**. A **proportion** is made up of two equal ratios. You can use a proportion to show that the ratios are equal.

Like fractions and decimals, **percents** are used to show part of a whole amount. Percents are often used to calculate discounts, taxes, interest, budgets, and so on. You will also learn how percents can be converted into fractions and decimals.

● What numbers or things have you compared today? Can you write these comparisons as a ratio? For example, 3 pairs of socks for $10 would be written as 3 to $10 or 3:10 or $\frac{3}{\$10}$.

● Have you noticed any percents in store windows that you pass? What information do the percents represent? Visit a store or look in a newspaper to find advertised sales and notice the percents.

SECTIONS

Reorganization

SALE

Incredible Savings

UP TO **70** % OFF

Ratio and Proportion

MEASUREMENT AND GEOMETRY

Finding Ratios and Rates

Ratios compare quantities or amounts. When you compare inches to inches or dollars to dollars, you are comparing *like* quantities. To compare 3 inches to 8 inches, you could write 3 to 8, 3:8, or $\frac{3}{8}$. The colon in 3:8 says this is a ratio.

EXAMPLE Tanya works for a photo lab. During an 8-hour day, Tanya works 3 hours at the counter helping customers and 5 hours in the lab. What is the ratio of the time Tanya works at the counter to the time she works in the lab?

(1) $\frac{3 \text{ hours}}{8 \text{ hours}}$

(2) $\frac{3 \text{ hours}}{5 \text{ hours}}$

(3) $\frac{5 \text{ hours}}{8 \text{ hours}}$

(4) $\frac{8 \text{ hours}}{5 \text{ hours}}$

(5) $\frac{8 \text{ hours}}{3 \text{ hours}}$

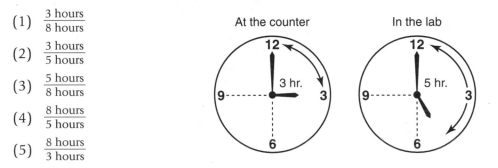

At the counter — 3 hr.   In the lab — 5 hr.

Answer **(2)** is correct. The ratio of hours working at the counter to hours working in the lab is 3:5 or $\frac{3}{5}$. The ratio shows that for every **3 hours** Tanya works at the counter, she works **5 hours** in the lab.

You can also compare *unlike* quantities or amounts. A ratio comparing unlike quantities is called a **rate**. A rate usually has a denominator of 1. It is very important to use the units of both the numerator and denominator in a rate.

EXAMPLE Dean does a lot of city driving. He finds that he drives 15 miles on 1 gallon of gasoline. What is the rate of miles to gallons? Remember the order in which you are asked to compare. Here you are comparing *miles* to *gallons*.

(1) $\frac{1 \text{ mile}}{15 \text{ gallons}}$

(2) $\frac{1 \text{ gallon}}{15 \text{ miles}}$

(3) $\frac{15 \text{ gallons}}{15 \text{ miles}}$

(4) $\frac{1 \text{ mile}}{1 \text{ gallon}}$

(5) $\frac{15 \text{ miles}}{1 \text{ gallon}}$

Answer **(5)** is correct. The rate of miles to gallons is 15:1 or $\frac{15}{1}$. This means Dean can drive **15 miles** for every **one gallon** of gasoline. When writing rates, it is common to use the word *per* instead of *for every*. In this case you could write that Dean can drive 15 miles per gallon.

Circle the best answer for each item.

1 Ned is making bread. The recipe calls for 1 cup of sugar and 6 cups of flour. Ned wants to know the ratio of the ingredients in case he decides to double the recipe. What is the ratio of cups of sugar to cups of flour?

(1) $\frac{1 \text{ cup sugar}}{3 \text{ cups flour}}$

(2) $\frac{1 \text{ cup sugar}}{6 \text{ cups flour}}$

(3) $\frac{1 \text{ cup sugar}}{12 \text{ cups flour}}$

(4) $\frac{1 \text{ cup sugar}}{36 \text{ cups flour}}$

(5) $\frac{6 \text{ cups sugar}}{1 \text{ cup flour}}$

2 The City Library has 1,055 fiction books and 1,267 nonfiction books checked out. What is the ratio of nonfiction books to fiction books checked out of the library?

(1) $\frac{1,055 \text{ fiction}}{1,055 \text{ nonfiction}}$

(2) $\frac{1,055 \text{ fiction}}{1,267 \text{ nonfiction}}$

(3) $\frac{1,055 \text{ nonfiction}}{1,267 \text{ fiction}}$

(4) $\frac{1,267 \text{ nonfiction}}{1,055 \text{ fiction}}$

(5) $\frac{1,267 \text{ nonfiction}}{1,267 \text{ fiction}}$

3 A 12-oz. package of wheat bran cereal costs $2.88, or $0.24 for each ounce. What is the rate of cost in dollars to weight in ounces?

(1) $\frac{\$0.24}{1 \text{ oz.}}$

(2) $\frac{\$0.24}{12 \text{ oz.}}$

(3) $\frac{\$2.88}{1 \text{ oz.}}$

(4) $\frac{1 \text{ oz.}}{\$0.24}$

(5) $\frac{12 \text{ oz.}}{\$0.24}$

4 During a 40-hour week, Eiko earns $18 an hour as a copyeditor. What is the rate of money earned per hour?

(1) $\frac{1 \text{ hr.}}{\$.18}$

(2) $\frac{40 \text{ hr.}}{\$18}$

(3) $\frac{40 \text{ hr.}}{1 \text{ hr.}}$

(4) $\frac{\$18}{1 \text{ hr.}}$

(5) $\frac{\$18}{40 \text{ hr.}}$

5 Daksha works a total of 45 hours a week as a baker making cakes and cookies. She spends about 19 hours baking cakes. What is the ratio of hours baking cakes to hours baking cookies?

(1) $\frac{19 \text{ hours}}{45 \text{ hours}}$

(2) $\frac{19 \text{ hours}}{26 \text{ hours}}$

(3) $\frac{26 \text{ hours}}{45 \text{ hours}}$

(4) $\frac{26 \text{ hours}}{19 \text{ hours}}$

(5) Not enough information is given.

6 **Reason** Refer to the second example on page 158. Suppose Dean has two cars. The car discussed in the example is listed below as Car 1. His second car uses 10 gallons to drive 100 miles. On a separate piece of paper, explain how you can find which car gives Dean the better rate of miles to gallons.

Car 1: 15 miles to 1 gallon
Car 2: ? miles to 1 gallon

Which car gets more miles per gallon?

Working with Ratios and Rates

A ratio is a way to compare two quantities. A ratio that compares two unlike quantities is called a rate. You can write ratios in lowest terms the same way you can reduce fractions. Remember that the <u>order</u> in which you compare quantities is very important. The amount mentioned <u>before</u> the word *to* is always the numerator. The amount <u>after</u> the word *to* is always the denominator.

EXAMPLE   Clyde worked 8 hours on Thursday and 10 hours on Friday. What is the ratio of hours worked on Thursday to hours worked on Friday?

Step 1:   Write the quantities in a ratio in the same order that you are asked to compare them. In this example, compare the hours worked on Thursday to the hours worked on Friday.

$\dfrac{\text{Thursday hours}}{\text{Friday hours}}$
$\dfrac{8}{10}$

Step 2:   Reduce the ratio to lowest terms the same way you reduce a fraction. Divide the numerator and denominator by the same number. The ratio is in lowest terms when no number except 1 will divide both numbers evenly.

$\dfrac{8}{10} = \dfrac{8 \div 2}{10 \div 2} = \dfrac{4}{5}$

The ratio of hours worked on Thursday to hours worked on Friday is $\dfrac{4}{5}$ or 4:5.

TIP

When a ratio is an improper fraction, do *not* change it to a mixed number or a whole number. A ratio is always expressed as a fraction. The fraction $\dfrac{10}{1}$ equals the whole number 10. However, as a ratio, keep the number in the form $\dfrac{10}{1}$.

EXAMPLE   Nancy earned $100 for 10 hours of work. What is the rate of her earnings to the hours she worked?

Step 1:   Write the quantities in a rate in the same order that you are asked to compare them. Include units when comparing unlike quantities (amounts with different units of measurement).

$\dfrac{\text{earnings}}{\text{hours}} = \dfrac{\$100}{10 \text{ hr.}}$

Step 2:   Reduce the rate to lowest terms.

$\dfrac{\$100}{10 \text{ hr.}} = \dfrac{\$100 \div 10}{10 \div 10} = \dfrac{\$10}{1 \text{ hour}}$

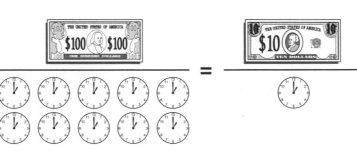

The rate of Nancy's earnings to the hours she worked is $10/1 hour or $10:1 hour.

A. Write each quantity as a ratio in lowest terms. Include units when comparing unlike quantities. The first one is done for you. The third one is started for you.

1 4 inches to 6 inches

$$\frac{4}{6} = \frac{2}{3}$$

2 16 employees to 21 employees

3 2 gallons of paint to 400 square feet

$$\frac{2 \; gallons}{400 \; square \; feet}$$

4 2 pounds of apples to $1.50

5 4 hours to 10 hours

6 8 grams to 20 grams

7 3 pairs of socks to $12

8 12 women to 8 men

9 $280 to 20 hours

10 40 books to 15 books

B. Write the quantities as a ratio in lowest terms. The first one is done for you.

11 It rained 9 days out of 31 days last month. What is the ratio of rainy days to total days in the month?

9 rainy days/31 days in the month

$$\frac{9}{31}$$

13 Midtown Motors sold 35 cars and 14 vans last week. What is the ratio of cars sold to vans sold?

12 Mary's softball team won 12 games out of 18 games played. What is the ratio of games won to games played?

14 Anwar drove 230 miles on 10 gallons of gasoline. What is the rate of miles to gallons?

C. Solve.

15 Today 15 of Ramona's customers paid with cash and 25 customers used a credit card. What is the ratio of cash customers to charge customers?

16 Kara earned $42 for 7 hours of work. What is the ratio of her earnings to the hours she worked?

Solving Proportions

A **proportion** is an equation comparing two equal ratios. The numbers in the proportion are called **terms.** A proportion has four terms. Sometimes one of the terms is not known. If you know three of the four terms, you can solve for the unknown term.

EXAMPLE Find the value of the unknown term in this proportion: $\frac{10}{2} = \frac{?}{5}$

Step 1: Cross multiply to solve. Start by cross multiplying the numerator and denominator you know.

$\frac{10}{2} \diagdown \frac{?}{5}$

$10 \times 5 = 50$

Step 2. Divide the result by the remaining number. In this case, divide 50 by the remaining term of 2. The answer, **25,** is the unknown term.

$50 \div 2 = 25$

Step 3: Check your answer by cross multiplying. If the results are equal, you know your answer is correct.

$\frac{10}{2} \diagtimes \frac{25}{5}$

$10 \times 5 = 50$

$2 \times 25 = 50$

Another way to check a solved proportion is to recognize that it has the same value as two equivalent fractions. In the example above, $\frac{10}{2} = \frac{25}{5}$ because they both can be reduced to $\frac{5}{1}$.

$$\frac{10}{2} = \frac{5}{1} \qquad \frac{25}{5} = \frac{5}{1}$$

> **TIP**
>
> If the numbers you are dividing do not divide evenly, write the remainder as a decimal. For example:
>
> $\frac{3}{4} = \frac{?}{6}$
>
> $3 \times 6 = 18$
>
> $18 \div 4 = 4.5$
>
> $\frac{3}{4} = \frac{4.5}{6}$

EXAMPLE Elena is a caterer. She uses 3 cups of ground coffee to serve 16 people. How many cups of ground coffee should she use to serve 80 people?

$$\frac{}{16 \text{ people}} = \frac{? \text{ cups}}{80 \text{ people}}$$

Step 1: Write the ratio that has two known quantities.

$\frac{3 \text{ cups}}{16 \text{ people}}$

Step 2: Write the second ratio with the unknown term.

$\frac{? \text{ cups}}{80 \text{ people}}$

Step 3: Write a proportion using the two ratios. Be sure both ratios are set up in the same order: cups/people = cups/people

$\frac{3 \text{ cups}}{16 \text{ people}} = \frac{? \text{ cups}}{80 \text{ people}}$

Step 4. Solve the proportion.

$3 \times 80 = 240$

$240 \div 16 = 15$

Step 5: Check your answer by cross multiplying. The cross products are equal: 240 = 240. Thus the ratios in the proportion are equal.

$\frac{3}{16} \diagtimes \frac{15}{80}$

$3 \times 80 = 240$

$16 \times 15 = 240$

A. **Solve each proportion. Show your work. The first one is done for you.**

1 $\frac{3}{4} = \frac{?}{12}$

$3 \times 12 = 36$

$36 \div 4 = 9$ Thus, $\frac{3}{4} = \frac{9}{12}$

Check: $\frac{3}{4} = \frac{9}{12}$

$3 \times 12 = 36$

$4 \times 9 = 36$

2 $\frac{4}{?} = \frac{12}{18}$

3 $\frac{30}{18} = \frac{10}{?}$

4 $\frac{7}{14} = \frac{?}{4}$

5 $\frac{20}{?} = \frac{4}{11}$

6 $\frac{?}{30} = \frac{10}{12}$

7 $\frac{?}{42} = \frac{10}{12}$

8 $\frac{5}{24} = \frac{4}{?}$

9 $\frac{5}{8} = \frac{?}{40}$

B. **Use proportions to solve these problems. Show your work. The first one is started for you.**

10 Toya gets paid an hourly wage. During a 40-hour week, Toya earns $240 in her job as a receptionist. This week she worked only 35 hours. How much did she earn?

$\frac{40 \text{ hr.}}{\$240} = \frac{35 \text{ hr.}}{?}$

11 Kelsey swims 35 laps in 25 minutes. How many laps does he swim in 15 minutes?

12 Sarah is a photographer. She uses 5 rolls of black-and-white film out of 15 rolls. If Sarah uses 51 rolls of film, how many rolls are in black-and-white?

13 Dixie walks 45 minutes a day. How many hours will she walk in 8 days?

C. **Solve. Show your work.**

14 One meter is 39.37 inches. How many inches is 2.5 meters?

15 Noriko is in charge of the airport parking lots. Lot A has 4,000 spaces. Today there are 2,500 cars in Lot A. Lot B has 2,000 spaces. If both lots have the same ratio of cars to spaces, how many cars are in Parking Lot B?

Finding Sale Prices

EXAMPLE Matt works in a hardware store. One morning Jack Prow, Matt's boss, gives Matt a list of a few items to put on sale.

*** PROW'S PAINT AND HARDWARE ***

Sale Prices
2-inch brushes   4 for $12.00
masking tape   2 rolls for $4.98
latex paint   3 quarts for $14.40

Which ratio compares the number of brushes on sale to the sale price?

(1) $\frac{4}{\$12}$

(2) $\frac{2}{\$12}$

(3) $\frac{\$12}{4}$

(4) $\frac{\$12}{2}$

(5) $\frac{4}{2}$

Answer **(1)** is correct. The ratio means that **every group of 4 brushes costs $12**.

EXAMPLE A few hours later Ms. Okano, an art teacher, calls the store. She wants to buy supplies for her students. Ms. Okano tells Matt she needs to buy ten 2-inch brushes. Matt wants to find the sale price for 10 brushes. Which is the correct proportion to find the price of 10 brushes?

(1) $\frac{4}{\$10} = \frac{12}{?}$

(2) $\frac{4}{\$12} = \frac{10}{?}$

(3) $\frac{10}{\$12} = \frac{?}{4}$

(4) $\frac{\$12}{4} = \frac{10}{?}$

(5) $\frac{\$12}{10} = \frac{4}{?}$

Answer **(2)** is correct. A proportion is an equation of two ratios. In both ratios, the number of brushes is the top number, and the price is the bottom number. The equal ratios make a proportion. Matt needs to compare the sale price of $\frac{4 \text{ brushes}}{\$12}$ to the ratio that includes the 10 brushes Ms. Okano needs. The proportion is $\frac{4}{\$12} = \frac{10}{\$?}$. Matt cross multiplies $12 and 10. Then he divides $120 by 4 to find the price of 10 brushes.

Set up the proportion.   Cross multiply.   Divide.

$\frac{4}{\$12} \diagup\!\!\!\diagdown \frac{10}{?}$ $\$12 \times 10 = \$120$ $\$120 \div 4 = \mathbf{\$30}$

Based on the sale price of 4 brushes for $12, **10 brushes will cost $30**.

Circle the best answer for each item.

Items 1–3 refer to the following information.

Kevin works at an office supply warehouse. Kevin's boss gives him this list of items on sale.

OFFICE SUPPLY WAREHOUSE		
Bond paper	10 reams for	$22.50
Steno pads	12 for	$3.96
Correction fluid (0.75 oz.)	3 bottles for	$2.49

1 Cindy bought 6 reams of paper. Which is the correct expression to find the price of 6 reams of paper?

(1) $\frac{6}{10} = \frac{\$22.50}{?}$

(2) $\frac{6}{\$22.50} = \frac{10}{?}$

(3) $\frac{10}{\$22.50} = \frac{6}{?}$

(4) $\frac{10}{\$22.50} = \frac{?}{6}$

(5) $\frac{\$22.50}{10} = \frac{6}{?}$

2 Carlos bought 18 steno pads. Which is the correct expression to find the price of 18 steno pads?

(1) $\frac{12}{2} = \frac{\$3.96}{?}$

(2) $\frac{12}{18} = \frac{?}{\$3.96}$

(3) $\frac{12}{\$3.96} = \frac{2}{?}$

(4) $\frac{12}{\$3.96} = \frac{18}{?}$

(5) $\frac{18}{\$3.96} = \frac{12}{?}$

3 Becky bought 4 bottles of correction fluid. What is the price of 4 bottles of correction fluid?

(1) $1.87
(2) $2.49
(3) $3.32
(4) $7.48
(5) $13.28

Items 4–6 refer to the following information.

Yolanda is a clerk at Green's Fruits and Vegetables. The list below shows the produce that is on sale this week.

GREEN'S PRODUCE		
Lettuce	2 heads for	$1.28
Potatoes	5 pounds for	$3.15
Carrots	3 bags for	$1.22

4 Marvin needs 5 heads of lettuce to make a salad. Which is the correct expression to find the price of 5 heads of lettuce?

(1) $\frac{2}{5} = \frac{?}{\$1.28}$

(2) $\frac{2}{\$1.28} = \frac{5}{?}$

(3) $\frac{2}{\$1.28} = \frac{?}{5}$

(4) $\frac{5}{2} = \frac{\$1.28}{?}$

(5) $\frac{5}{\$1.28} = \frac{2}{?}$

5 Daksha needs 2 bags of carrots for a stew she is making for an office party. To the nearest cent, what is the price of 2 bags of carrots?

(1) $0.41
(2) $0.81
(3) $1.83
(4) $3.66
(5) Not enough information is given.

6 **Communicate** Vernon has a catering business. He spent $12.60 on potatoes at Green's Produce. On a separate piece of paper, explain how to find the number of pounds of potatoes Vernon bought. Then solve the problem.

MEASUREMENT AND GEOMETRY

Using a Map Scale

EXAMPLE Luis is a truck driver for Brown's Bakery. He needs to drive east from Lake City to Mt. Vernon, then south from Mt. Vernon to Hartford.

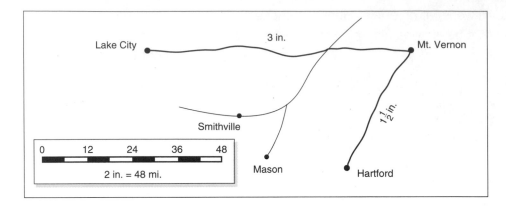

The map has a *scale* that shows you how to figure out distances. The scale shows the ratio of inches on the map to miles of actual distance.

On Luis' map, the scale is 2 inches equals 48 miles. The distance from Lake City to Mt. Vernon on the map is 3 inches. What is the actual distance in miles?

(1) 56
(2) 68
(3) 72
(4) 78
(5) 94

Answer **(3)** is correct. Set up a proportion comparing the map scale and a ratio with the distance shown on the map from Lake City to Mt. Vernon. Luis writes a proportion to solve for the unknown. The proportion is $\frac{2 \text{ in.}}{48 \text{ mi.}} = \frac{3 \text{ in.}}{? \text{ mi.}}$.

Set up the proportion.	Cross multiply.	Divide.
$\frac{2 \text{ in.}}{48 \text{ mi.}} \diagup \frac{3 \text{ in.}}{? \text{ mi.}}$	$48 \times 3 = 144$	$144 \div 2 = \textbf{72 miles}$

EXAMPLE Mt. Vernon and Hartford are $1\frac{1}{2}$ inches apart on the map. What is the actual distance in miles from Mt. Vernon to Hartford?

(1) 12
(2) 24
(3) 25
(4) 36
(5) 48

Answer **(4)** is correct. Luis sets up a proportion and solves.

Set up the proportion.	Cross multiply.	Divide.
$\frac{2 \text{ in.}}{48 \text{ mi.}} \diagup \frac{1\frac{1}{2} \text{ in.}}{? \text{ mi.}}$	$48 \times 1\frac{1}{2} = 72$	$72 \div 2 = \textbf{36 miles}$

Circle the best answer for each item.

Items 1–3 refer to the following information.

Cheryl is a saleswoman for a computer company. She leaves her home in Riverton to drive to her first customer in Plainview. Then she continues on to Rock Falls.

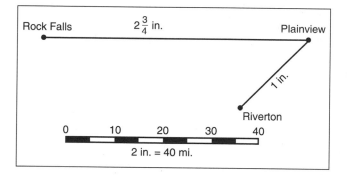

1 What is the actual distance from Riverton to Plainview in miles?

(1) 1
(2) 10
(3) 20
(4) 40
(5) 80

2 What is the actual distance from Plainview to Rock Falls in miles?

(1) 50
(2) 55
(3) 60
(4) 110
(5) 220

3 One of Cheryl's customers is in Hillsdale. Hillsdale is 60 miles north of Plainview. If Hillsdale were shown on this map, how many inches north of Plainview would it be?

(1) $1\frac{1}{3}$
(2) $1\frac{1}{2}$
(3) 3
(4) 6
(5) 60

Items 4–6 refer to the following information.

Jerome works for a recycling company. He drives from Mesa to Canyon City to Bluffton.

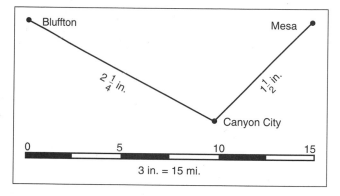

4 What is the actual distance from Mesa to Canyon City in miles?

(1) $3\frac{1}{2}$
(2) $5\frac{1}{2}$
(3) $6\frac{1}{2}$
(4) $7\frac{1}{2}$
(5) $10\frac{1}{2}$

5 What is the actual distance from Canyon City to Bluffton in miles?

(1) $4\frac{1}{2}$
(2) $6\frac{2}{3}$
(3) $6\frac{3}{4}$
(4) $11\frac{1}{4}$
(5) Not enough information is given.

6 **Reason** Assume Jerome is driving at a constant rate of 40 miles per hour. (In other words, he is driving 40 miles for each hour he drives.) Explain about how long (the number of minutes) it would take Jerome to drive from Mesa to Bluffton through Canyon City.

Section 11 Review

Write each ratio in lowest terms.

1 At Quality Products 20 employees work in the office and 28 work in the warehouse. What is the ratio of employees who work in the office to employees who work in the warehouse?

2 During a football game, the Greyhounds scored 36 points. Of the total points, 15 points were scored by kicking field goals. What is the ratio of points scored with field goals to total points scored?

Solve each proportion. Show your work.

3 $\frac{6}{7} = \frac{?}{42}$

4 $\frac{12}{8} = \frac{30}{?}$

5 Andre drove 106 miles in 2 hours. If he can continue this speed, how far can he drive in 5 hours?

6 Of every 400 parts Tri-State Assembly makes, an average of 3 parts are defective. On Monday Tri-State Assembly made 1,600 parts. How many of these parts are likely to be defective?

Item 7 refers to the following ad.

CRAFTER'S WORLD		
Fabric paints	5 bottles for	$2.25
Plastic canvas	3 sheets for	$0.99
Yarn, 4 oz.	2 skeins for	$2.86

7 Eva works at Crafter's World. A customer needs 5 sheets of plastic canvas to make holiday decorations. What price should Eva charge the customer?

Check your answers on pages 300–301.

Use proportions to solve each problem. Show your work.

8 It takes Manuel 2 hours to mat and frame 5 pictures. How long will it take him to mat and frame 16 pictures?

9 Michela is planning to paint her house. The color she wants requires 3 parts blue to 4 parts gray paint. How many gallons of blue paint does she need to mix with 10 gallons of gray paint?

10 Darius drove 300 miles on 12 gallons of gasoline. His tank holds 16 gallons. How far could Darius drive on a full tank of gasoline?

11 At Pioneer Insurance 5 out of 9 employees are women. Pioneer Insurance employs 432 people. How many women work at Pioneer Insurance?

Items 12–14 refer to the following map.

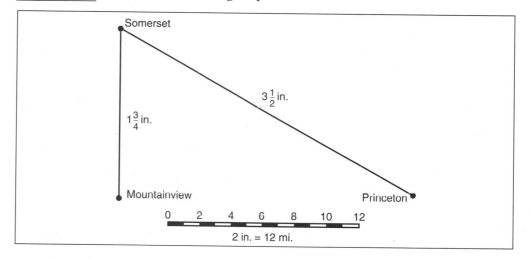

12 What is the actual distance from Mountainview to Somerset in miles?

13 What is the actual distance from Somerset to Princeton in miles?

14 Pine Ridge is 30 miles north of Somerset. If Pine Ridge were shown on this map, how many inches north of Somerset would it be?

Percent Basics

The Meaning of Percent

Percent is another way to show part of a whole. The sign for percent is %.

EXAMPLE Lana, a store clerk, is selling an AM/FM radio with a CD player for $100. The customer wants to make a down payment. Lana tells the customer that he must make a 25% down payment. She figures out that 25% of $100 is $25.

● **Percent** means "for every 100" or "out of 100."

The figure is divided into 100 equal parts. Since 25 out of 100 parts are shaded, the shaded portion is 25% of the whole figure.

● **Percents** can also be written as fractions or decimals.

The customer gives Lana $25 for the down payment. The customer was paying what fraction of the total cost for the radio?

(1) $\frac{1}{4}$ (4) $\frac{2}{3}$

(2) $\frac{1}{3}$ (5) $\frac{3}{4}$

(3) $\frac{1}{2}$

Answer **(1)** is correct. You can describe a percent as a fraction in lowest terms. $\frac{25}{100} = \frac{25 \div 25}{100 \div 25} = \frac{1}{4}$, so the customer was paying $\frac{1}{4}$ of $100, or $25.

EXAMPLE You can also write a percent as a decimal. What is 25% written as a decimal? (Hint: 25% is also $\frac{25}{100}$, or 25 hundredths.)

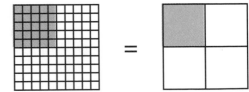

(1)   0.0025
(2)   0.025
(3)   0.25
(4)   25.0
(5)   250.0

Answer **(3)** is correct. Since you know **25%** equals $\frac{25}{100}$, you can write the decimal **0.25** for 25%.

PRACTICE

Circle the best answer for each item.

Items 1–3 refer to the following information.

Francisco is a bank teller. He counted the number of pennies in his bank drawer. The drawer held 30% of a dollar in pennies. The figure below shows this percent.

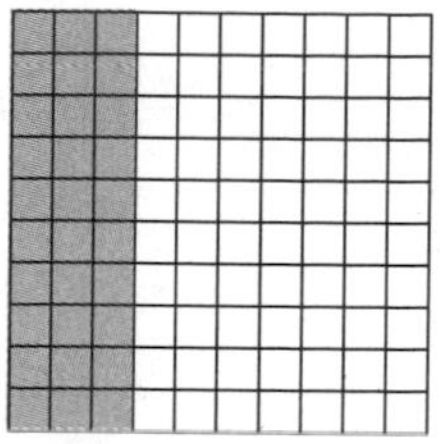

1. How many pennies does Francisco have in the drawer?

(1) 0.03
(2) 0.3
(3) 3
(4) 30
(5) 300

2. What fraction of a dollar does Francisco have in pennies?

(1) $\frac{1}{100}$
(2) $\frac{3}{100}$
(3) $\frac{3}{10}$
(4) $\frac{1}{3}$
(5) $\frac{3}{1}$

3. What dollar amount does Francisco have in pennies?

(1) $0.30
(2) $3.00
(3) $3.30
(4) $30.00
(5) $300.00

Items 4–6 refer to the following information.

Gina is a mathematics teacher. She gave a test that had 100 questions. The highest score on the test was 90%. The figure at the right shows this percent.

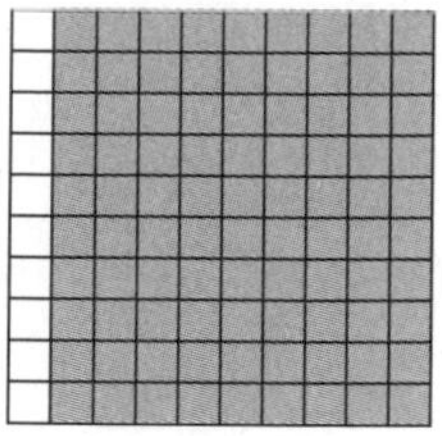

4. Marcus received the highest score. Gina wrote the scores on the tests as fractions. What fraction would you see on Marcus' test?

(1) $\frac{9}{100}$
(2) $\frac{90}{100}$
(3) $\frac{100}{90}$
(4) $\frac{10}{9}$
(5) $\frac{9}{1}$

5. Leah keeps track of all her tests in her notebook. She writes her scores as decimals so that she can average them at the end of the semester. Leah saw the fraction $\frac{85}{100}$ on her test. What decimal does Leah write in her notebook?

(1) 0.085
(2) 0.80
(3) 0.85
(4) 8.50
(5) Not enough information is given.

6. **Communicate** On a separate piece of paper, explain how you would find a fraction, a decimal, and a percent to describe the colored portion of the figure to the right.

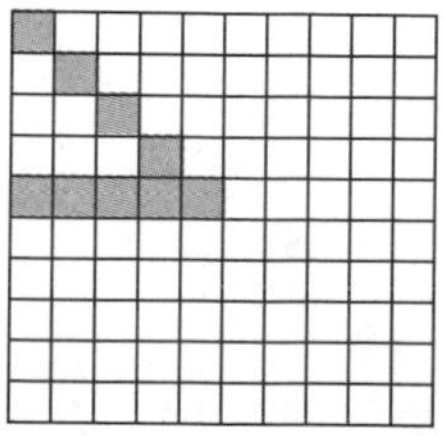

Check your answers on page 301.

Relating Percents, Fractions, and Decimals

TIP

Percents, fractions, and decimals are different ways to show parts of a whole.

You can write any percent as a decimal or a fraction. Since these numbers represent the same amount only in different forms, we can say they are **equivalent.**

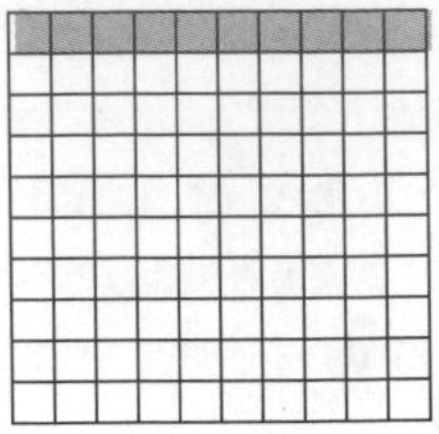

EXAMPLE Write a fraction, a decimal, and a percent to describe the colored portion of the figure at the right.

Step 1: Count the number of colored parts. 10

Step 2: Count the total number of parts. 10 out of 100 parts are colored. Write the fraction for "10 out of 100." Reduce. $\frac{10}{100} = \frac{1}{10}$

Step 3: Write the decimal. Hundredths is two places to the right of the decimal point. 0.10

Step 4: Write the percent. Percent means "out of 100," so 10 out of 100 would be 10%. 10%

PRACTICE

Write a fraction, a decimal, and a percent to describe the colored portion of each figure. The first one is done for you.

1.

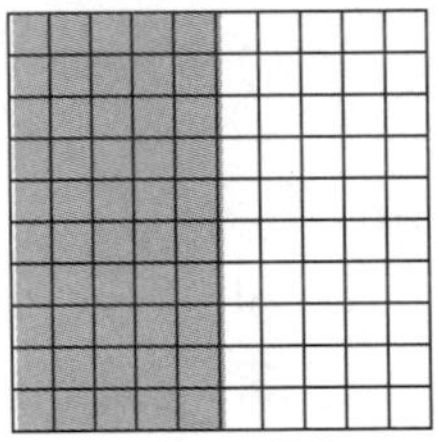

Fraction $\frac{50}{100} = \frac{1}{2}$

Decimal 0.50

Percent 50%

2.

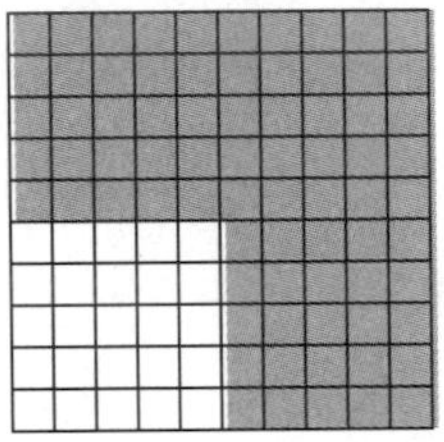

Fraction ________

Decimal ________

Percent ________

3.

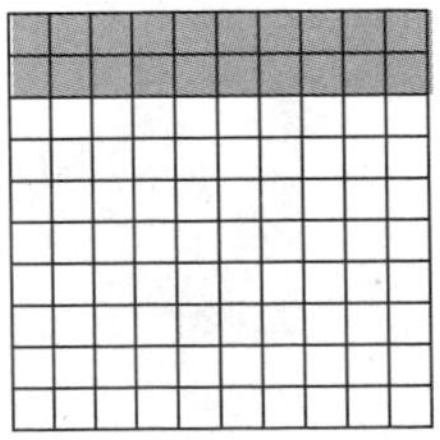

Fraction ________

Decimal ________

Percent ________

4.

Fraction ________

Decimal ________

Percent ________

Check your answers on page 301.

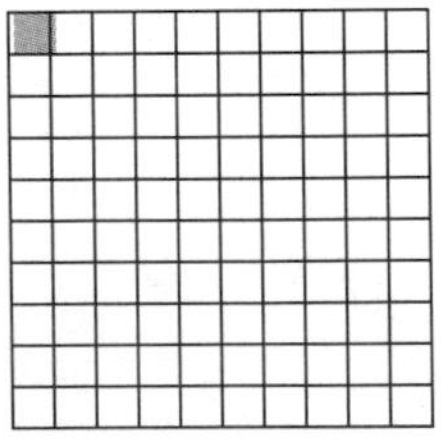

Changing Percents and Decimals

You can change a percent to a decimal and a decimal to a percent.

EXAMPLE Change 150% to a decimal.

TIP

To change a percent with a fraction to a decimal, change the fraction to its decimal form, drop the percent sign, then move the decimal point. For example, change $5\frac{3}{4}$% to a decimal.

First: $5\frac{3}{4}$% = 5.75%

Then: 5.75% = 0.0575

Step 1: Drop the percent sign. $\quad$ 150

Step 2: Write the number as a fraction with 100 as the denominator (since a percent is "out of 100"). Move the decimal point two places to the left. Add a zero as a placeholder if necessary. If a percent does not have a decimal point, it is understood to have one to the right of the ones place.

$\dfrac{150}{100}$

$1.50. = 1.5$

EXAMPLE Change 0.875 to a percent.

Step 1: Multiply by 100. Move the decimal point two places to the right. $\quad$ 0.87.5

Step 2: Write the percent sign after the number. $\quad$ 87.5%

PRACTICE

Change each percent to a decimal. The first one is done for you.

❶ 32% = _0.32_ ❷ 4% ❸ 250% ❹ $6\frac{1}{2}$%

❺ 5.25% ❻ 0.14% ❼ $97\frac{1}{4}$% ❽ 100%

Change each decimal to a percent. The first one is done for you.

❾ 2.60 = _260%_ ❿ 0.03 ⓫ 0.625 ⓬ 0.0008

⓭ 4.635 ⓮ 0.2 ⓯ 5.86 ⓰ $0.34\frac{1}{2}$

Changing Percents and Fractions

You can change a percent to a fraction and a fraction to a percent.

EXAMPLE   Change 75% to a fraction.

Step 1:   Drop the percent sign.   75

Step 2:   Write the number as a fraction with a $\frac{75}{100}$
          denominator of 100 (since a percent is
          "out of 100").

Step 3:   Reduce the fraction to lowest terms. $\frac{75}{100} = \frac{75 \div 25}{100 \div 25} = \frac{3}{4}$

EXAMPLE   Change $\frac{1}{8}$ to a percent.

Step 1:   Change the fraction to a decimal. To do this,
          divide the denominator into the numerator.

$$\begin{array}{r} 0.125 \\ 8\overline{)1.000} \\ -8 \\ \hline 20 \\ -16 \\ \hline 40 \\ -40 \\ \hline 0 \end{array}$$

Step 2:   Change the decimal to a percent. $0.12.5 = 12.5\%$

PRACTICE

Change each percent to a fraction in lowest terms. The first one is started for you.

1   60% = $\frac{60}{100}$ **2**   8%   **3**   52%   **4**   1%

5   250%   **6**   16%   **7**   100%   **8**   42%

Change each fraction to a percent. The first one is started for you.

9 $\frac{1}{2} = 2\overline{)1.0}^{\,0.5\,=} \atop {\underline{-1.0} \atop 0}$ **10** $\frac{3}{5}$ **11** $\frac{7}{10}$ **12** $\frac{3}{8}$

For more practice changing percents and fractions, see Calculator Handbook page 252.

13 $\frac{1}{4}$ **14** $\frac{5}{8}$ **15** $6\frac{5}{10}$ **16** $\frac{15}{5}$

Section 12 Review

Change each percent to a decimal.

1 37%

_____

2 4%

_____

3 225%

_____

4 $6\frac{1}{2}\%$

_____

Change each decimal to a percent.

5 0.46

_____

6 0.08

_____

7 2.5

_____

8 0.375

_____

Change each percent to a fraction.

9 50%

_____

10 40%

_____

11 25%

_____

12 90%

_____

Change each fraction to a percent.

13 $\frac{7}{10}$

_____

14 $\frac{3}{4}$

_____

15 $\frac{4}{5}$

_____

16 $\frac{7}{8}$

_____

Check your answers on page 302.

Solving for the Part *(p)*

The Percent Formula

Every percent problem has three elements: the base, the rate, and the part. For example: "30 is 50% of 60" or "50% of 60 is 30."

- The **base** *(b)* is the whole amount. The base (here, 60) represents 100%. The base usually follows the word *of*.

- The **rate** *(r)* is easy to find because it is always followed by the percent sign (%) or the word *percent* (here, 50%).

- The **part** *(p)* is part of the whole amount. The part (here, 30) usually precedes or follows the word *is*.

EXAMPLE Chris earns $240 a week as a receptionist. Each week he puts $9.60 or 4% of his earnings into a savings account. Which number is the base?

(1) 4%   (4) $9.60
(2) 0.04   (5) $240
(3) $3.84

Answer **(5)** is correct. The base, or the whole amount, is **$240**.

You can see the relationship of the three elements in the following formula. The triangle on the left will help you remember how these elements are related.

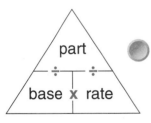

base × rate = part   **$240 × 4% = $9.60**

In percent problems, you will know two of the three elements. To solve, first identify the elements that you know. Then solve for the unknown element.

EXAMPLE Karl makes repairs for a large apartment building. Velma, Karl's supervisor, estimates he will spend 20% of his 10-hour day fixing the garbage disposal in Apartment 3B. Karl wants to find how many hours the job should take. Which element does Karl need to find?

(1) the base
(2) the rate
(3) the part
(4) the proportion
(5) Not enough information is given.

Answer **(3)** is correct. Karl knows the base (his whole 10-hour day) and the rate (20%). He needs to find the **part** (hours to fix the garbage disposal). To solve, he needs to multiply the base (10) by the rate (20%).

Circle the best answer for each item.

1 Hector works 8 hours a day in an office. He spends 20% of his time photocopying and filing. To find how many hours Hector spends on these tasks, which element do you need to find?

(1) the base
(2) the rate
(3) the part
(4) the proportion
(5) Not enough information is given.

2 Margaret earns $550 a week. Her employer deducts 8% or $44 for social security. Which number is the rate?

(1) $44
(2) 8%
(3) 8
(4) $352
(5) $550

3 Arturo's company employs 64 people. Of the employees, 16 are women. Arturo wants to know what percent of the employees are women. Which element does Arturo need to find?

(1) the base
(2) the rate
(3) the part
(4) the whole amount
(5) Not enough information is given.

4 Joyce works 6 hours a day at a school cafeteria. She spends 50%, or 3 hours, of her workday preparing food. Which number is the base?

(1) 6
(2) 3
(3) 50%
(4) 18
(5) 150

5 Mathew saw a sweatshirt originally priced at $24 on sale for 20% off the original price. Mathew wants to know how much he will save by buying the sweatshirt on sale. Which element does Mathew need to find?

(1) the base
(2) the rate
(3) the part
(4) the percent
(5) Not enough information is given.

6 Fashions Plus has a 25% discount on sweaters. A customer saves $10 on one sweater. To find the regular price of the sweater, what element do you need to find?

(1) the base
(2) the rate
(3) the part
(4) the interest
(5) Not enough information is given.

7 The Bluejays won 75%, or 12 games, of the 16 games they played this season. Which number is the part?

(1) 0.75
(2) 75%
(3) 16
(4) 12
(5) Not enough information is given.

8 **Communicate** Given the following numbers, $500, 20%, and $100, write a percent word problem. Use the triangle on page 176 to help you label the three elements. Then show the relationship of the three numbers using the percent formula.

Finding the Part (*p*)

Recall that the **part** is a portion of the whole amount. You find the part by multiplying the whole amount (base) by the percent (rate).

base × **rate** = **part**

EXAMPLE   What is 25% of 800?

Step 1: Identify the rate and the base. You know that 25% is the rate because 25 is followed by the percent sign. The number 800 is the base. The base often follows the word *of*.

The rate is 25%.
The base is 800.

Step 2: Change the percent to a decimal. Drop the percent sign (%). Then move the decimal point two places to the *left*.

$25\% = .25. = 0.25$

Step 3: Use the percent formula, *part = base × rate*, to find the part. Multiply the base (800) by the rate (0.25) to solve for the part.

$$\begin{array}{r} 800 \\ \times 0.25 \\ \hline 4000 \\ +1600 \\ \hline 200.00 \end{array}$$ ← 2 decimal places

← 2 decimal places

200 is 25% of 800.

> **TIP**
>
> When the rate is greater than 100%, the part will be greater than the base. Change the rate to a decimal as you would with rates less than 100%. For example, 150% becomes 1.5.

Use the percent triangle to help you remember the formula. Cover the letter *p* because you need to solve for the part. The other elements, the base and the rate, are connected by the multiplication sign. You can solve the problem by multiplying the base times the rate.

Sometimes the rate contains a fraction.

EXAMPLE   What is $5\frac{1}{2}\%$ of $60?

Step 1: Identify the rate and the base.

The rate is $5\frac{1}{2}\%$.
The base is $60.

Step 2: Change the fraction percent to a decimal percent.

$5\frac{1}{2}\% = 5.5\%$

Step 3: Change the percent to a decimal.

$5.5\% = 0.05.5 = 0.055$

Step 4: Multiply the base ($60) by the rate (0.055) to find the part.

$$\begin{array}{r} \$60 \\ \times 0.055 \\ \hline \$3.300 \end{array}$$ ← 3 decimal places

← 3 decimal places

$3.30 is $5\frac{1}{2}\%$ of $60.

A. **Solve. Show your work. The first one is done for you.**

1 What is 35% of 220?

$$35\% = 0.35$$

$$
\begin{array}{r}
220 \\
\times 0.35 \\
\hline
1100 \\
+660 \\
\hline
77.00
\end{array}
$$

2 What is 40% of $45?

3 What is 8% of 16?

4 What is 250% of 48?

5 What is $4\frac{1}{2}$% of $50?

6 What is $7\frac{1}{4}$% of 144?

B. **Solve. Show your work. The first one is done for you.**

7 Laura bought a pair of jeans that regularly cost $32 for 25% off the regular price. How much did Laura save?

$$25\% = 0.25$$

$$
\begin{array}{r}
\$32 \\
\times 0.25 \\
\hline
160 \\
+64 \\
\hline
\$8.00
\end{array}
$$

8 The Jaguars won 80% of the 15 games they played this season. How many games did they win?

9 Of the 8,000 people at the baseball game, 5% were season ticket holders. How many people at the game were season ticket holders?

10 Electronic Buys sold 240 items last week. Of the items sold, 45% were kitchen appliances. How many kitchen appliances did Electronic Buys sell last week?

C. **Solve.**

11 Rick earns $32,500 a year. He spends 16% of his salary on rent. How much does Rick spend on rent a year?

12 A gallon of gasoline costs $1.32. Of that amount, 27% goes to state and local taxes. To the nearest cent, how much money per gallon goes to taxes?

Simple Interest Formula

Interest is a charge that someone pays to borrow or to use someone else's money. The amount of interest charged is based on the **principal** or amount of money borrowed. An **interest rate** is used to figure out how much interest the borrower has to pay.

You can use a formula (similar to the percent formula) to find the amount of simple interest owed.

$i = p \times r \times t$, where i is the amount of **interest** (similar to the **part**), p is the **principal** (similar to the **base**), r is the **interest rate** (the rate), and t is the **time** in years over which the money is borrowed (time must be included because the rate includes time, usually % per year).

EXAMPLE Gordon wants to buy a refrigerator. J & B Appliances has advertised a used refrigerator for $388. Gordon wants to pay $100 down on the refrigerator and the rest in monthly payments over 2 years. The store would charge an 18% interest rate on the remaining $288. Which is the correct expression to find how much simple interest Gordon will pay on the money he owes (the principal)?

(1) $388 + 0.18 + 2
(2) $388 × 0.18 × 2
(3) $388 × 100 × 2
(4) $288 × 0.18 × 2
(5) $288 + 0.18 + 2

Answer **(4)** is correct. Before finding the interest you must identify the principal. The principal is the amount owed. Gordon still owes $288 (he paid $100 when he bought the refrigerator). So $288 is the principal. The rate is 18%. The time is 2 years. Use the simple interest formula to find the interest Gordon owes. To multiply by the rate, first change the percent to a decimal: 18% = 0.18. = 0.18

$288 × 0.18 × 2 = $103.68, so Gordon will pay **$103.68** in interest in order to pay in installments over 2 years.

EXAMPLE If Gordon buys the refrigerator, how much will he end up paying for the refrigerator including the original price and the interest?

(1) $184.32
(2) $339.84
(3) $391.68
(4) $443.52
(5) $491.68

Answer **(5)** is correct. Add the cost of the refrigerator ($388.00) and the interest ($103.68). **$388.00 + $103.68 = $491.68**, so the total cost will be **$491.68**.

Circle the best answer for each item.

1 Dan borrowed $650 for 1 year at a rate of 12% to make home improvements. Which is the correct expression to find the interest Dan paid?

(1) $650 + 0.12 + 1
(2) $650 × 0.12 × 1
(3) $650 × 0.12 × 0.01
(4) $650 × 12 × 1
(5) $650 × 0.12 + 1

2 The Davis family took out a loan of $420 for 1 year at a rate of 18%. How much interest will the family owe?

(1) $75.60
(2) $76.60
(3) $110.60
(4) $326.00
(5) $336.00

3 Donna took out a $1,240 loan for college for 2 years. If the interest rate is $8\frac{1}{2}$%, how much will Donna owe in interest?

(1) $52.70
(2) $105.40
(3) $198.40
(4) $210.80
(5) $2,108.00

4 Mr. Martinez deposited $1,600 in his savings account. If his savings account earns $5\frac{1}{4}$% interest, how much will Mr. Martinez earn in interest after 3 years?

(1) $240
(2) $252
(3) $320
(4) $400
(5) $533

5 Morris bought a stove for $560. He paid $150 down. He will pay the rest at the end of 2 years at 16% interest. Which is the correct expression to find how much interest Morris will owe after 2 years?

(1) $150 × 0.16 × 2
(2) $410 × 0.16 × 2
(3) $410 × 16 × 2
(4) $560 × 0.16 × 2
(5) $710 × 0.16 × 2

6 Recall the amount paid back is the total of the principal plus the interest. How much will Morris have to pay back?

(1) $410.00
(2) $524.10
(3) $541.20
(4) $608.00
(5) Not enough information is given.

7 Freddie got a personal loan of $2,100 from a friend. He will pay the money back over 1 year at 12% interest. How much will Freddie pay back to his friend?

(1) $252
(2) $2,100
(3) $2,112
(4) $2,352
(5) Not enough information is given.

8 **Reason** Gina bought a used car for $12,000. Gina paid $2,000 as a down payment. She will pay the rest in monthly payments over 2 years at 15% interest. Explain how you would estimate the monthly payment. Then do the actual calculations. Show your work.

DATA ANALYSIS — Reading and Making a Circle Graph

EXAMPLE Sonia Velez wants to make a circle graph of the Velez family budget. She gathers information and writes it in a table. The percents in the table total 100% and represent the amount of money the family has for expenses.

Rent	30%
Food	25%
Personal needs	20%
Transportation	15%
Other	10%

A circle graph is a good way to compare percents. The whole circle represents 100%. The sections of the circle show parts of the whole.

  The sections of a circle graph are formed with central angles. The central angles of a circle add up to 360°. The angle shown has a measure of 45°.

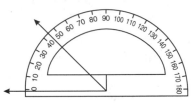

To make a circle graph, Sonia first draws a circle. To make her first section, Sonia wants to find how many degrees are in the angle representing 30%. Which is the correct expression to find how many degrees are in the angle representing 30%?

(1) $360° \times 30\%$ (4) $360° - 30$
(2) $360° \div 30\%$ (5) $30\% \div 360°$
(3) $360° + 30\%$

Answer **(1)** is correct. The entire circle ($360°$) is the base. The rate is 30%. The degrees representing 30% is the part. Use the percent formula to solve the problem. *part = base × rate* = $360° \times 30\%$

Change the percent to a decimal before multiplying. 30% = 0.30
$360 \times 0.30 = 108$, so Sonia draws an angle that measures **108°**. Using this same process for each expense, Sonia completes the circle graph.

EXAMPLE Sonia knows the family's take-home pay for one month is $1,800. Which is the correct expression to find out what dollar amount is spent on rent?

(1) $\$1,800 \times 30$ (4) $\$1,800 \div 30$
(2) $\$1,800 \times 0.30$ (5) $30 \div \$1,800$
(3) $\$1,800 \div 0.30$

Answer **(2)** is correct. The family's take-home pay ($1,800) is the base. The rate is 30%. The dollar amount spent on rent is the part. Use the percent formula to solve the problem.
part = base × rate = $\$1,800 \times 30\%$

Change the percent to a decimal before multiplying. 30% = 0.30
$\$1,800 \times 0.30 = \$540$, so the Valdez family spends **$540** on rent each month.

PRACTICE

Circle the best answer for each item.

Items 1–3 refer to the following table.

Paycheck Deductions	Percent of Total Deductions
Federal income tax	50%
State income tax	20%
Health insurance	15%
Credit union	10%
Retirement fund	5%

1. Marlo wants to make a circle graph of the deductions from her paycheck. Which is the correct expression to find the number of degrees in the angle representing 15%?

   (1) 360° ÷ 15
   (2) 360° × 15
   (3) 360° × 0.15
   (4) 360° ÷ 0.15
   (5) 15 ÷ 360°

2. The total deducted from Marlo's check this pay period is $120. What amount went for health insurance?

   (1) $12.00
   (2) $15.00
   (3) $18.00
   (4) $54.00
   (5) $120.15

3. For which item is the greatest amount deducted?

   (1) federal income tax
   (2) state income tax
   (3) health insurance
   (4) credit union
   (5) retirement fund

Items 4–6 refer to the following table.

Use of Electricity for Quan Household

Heating	40%
Refrigerator	20%
Water heater	10%
Cooking	5%
Other	25%

4. Mr. Quan wants to make a circle graph of the household use of electricity. Which is the correct expression to find the number of degrees representing the angle for heating (40%)?

   (1) 40 ÷ 360°
   (2) 360° ÷ 40
   (3) 360° × 40
   (4) 360° × 0.40
   (5) 360° ÷ 0.40

5. Mr. Quan wants to find out how much of his electric bill is spent on heat. Which is the correct expression to find the cost of heat each month?

   (1) $32 × 0.40
   (2) $32 ÷ 40
   (3) $32 × 40
   (4) $32 ÷ 0.40
   (5) Not enough information is given.

6. **Communicate** On a separate piece of paper, explain how to find the number of degrees in each section of the circle graph. Include calculations with your answer.

Check your answers on page 304.

Section 13 Review

Solve. Show your work.

1 What is 64% of 300?

2 What is 6% of 72?

3 What is 90% of $96?

4 What is 400% of $56?

5 What is 125% of 36?

6 What is $3\frac{1}{2}$% of 150?

Solve. Show your work.

7 The Mortons' health insurance covers 80% of the cost of medical care. John Morton had a series of lab tests that cost $145. How much of this cost was covered by his insurance?

9 In a survey, 3,000 adults were asked which newspaper they prefer to read. Fifty-five percent said they prefer a morning paper. How many adults prefer a morning paper?

8 Jill buys a coat that regularly sells for $125 for 33% off. How much does Jill save?

10 Miguel puts 12% of his earnings in a savings account. This week Miguel earns $310. How much will he put in his savings account?

Check your answers on page 304.

11 Seth borrows $1,500 from his brother for 1 year and agrees to pay him $8\frac{1}{2}$% interest. How much interest will Seth owe his brother?

12 Mayeta borrows $2,000 from a friend for 2 years at 14% interest to start her own cleaning service. How much interest will Mayeta pay on this loan?

13 Raghib lends his brother $1,200 for 1 year at $12\frac{1}{2}$% interest. How much interest will Raghib be paid on the loan?

14 Aretha is recarpeting her house for $1,750. She pays $250 down. She will pay the rest in monthly payments over 2 years at 18% interest. Estimate how much Aretha will have to pay back.

15 Jackie borrows $1,500 from her father for classes at the Technical Institute. She agrees to pay the money back at the end of 2 years at 15% interest. What is the amount Jackie will pay back?

16 Jesse buys furniture for $990. He will pay for it in monthly payments over 2 years at 18% interest. Estimate Jesse's monthly payment for the furniture.

17 Carol borrowed $9,350 at 12% interest to start a business. She will pay back the loan at the end of 5 years. How much will Carol pay back at the end of 5 years?

18 The Tallchiefs bought a refrigerator for $475 and a stove for $680. They paid $150 down. They will pay the rest in monthly payments over 1 year at 16% interest. Estimate the Tallchiefs' monthly payment.

Circle the best answer for each item.

Items 19–21 refer to the following information.

Dayna is the manager of the clothing departments at Shop-Wise. She is scheduled to work 40 hours each week. The table shows what parts of her total week she spends working at different tasks.

Task	Percent of Time
Supervising	40%
Planning	25%
Selling	20%
Customer relations	15%

19 Which is the correct expression to find how many hours Dayna spends on customer relations each week?
   (1) 15 × 40%
   (2) 15 ÷ 40%
   (3) 40% × 0.15
   (4) 40 × 15%
   (5) 40 ÷ 15%

20 Which is the correct expression to find the total rate of time Dayna spends on supervising and planning each week?
   (1) 40% × 25%
   (2) 40% + 25%
   (3) 40% − 25%
   (4) 40 × 40% × 25%
   (5) 40 × 40% + 25%

21 How many hours does Dayna spend selling each week?
   (1)   8
   (2) 10
   (3) 20
   (4) 40
   (5) 80

Items 22–24 refer to the following table.

Wagner Family Budget

Rent and utilities	35%
Food	25%
Transportation	20%
Clothes	15%
Other	5%

22 Mrs. Wagner wants to make a circle graph of the family budget. Find the number of degrees in each section of the graph.

   Rent and utilities _____

   Food _____

   Transportation _____

   Clothes _____

   Other _____

Circle the best answer for each item.

23 For which item do the Wagners budget the least amount?
   (1) Rent and utilities
   (2) Food
   (3) Transportation
   (4) Clothes
   (5) Other

24 The Wagners' total monthly income is $2,200. How much do they budget for rent and utilities?
   (1) $35
   (2) $350
   (3) $550
   (4) $570
   (5) $770

  Check your answers on page 305.

**ALGEBRA
TOPICS**

The Percent Formula

You can rewrite the percent formula to solve for the **rate** in percent problems.

EXAMPLE Jan traveled a distance of 780 miles on her vacation. The first day of her trip she drove 234 miles. What percent of the total distance did she drive the first day?

Before you can solve for the rate, *r*, you must identify the base and the part. The base, *b*, is the total distance, 780 miles. The distance Jan drove the first day is part of the total distance. The part, *p*, is 234 miles.

Using the percent triangle, you can rewrite the percent formula to solve for the rate. Cover the letter *r*. The other elements, the part and the base, are separated by the division sign.

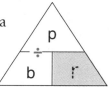

 part ÷ base = rate

Which is the correct expression to find the percent of the total distance Jan drove the first day?
(1) 234 ÷ 780
(2) 234 × 780
(3) 780 ÷ 234
(4) 780 − 234
(5) 780 + 234

Answer **(1)** is correct. The total distance (780) is the base. The distance Jan drove on the first day (234) is the part. To find the rate, divide the part by the base. **234 ÷ 780 = 0.30 = 30%,** so Jan drove **30%** of the total distance the first day.

EXAMPLE The Hillside Neighborhood Association has 450 members. In May 360 members attended the annual meeting. Which is the correct expression to find the percent of the association members who attended the meeting?
(1) 360 ÷ 450
(2) 450 ÷ 360
(3) 450 − 360
(4) 450 + 360
(5) 450 × 360

Answer **(1)** is correct. The total membership (450) is the base. The members at the meeting (360) is the part. Divide the part by the base to find the rate. **360 ÷ 450 = 0.8 = 80%,** so **80%** of the members attended the meeting.

PRACTICE

Circle the best answer for each item.

1 Mr. Cruz earns $1,150 a month. He pays $299 a month for rent. Which is the correct expression to find what percent of his income he pays for rent?

(1) $1,150 ÷ $299
(2) $1,150 − $299
(3) $299 ÷ $1,150
(4) $299 + $1,150
(5) $299 × $1,150

2 Chiang bought a used car that cost $4,500. She made a down payment of $500. Which is the correct expression to find what percent of the total cost she put down on the car?

(1) $500 ÷ $4,500
(2) $500 × $4,500
(3) $500 + $4,500
(4) $4,500 − $500
(5) $4,500 ÷ $500

3 A set of tires regularly sells for $200. Lamont saved $30 by buying the tires on sale. Which is the correct expression to find what percent of the regular price Lamont saved by buying the tires on sale?

(1) $200 − $30
(2) $200 ÷ $30
(3) $30 ÷ $200
(4) $30 × $200
(5) $30 + $200

4 Computer Network employs 110 people. Of the employees, 65 are women. Which is the correct expression to find what percent of the employees are women?

(1) 110 ÷ 65
(2) 110 − 65
(3) 65 ÷ 110
(4) 65 + 110
(5) 65 × 110

5 Ms. Chu earns $380 a week. Her employer deducts $45 out of her paycheck each week. Which is the correct expression to find what percent of Ms. Chu's earnings are taken out of her paycheck for deductions?

(1) $380 − $45
(2) $380 ÷ $45
(3) $45 + $380
(4) $45 ÷ $380
(5) $45 × $380

6 Flavia earns $420 a week. She shops for food once a week. Which is the correct expression to find what percent of Flavia's earnings is spent on food?

(1) $50 ÷ $420
(2) $50 × $420
(3) $420 − $50
(4) $420 ÷ $50
(5) Not enough information is given.

7 There are 300 calories in a box of macaroni and cheese. Of the 300 calories, 117 calories come from fat. Which is the correct expression to find what percent 117 calories is of 300 calories?

(1) 300 ÷ 117
(2) 300 − 117
(3) 117 + 300
(4) 117 ÷ 300
(5) Not enough information is given.

8 **Reason** To find the *part* in a percent problem you used the percent formula, *part = base × rate.* To find the *rate* in a percent problem, you rewrote the formula so that *rate = part ÷ base.* On a separate piece of paper, explain in a sentence or two how the two formulas are related.

Reading a Pay Stub

EXAMPLE Bill earns $1,100 per month. The amount he earns before taxes and other deductions are taken out of his paycheck is called his gross pay.

Bill can see from this payroll check stub that amounts are taken out of each check to pay federal and state income taxes. Bill wants to find what percent of his gross pay is used to pay federal income tax.

Bill's gross pay is $1,100. Since the gross pay is the whole amount he earns, $1,100 is the base. The part is $132, the amount deducted for federal income tax.

Bill Mitchell	
Gross pay	$1,100.00
Deductions:	
Federal income tax	132.00
State income tax	15.40
Other	84.13
Total deductions	$ 231.53
Net pay	$ 868.47

Which is the correct expression to find the percent of Bill's gross pay used to pay federal income tax?
(1) $132 + $1,100
(2) $132 ÷ $1,100
(3) $1,100 ÷ $132
(4) $1,100 × $132
(5) $1,100 − $132

Answer **(2)** is correct. Bill uses the percent triangle. He covers the letter r because he needs to solve for the rate. He needs to divide the part ($132) by the base ($1,100). **$132 ÷ $1,100 = 0.12 = 12%**, so Bill finds that **12%** of his gross pay is taken out of each check to pay federal income tax.

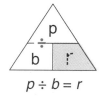

$p \div b = r$

EXAMPLE Next, Bill wants to find what percent of his gross pay is used to pay state income tax. Which is the correct expression to find the percent of Bill's gross pay used to pay state income tax?
(1) $15.40 ÷ $1,100.00
(2) $15.40 + $1,100.00
(3) $1,100.00 × $15.40
(4) $1,100.00 ÷ $15.40
(5) $1,100.00 − $15.40

Answer **(1)** is correct. To solve for the rate, divide the part ($15.40) by the base ($1,100.00). **$15.40 ÷ $1,100.00 = 0.014 = 1.4%**, so Bill finds that **1.4%** of his gross pay is used to pay state income tax.

Circle the best answer for each item.

Items 1–3 refer to the following payroll check stub.

```
Leona Hill

Gross pay                    $600.00

Deductions:

 Federal income tax            84.00
 State income tax              12.60
 Health insurance              30.00
 Total deductions            $126.60

Net pay                      $473.40
```

Items 4–6 refer to the following payroll check stub.

```
Sylvia Sanchez

Gross pay                  $1,250.00

Deductions:

 Federal income tax          143.75
 State income tax             23.75
 Retirement fund              62.50
 Insurance                    25.00
 Total deductions          $ 255.00

Net pay                    $ 995.00
```

1 Which is the correct expression to find the percent of Leona's gross pay used for health insurance?

(1) $30.00 ÷ $473.40
(2) $30.00 ÷ $600.00
(3) $600.00 − $30.00
(4) $600.00 × $30.00
(5) $600.00 ÷ $30.00

2 Which is the correct expression to find the percent of Leona's gross pay used for taxes?

(1) $84.00 ÷ $600.00
(2) $84.00 + $12.60 ÷ $600.00
(3) ($84.00 + $12.60) ÷ $600.00
(4) $600.00 − ($84.00 + $12.60)
(5) $600.00 ÷ ($84.00 + $12.60)

3 Leona plans to contribute 3% of her gross pay to a retirement fund. Which is the correct expression to find how much she will contribute each pay period?

(1) 3 × $600.00
(2) 97% × $600.00
(3) $600.00 × 3%
(4) $600.00 ÷ 3%
(5) $600.00 ÷ 300

4 Which is the correct expression to find the percent of Sylvia's gross pay used for federal income tax?

(1) $143.75 ÷ $1,250.00
(2) $143.75 × $1,250.00
(3) $230.00 ÷ $1,250.00
(4) $1,250.00 ÷ $143.75
(5) $1,250.00 − $143.75

5 Sylvia saves 5% of her gross pay. Which is the correct expression to find how much she saves each pay period?

(1) 5% ÷ $995.00
(2) 5% ÷ $1,250.00
(3) $995.00 × 5%
(4) $1,250.00 × 5%
(5) Not enough information is given.

6 **Connect** On a separate piece of paper, explain how to find the percent of Sylvia's gross pay used for deductions other than taxes. Then find the percent.

Finding the Rate

Use the percent triangle to help you find the rate, r, (percent). Cover the letter r because you need to solve for the rate. The other two elements, the part and the base, are separated by the division sign.

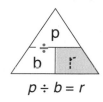

$p \div b = r$

EXAMPLE What percent of 62 is 31?

Step 1: Identify the base and the part.

The base is 62.
The part is 31.

Step 2: Rewrite the percent formula to find the rate.

$p \div b = r$

Step 3: Divide the part (31) by the base (62) to find the rate.

$31 \div 62 = r$

$$\begin{array}{r} 0.5 \\ 62\overline{)31.0} \\ -310 \\ \hline 0 \end{array}$$

Step 4: Change the decimal to a percent. Move the decimal point two places to the right.

$0.5 = 0.50. = 50\%$

31 is 50% of 62.

EXAMPLE To the nearest whole percent, what percent of 16 is 9?

Step 1: Identify the base and the part.

The base is 16.
The part is 9.

Step 2: Rewrite the percent formula to find the rate.

$p \div b = r$

Step 3: Divide the part (9) by the base (16) to find the rate. Carry the division out to three decimal places.

$9 \div 16 = r$

$$\begin{array}{r} 0.562 \\ 16\overline{)9.000} \\ -80 \\ \hline 100 \\ -96 \\ \hline 40 \\ -32 \\ \hline 8 \end{array}$$

Step 4: Round the decimal to the nearest hundredth.

0.562 rounds to 0.56

Step 5: Change the decimal to a percent. Move the decimal point two places to the right.

$0.56 = 0.56. = 56\%$

9 is about 56% of 16.

A. **Solve. Show your work. Round your answers to the nearest whole percent. The first one is done for you.**

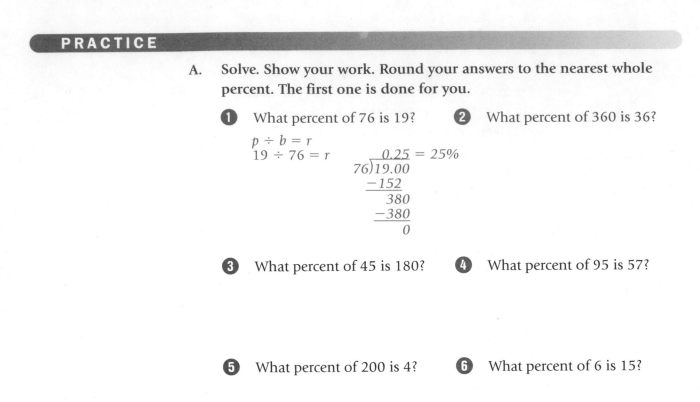

1 What percent of 76 is 19?

$p \div b = r$
$19 \div 76 = r$

$$\begin{array}{r} 0.25 = 25\% \\ 76\overline{)19.00} \\ -152 \\ \hline 380 \\ -380 \\ \hline 0 \end{array}$$

2 What percent of 360 is 36?

3 What percent of 45 is 180?

4 What percent of 95 is 57?

5 What percent of 200 is 4?

6 What percent of 6 is 15?

B. **Solve. Show your work. Round your answers to the nearest whole percent. The first one is started for you.**

7 Lyle answered 32 out of 40 questions on a test correctly. What percent of the questions did he answer correctly?

$32 \div 40 = r$

8 Disk Company employs 74 people. Of the employees, 42 are women. What percent of the employees are women?

9 Bev saved $9 on a sweater that regularly sells for $36. What percent of the regular price did she save?

10 Dwayne's weekly gross pay is $460.00. Deductions of $73.60 are taken out of his paycheck each week. What percent of Dwayne's gross pay is taken out of his paycheck for deductions?

C. **Solve. Round your answers to the nearest whole percent.**

11 The Goto family earned a total annual income of $46,000. The family saved $1,100 for the year. What percent of their total income did they save?

12 Furniture Buys was having a sale on couches. One of the couches was marked down from $950 to $570. What percent of the original price was the sale price?

Using Proportions to Solve Percent Problems

Up until now, you have solved percent problems using the percent formula working with the rate as a decimal. You can also solve a percent problem using the percent formula as a proportion. To solve a percent problem using a proportion, think of the rate (percent) as a ratio comparing an amount out of 100, or $\frac{?}{100}$.

● **The rate (percent) can be written as a fraction $= \frac{?}{100}$**

● **The percent formula, $p \div b = r$, can be written as a proportion: $\frac{part}{base} = \frac{rate}{100}$.**

TIP

Recall the steps for solving a proportion:

Step 1: Cross multiply

Step 2: Divide

EXAMPLE Andrew's doctor told him to avoid foods in which more than 30% of the calories come from fat. Andrew decides to compare the fat content of a bag of microwave popcorn and a bag of potato chips. The labels from the products are shown below. Andrew knows that there are 9 calories in every gram of fat. He can find how many calories come from fat by multiplying the grams of fat by 9: $5 \times 9 = 45$.

MICROWAVE *LITE* POPCORN
SERVING SIZE ....4 CUPS
CALORIES 150
PROTEIN, grams 5
FAT, grams 5
CARBOHYDRATE, grams 21
SODIUM, milligrams 180

RANCH STYLE LIGHT CHIPS
SERVING SIZE ....14 CHIPS
CALORIES 130
PROTEIN, grams 1
FAT, grams 6
CARBOHYDRATE, grams 18
SODIUM, milligrams 110

Which is the correct proportion to find what percent 45 calories is of 150 calories?

(1) $\frac{150}{100} = \frac{45}{?}$ (4) $\frac{45}{150} = \frac{100}{?}$

(2) $\frac{150}{45} = \frac{?}{100}$ (5) $\frac{45}{100} = \frac{?}{150}$

(3) $\frac{45}{150} = \frac{?}{100}$

Answer **(3)** is correct. 45 calories is the part, and 150 calories is the base. The rate is unknown. Set up the proportion and solve for the unknown.
$\frac{45}{150} = \frac{?}{100} = 4{,}500 \div 150 = 30$, so 45 calories is **30%** of 150 calories.

EXAMPLE Andrew wants to find the percent of calories that comes from fat in potato chips. First, he multiplies the 6 grams of fat in a serving by 9 to find 54 of the 130 calories come from fat. Using a proportion, what percent of 130 calories is 54 calories?

(1) 30%   (4) 42%

(2) 36%   (5) 43%

(3) 40%

Answer **(4)** is correct. Set up the proportion and solve for the unknown.
$\frac{54}{130} = \frac{?}{100} = 5{,}400 \div 130 = 41.5$, so about **42%** of the 130 calories is from fat.

Circle the best answer for each item.

Items 1–3 refer to the following label.

SPAGHETTI	
SERVING SIZE ....2 OUNCES (DRY)	
CALORIES 200	
PROTEIN, grams 7	
FAT, grams 1	
CARBOHYDRATE, grams 41	
SODIUM, milligrams 0	

Items 4–6 refer to the following label.

OAT CEREAL	
SERVING SIZE ....1 OUNCE	
CALORIES 120	
PROTEIN, grams 4	
FAT, grams 2	
CARBOHYDRATE, grams 21	
SODIUM, milligrams 290	

1 What two amounts do you need to know to find how many calories are in a 2-ounce serving of spaghetti?

(1) grams of fat and serving size

(2) grams of fat and number of calories in a gram of fat

(3) grams of fat and grams of protein

(4) serving size and number of calories in a gram of fat

(5) grams of protein and number of calories in a gram of fat

2 Recall there are 9 calories in 1 gram of fat. Which is the correct proportion to find what percent 9 calories is of 200 calories?

(1) $\frac{200}{9} = \frac{?}{100}$

(2) $\frac{63}{200} = \frac{?}{100}$

(3) $\frac{9}{200} = \frac{?}{100}$

(4) $\frac{9}{200} = \frac{100}{?}$

(5) $\frac{200}{63} = \frac{?}{100}$

3 Using a proportion, find what percent of 200 calories is 9 calories.

(1) 0.045%

(2) 4.5%

(3) 9%

(4) 22.2%

(5) 222%

4 Which is the correct proportion to find what percent 18 calories is of 120 calories?

(1) $\frac{120}{18} = \frac{?}{100}$

(2) $\frac{18}{120} = \frac{?}{100}$

(3) $\frac{18}{100} = \frac{?}{120}$

(4) $\frac{18}{120} = \frac{100}{?}$

(5) $\frac{18}{100} = \frac{120}{?}$

5 Using a proportion, what percent of 120 calories is 18 calories?

(1) 15%

(2) 21%

(3) 64%

(4) 66%

(5) Not enough information is given.

6 **Communicate** Andrew was told to avoid foods in which over 30% of the calories come from fat. Should Andrew avoid spaghetti or oat cereal? Look at the labels above. On a separate piece of paper, explain your answer in a sentence or two. Use percents in your explanation.

DATA ANALYSIS **Budgets and Circle Graphs**

EXAMPLE Ella works for the Conrad Chemical Company. She is helping her boss, Paula, collect information on the company's yearly budget. Paula gives Ella the following circle graph.

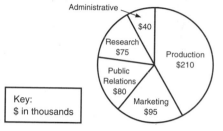

Conrad Chemical Company Yearly Budget

Key:
$ in thousands

Which is the correct expression to find what percent of the total budget is spent on marketing?
(1) $95 × $500
(2) $95 ÷ $500
(3) $500 ÷ $95
(4) $500 − $95
(5) $500 + $95

Answer **(2)** is correct. First, find the total dollar amount for all the areas on the graph: $210 + $95 + $80 + $75 + $40 = $500. To find the rate (percent), divide the part ($95) by the base ($500).

$95 ÷ $500 = 0.19 = 19%, so the amount spent on marketing is **19%** of the company budget.

Note: Because the key states that the dollar amounts are in thousands, you would multiply each amount by $1,000 to find the actual dollar amounts. The company's yearly budget is $500,000, and $95,000 is spent on marketing. This is still 19% of the total budget.

EXAMPLE The amount in the section labeled *Production* is $210. The amount in the *Research* section is $75. What is the total percent spent on research and production?
(1) 47%
(2) 54%
(3) 57%
(4) 63%
(5) 75%

Answer **(3)** is correct. First, add the amounts from the two sections on the graph: $210 + $75 = $285. To solve for the rate, divide the part ($285) by the base ($500). **$285 ÷ $500 = 0.57 = 57%**, so the amount spent on research and production is **57%** of the total budget.

Circle the best answer for each item.

Items 1–3 refer to the following graph.

The Morenos' Monthly Budget

Items 4–6 refer to the following graph.

Annual Sales of Team-Wear, Inc., 1999

Key:
$ in thousands

1 What is the total amount of the Morenos' monthly budget?

(1) $100
(2) $360
(3) $1,500
(4) $1,600
(5) $5,760

2 To the nearest whole percent, what percent of the Morenos' budget is for rent and utilities?

(1)   3%
(2)   17%
(3)   38%
(4)   60%
(5)   600%

3 To the nearest whole percent, what percent of the budget is for personal expenses and other expenses?

(1)   4%
(2)   22%
(3)   28%
(4)   45%
(5)   450%

4 What were Team-Wear, Inc.'s total sales in 1999?

(1) $65
(2) $135
(3) $200
(4) $2,000
(5) $200,000

5 To the nearest percent, what percent of the sales came from T-shirts and sweatshirts?

(1)   25%
(2)   43%
(3)   50%
(4)   68%
(5)   Not enough information is given.

6 **Reason** How much larger is the percent of sales from caps than from jackets? On a separate piece of paper, explain how you would find this difference. Then answer the question.

Section 14 Review

Solve. Show your work.

1 What percent of 10 is 16?

2 What percent of 72 is 54?

3 What percent of 300 is 12?

4 What percent of 15 is 75?

5 Sedika took a 600-mile trip. She drove 240 miles before stopping for lunch. What percent of the total trip did she drive before lunch?

6 Curtis paid $80 for a jacket that originally cost $96. What percent of the original price did Curtis pay?

7 Of the 220 employees at Adams Department Store, 125 work part-time. What percent of the employees work part-time?

8 Sarah works 40 hours per week as assistant manager of a hotel. She spends an average of 12 hours each week responding to guest comments and complaints. What percent of her time does Sarah spend responding to guest comments and complaints?

Items 9–11 Refer to the following graph.

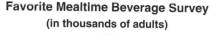

Favorite Mealtime Beverage Survey
(in thousands of adults)

Other 2
Soda 4
Coffee 7
Milk 5
Tea 2

10 What percent of those surveyed prefer milk?

9 What is the total number of adults surveyed?

11 What percent of those surveyed prefer coffee or tea?

Circle the best answer for each item.

Items 12–14 refer to the following label.

```
              YOGURT
SERVING SIZE ..................8 OUNCES
CALORIES .........................180
PROTEIN, grams .............................11
FAT, grams ...........................................5
CARBOHYDRATE, grams ................23
SODIUM, milligrams .......................125
```

12 There are two amounts you need to know to find how many calories that come from fat are in an 8-ounce serving of yogurt. One amount you need is grams of fat. What is the other amount?
    (1) milligrams of sodium
    (2) grams of protein
    (3) number of calories in a gram of fat
    (4) number of servings
    (5) number of calories in a serving

13 Recall that there are 9 calories in 1 gram of fat. Which is the correct proportion to find what percent 45 calories is of 180 calories?

    (1) $\frac{180}{45} = \frac{?}{100}$

    (2) $\frac{45}{100} = \frac{?}{180}$

    (3) $\frac{45}{180} = \frac{100}{?}$

    (4) $\frac{45}{180} = \frac{?}{100}$

    (5) $\frac{45}{100} = \frac{180}{?}$

14 Using a proportion, what percent of 180 calories is 45 calories?
    (1) 15%
    (2) 25%
    (3) 45%
    (4) 65%
    (5) Not enough information is given.

Items 15–17 refer to the following payroll check stub.

```
Tony Marquez

Gross pay              $500.00

Deductions:

  Federal income tax       63.40
  State income tax          6.60
  Credit union             25.00
  Total deductions      $ 95.00

Net pay                $405.00
```

15 Which is the correct expression to find the percent of Tony's gross pay deducted for his credit union account?
    (1) $\$25.00 \times \$500.00$
    (2) $\$25.00 \div \$500.00$
    (3) $\$500.00 - \$25.00$
    (4) $\$500.00 \times \$25.00$
    (5) $\$500.00 \div \$25.00$

16 What percent of Tony's gross pay is deducted for taxes?
    (1) 7%
    (2) 12%
    (3) 14%
    (4) 35%
    (5) 70%

17 Tony wants to contribute 4% of his gross pay to his company's savings plan. How much would Tony contribute each pay period?
    (1) $4.00
    (2) $20.00
    (3) $16.20
    (4) $125.00
    (5) $200.00

15 Solving for the Base and Percent of Change

ALGEBRA TOPICS

The Percent Formula

You can rewrite the percent formula to solve for the **base** in percent problems.

EXAMPLE Ellen works at a furniture store. She needs to price a new sofa 45% higher than the original cost. But Ellen has a problem. She does not know what the store paid for the sofa. Her boss tells her that she should add $292.50 to the cost. If $292.50 represents 45% of the cost, how can Ellen find the cost?

In this example, Ellen knows that the rate is 45% and that $292.50 is part of the original cost. The cost is the whole amount, or base. She needs to solve for the base. Ellen covers the letter *b* on the percent triangle, which shows to divide the part by the rate to solve for the base.

 part ÷ rate = base

Which is the correct expression to find the cost of the sofa?
(1) $292.50 ÷ 45
(2) $292.50 + 45
(3) $292.50 ÷ 45%
(4) $292.50 × 45%
(5) $292.50 − 45%

Answer **(3)** is correct. To find the base, divide the part ($292.50) by the rate (45%). First change the rate from a percent to a decimal: **45% = 0.45** **$292.50 ÷ 0.45 = $650,** so the base or the original price (cost) of the sofa is $650.

EXAMPLE A bookstore manager adds 20% to the price she pays for each hardcover book sold in the store. If the added cost to a book is $5, which is the correct expression to find the cost of the book?
(1) $5 × 0.20
(2) $5 ÷ 0.20
(3) $5 − 20
(4) $5 ÷ 20
(5) $5 + 20

Answer **(2)** is correct. Divide the part ($5) by the rate (20%) to find the base. Remember to first change the percent to a decimal (20% = 0.20). **$5 ÷ 0.20 = $25,** so $25 is the base, the original cost of the book.

Circle the best answer for each item.

1 Clara puts 6% of her gross pay in the credit union each pay period. This pay period she put $17.52 in the credit union. Which is the correct expression to find Clara's gross pay?

(1) $0.06 \div \$17.52$
(2) $\$17.52 \div 0.06$
(3) $\$17.52 - 6$
(4) $\$17.52 + 6$
(5) $\$17.52 \times 6$

2 Michael found that 75% of the people who come into his hardware store make a purchase. Today Michael made 270 sales. Which is the correct expression to find how many people came into the hardware store?

(1) $\$270 \times 0.75$
(2) $\$270 \div 0.75$
(3) $\$270 + 75$
(4) $\$270 - 75$
(5) $\$270 \div 75$

3 Dustin answered 22 questions correctly on a test. His score was 88%. Which is the correct expression to find how many questions were on the test?

(1) $22 + 88$
(2) 22×0.88
(3) $22 \div 0.88$
(4) $0.88 \div 22$
(5) $88 - 22$

4 This week Curtis spent 45% or 18 hours of his time taking inventory. Which is the correct expression to find how many hours Curtis worked this week?

(1) $18 \div 45$
(2) $18 \div 0.45$
(3) 18×0.45
(4) $45 - 18$
(5) $45 + 18$

5 Lenora made 20% of the sales at Super Sporting Goods this week. She made 85 sales. Which is the correct expression to find how many sales Super Sporting Goods had this week?

(1) $85 - 20$
(2) $85 \div 20$
(3) $85 + 20$
(4) $85 \div 0.20$
(5) 85×0.20

6 Transport Movers need to load boxes onto moving vans. They have 3 vans. By noon, the movers loaded 129 boxes. Which is the correct expression to find the number of boxes the movers need to load?

(1) $129 \div 0.3$
(2) $129 - 3$
(3) 129×0.33
(4) $129 \div 3$
(5) Not enough information is given.

7 Carmen attached the following sale tag to a pair of pants. Which is the correct expression to find the regular price for the pair of pants?

(1) $\$13.60 \times 0.40$
(2) $\$13.60 \div 40$
(3) $\$13.60 - 0.40$
(4) $\$13.60 \div 0.40$
(5) Not enough information is given.

Discount Fashions
Pants
Discount   40%
You save $13.60

8 **Reason** To find the *part* in a percent problem, you used the percent formula, *part = base × rate*. To find the *rate* in a percent problem, you rewrote the formula so that *rate = part ÷ base*. On a separate piece of paper, explain in a sentence or two how to rewrite the percent formula to find the base.

Selling at a Discount

EXAMPLE Paul works at Discount Fashions, a clothing store. The store buys clothes in large quantities at low prices and sells the clothes at a discount. Sometimes customers return clothes to the store. If the tags have been taken off, Paul must print out new tags and attach them to the clothing.

Paul needs to print a tag for a jacket. From the top of the tag, he knows that the discount rate is 25%, and the amount of customer savings is $15.50. Paul needs to find the regular price.

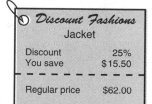

Discount Fashions
Jacket

Discount 25%
You save $15.50

Which is the correct expression to find the regular price of the jacket?
(1) 75% ÷ $15.50
(2) 25% ÷ $15.50
(3) $15.50 × 75%
(4) $15.50 × 25%
(5) $15.50 ÷ 25%

Answer **(5)** is correct. The regular price of the jacket is the base. The customer savings, $15.50, is the part of the base. The rate of discount is 25%. Using the percent formula, Paul divides the part ($15.50) by the rate (25%).

First change the percent to a decimal. 25% = 0.25
$15.50 ÷ 0.25 = $62.00, so the regular price of the jacket is **$62.00.**

EXAMPLE Next, Paul needs to find the sale price of the jacket. Which is the correct expression to find the sale price of this jacket?
(1) $62.00 + $15.50
(2) $62.00 − $15.50
(3) $62.00 × 25%
(4) $62.00 ÷ 25%
(5) $15.50 − 25%

Discount Fashions
Jacket

Discount 25%
You save $15.50

Regular price $62.00

Answer **(2)** is correct. To find the sale price of the jacket, subtract the amount of customer savings ($15.50) from the regular price ($62.00).
$62.00 − $15.50 = $46.50, so the sale price of the jacket is **$46.50.**

Circle the best answer for each item.

 1 Which is the correct expression to find the regular price of this sweater?

Discount Fashions
Sweater

Discount   30%
You save $7.20

(1) $7.20 × 30%
(2) $7.20 ÷ 30%
(3) $7.20 × 70%
(4) $7.20 + 30%
(5) 70% ÷ $7.20

2 Which is the correct expression to find the regular price of this coat?

Discount Fashions
Coat

Discount   45%
You save $54.00

(1) 45% × $54.00
(2) 45% ÷ $54.00
(3) $54.00 ÷ 45%
(4) $54.00 − 45%
(5) $54.00 ÷ 55%

3 Which is the correct expression to find the sale price of these jeans?

Discount Fashions
Jeans

Discount   35%
You save $12.60

Regular price $36.00

(1) $12.60 × 35%
(2) $12.60 ÷ 35%
(3) $36.00 ÷ 35%
(4) $36.00 + $12.60
(5) $36.00 − $12.60

4 Which is the correct expression to find the regular price of this sweatshirt?

Discount Fashions
Sweatshirt

Discount   40%
You save $5.60

(1) $5.60 × 40%
(2) $5.60 ÷ 40%
(3) $5.60 × 60%
(4) $5.60 ÷ 60%
(5) 40% ÷ $5.60

5 Which is the correct expression to find the amount you save on this dress?

Discount Fashions
Dress

Discount   25%
You save

Regular price $64.00

(1) 25% ÷ $64.00
(2) $64.00 × 25%
(3) $64.00 ÷ 25%
(4) $64.00 × 75%
(5) Not enough information is given.

6 **Communicate** On a separate piece of paper, explain how to find the discount rate for this skirt. Next, explain how you would find the sale price of the skirt. No calculations are necessary.

Discount Fashions
Skirt

Discount
You save $6.50

Regular price $32.50

Finding the Base

So far, you have learned to find the **part** and the **rate** in a percent problem. Now you will learn how to find the third element, the **base**.

EXAMPLE   The number 28 is 20% of what number?

Step 1: Identify the elements. You know that 20% is the rate because of the percent sign. Remember, the base often comes after the word *of*. In this problem, the word *of* comes before the words *what number*. The part often precedes or follows the word *is*. The number 28 is the part.

The rate is 20%.
The part is 28.
The unknown number is the base.

Step 2: Use the percent triangle to find the base. Cover the letter *b*, because you need to solve for the base. The other two elements, the part and the rate, are separated by a division sign.

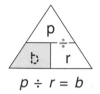

$$p \div r = b$$

Step 3: Divide the part (28) by the rate (20%) to find the base. First change the percent to a decimal.

$20\% = .20. = 0.20$

$$0.20\overline{)28.00.} \;\; \substack{140. \\ -20 \\ \hline 80 \\ -80 \\ \hline 0}$$

> **TIP**
>
> To divide by a decimal, move the decimal point to the right to make the decimal into a whole number. Then move the decimal point the same number of places to the right in the number you are dividing.
>
> $.4\overline{).32} = 4\overline{)3.2}^{\;.8}$

The number 28 is 20% of 140.

You can also solve these types of percent problems by using a proportion.

EXAMPLE   The number 50 is 125% of what number?

Step 1: Identify the elements.

The part is 50.
The rate is 125%.
The base is unknown.

Step 2: Write the percent as a ratio.

$125\% = \dfrac{125}{100}$

Step 3: Set up the proportion.

$\dfrac{part}{base} = \dfrac{rate}{100}$

$\dfrac{50}{?} = \dfrac{125}{100}$

Step 4: Cross-multiply.

$50 \times 100 = 5,000$

Step 5: Divide by the remaining number.

$5,000 \div 125 = 40$

The number 50 is 125% of 40.

Remember, when the rate is greater than 100%, the part will be greater than the base.

A. Solve. Show your work. The first one is done for you.

1 24 is 25% of what number?

$25\% = 0.25$

$$0.25\overline{)24.00}$$
$$\begin{array}{r} 96. \\ 0.25\overline{)24.00} \\ -225 \\ \hline 150 \\ -150 \\ \hline 0 \end{array}$$

2 36 is 40% of what number?

3 $9 is 75% of what number?

4 57 is 300% of what number?

5 40 is 8% of what number?

6 35 is 175% of what number?

B. Solve. Show your work. The first one is started for you.

7 For a sale, Earl's Furniture has discounted all items by 30%. The tag on a chair shows that a customer can save $22.50. What is the regular price of this chair?

$30\% = 0.30$

$$\begin{array}{r} \$7. \\ 0.30\overline{)\$22.50} \\ -21\ 0 \\ \hline 1\ 50 \end{array}$$

8 Ladonna's softball team won 14 games this season. The team won 70% of the games played. How many games did Ladonna's team play this season?

9 Vern spent 35% of his income last month for rent and utilities. He spent $490 for rent and utilities. What was Vern's income last month?

10 Elvia deposited money in a money market fund that pays 5% annual interest. She earned $34.25 interest in one year. How much did she deposit?

C. Solve these problems using a calculator.

11 Attendance at this year's opening game was 16,313. This number is 110% of last year's attendance. What was last year's attendance?

12 At CR Fashions, 31% of the weekly sales are of children's clothes. This week's sales of children's clothes is $2,356. What are the total sales for the week?

Earning a Commission

EXAMPLE Kim works in the office at Franklin Auto. Franklin Auto sells new and used cars. The salespeople at Franklin Auto earn a percent of the profits on the cars they sell. This percent is called the **rate of commission**. The amount a salesperson earns on a sale is called a **commission**.

Kim needs to complete the payroll records for three salespeople. He writes what he knows in a chart.

Salesperson	Base (Profit)	Rate	Part (Commission)
J. Carlson	$4,800	25%	
B. Ivey		20%	$1,350
M. Pace	$6,700		$1,675

Now Kim is ready to use the percent formula to complete the records. He starts with Janet Carlson's record. He knows the base and the rate. He needs to find the part. Which is the correct expression to find the part (commission)?

(1) $4,800 × 25
(2) $4,800 × 0.25
(3) $4,800 ÷ 0.25
(4) $4,800 ÷ 25
(5) $4,800 + 0.25

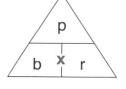

Answer **(2)** is correct. Multiply the base ($4,800) by the rate (25% = 0.25) to find the part. **$4,800 × 0.25 = $1,200,** so Janet's commission is **$1,200.**

EXAMPLE   Next, Kim goes to Ben Ivey's record. Kim knows the rate and the part. He needs to find the base. Which is the correct expression to find the base (profit)?

(1) $1,350 ÷ 20   (4)   20 ÷ $1,350
(2) $1,350 × 0.20   (5)   0.20 ÷ $1,350
(3) $1,350 ÷ 0.20

Answer **(3)** is correct. Divide the part ($1,350) by the rate (20% = 0.20) to find the base. **$1,350 ÷ 0.20 = $6,750,** so Ben brought in **$6,750** in profits during the pay period.

EXAMPLE   Last, Kim needs to find the percent or **rate of commission** that Max Pace earns. Which is the correct expression to find the rate of commission (percent)?

(1) $6,700 + $1,675   (4) $1,675 ÷ $6,700
(2) $6,700 − $1,675   (5) $1,675 × $6,700
(3) $6,700 ÷ $1,675

Answer **(4)** is correct. Divide the part ($1,675) by the base ($6,700) to find the rate. **$1,675 ÷ $6,700 = 0.25 = 25%,** so Max earns a 25% rate of commission.

PRACTICE

Circle the best answer for each item.

Items 1–4 refer to the following chart.

Salesperson	Base (Profit)	Rate	Part (Commission)
R. Montoya	$1,320.00	24%	
M. Quan		20%	$425.00
T. Simon	$1,320.00		$290.40

1. What is the amount of commission Roberto Montoya earned?
   (1) $31.68
   (2) $216.00
   (3) $316.80
   (4) $3,168.00
   (5) $5,500.00

2. What amount of profit did Mary Quan make for her company?
   (1) $85.00
   (2) $855.00
   (3) $1,855.00
   (4) $2,125.00
   (5) $2,500.00

3. What is Tara Simon's rate of commission?
   (1) 22%
   (2) 24%
   (3) 25%
   (4) 26%
   (5) 30%

4. What is the difference between the highest and lowest commission?
   (1) $26.40
   (2) $108.20
   (3) $134.60
   (4) $805.00
   (5) $895.00

Items 5–7 refer to the following chart.

Salesperson	Base (Profit)	Rate	Part (Commission)
L. Morgan	$4,350.00		$870.00
B. Whitefur		25%	$1,891.00
S. Chapa	$5,427.00	23%	

5. What is Lou Morgan's rate of commission?
   (1) 15.5%
   (2) 20%
   (3) 35%
   (4) 38%
   (5) 40%

6. What amount of profit did Bill Whitefur make for his company?
   (1) $75.64
   (2) $756.40
   (3) $7,564.00
   (4) $9,455.00
   (5) Not enough information is given.

7. What is the amount of commission Sue Chapa earned?
   (1) $236.00
   (2) $1,248.21
   (3) $2,137.21
   (4) $2,359.00
   (5) Not enough information is given.

8. **Reason** Ms. Montgomery sells furniture on commission. She receives a 6% rate of commission for new furniture she sells and 5% for used furniture. Today she sold a new couch for $630 and a used dining room set for $820. Which commission was higher, the commission for the new furniture or the old furniture? How much higher?

Check your answers on page 309.

Finding the Percent of Change

Finding the Percent of Increase

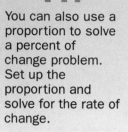

TIP

You can also use a proportion to solve a percent of change problem. Set up the proportion and solve for the rate of change.

$$\frac{\text{amt. of change}}{\text{original amt.}} = \frac{?}{100}$$

A **percent of change** problem is a type of percent problem. To solve the problem, identify the base as the original amount and find the part as the amount of change (or the difference between the new amount and the original amount). Then solve for the rate—the percent of increase.

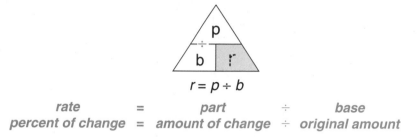

$$r = p \div b$$

rate	=	part	÷	base
percent of change	=	amount of change	÷	original amount

EXAMPLE In January, Warren's landlord raised the rent from $450 to $486 per month. What is the percent of increase in Warren's rent?

Step 1: Identify the original amount (the base). The base is $450.

Step 2: Identify the new amount. The new amount is $486.

Step 3: Find the amount of change (the part) by subtracting the original amount ($450) from the new amount ($486). $486 − $450 = $36 = part

Step 4: Use the percent formula. Divide the part ($36) by the base ($450) to find the rate of change. $36 ÷ $450 = 0.08

Step 5: Change the decimal to a percent. 0.08 = 0.08. = 8%

There was an 8% increase in Warren's monthly rent.

Finding the Percent of Decrease

EXAMPLE Chris's Deli lowered the price of its fruit cup from $4.00 to $3.20. What is the percent of decrease?

Step 1: Identify the original price (the base). The base is $4.00.

Step 2: Identify the new amount. The new amount is $3.20.

Step 3: Find the amount of change (the part) by subtracting the new amount ($3.20) from the original amount ($4.00). $4.00 − $3.20 = $0.80 = part

Step 4: Use the percent formula. Divide the part ($0.80) by the base ($4.00) to find the rate. $0.80 ÷ $4.00 = 0.20

Step 5: Change the decimal to a percent. 0.20 = 0.20. = 20%

There was a 20% decrease in the price of the fruit cup.

A. **Use the percent formula to find the percent of increase or decrease. Show your work. The first one is done for you.**

Original Amount	New Amount	Percent of Change
1 $575	$690.00	_20%_

$$\begin{array}{r} \$690.00 \\ -575.00 \\ \hline \$115.00 \end{array}$$

$$\begin{array}{r} 0.2 \\ \$575 \overline{)\$115.00} \\ -115.00 \\ \hline 0 \end{array}$$

2 $8.80	$6.60	_____
3 25 pounds	15 pounds	_____

B. **Use proportions to find the percent of increase or decrease. Show your work. The first one is started for you.**

Original Amount	New Amount	Percent of Change
4 4,800 people	5,376 people	_____

$$\frac{amount\ of\ change}{original\ amount} = rate\ of\ change$$

$$\frac{576}{4,800} = \frac{?}{100}$$

5 70 gallons	63 gallons	_____
6 $24.00	$22.56	_____

C. **Solve these problems using a calculator.**

7 Last year Michael's car insurance premium was $750. His car insurance premium this year is $1,125. What is the percent of increase in the premium?

8 A bus ticket from Mesa to Beaverton was $28.00. To attract more customers, the bus company reduced the fare to $23.80. What is the percent of decrease in the fare?

DATA ANALYSIS   Using a Double Bar Graph

EXAMPLE   Acorn Business Systems has offices in four states. The company sells office machines to other businesses. The following graph shows company sales by state for the first and second quarters. A quarter is three months.

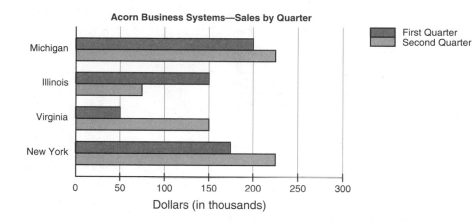

Acorn Business Systems—Sales by Quarter

First Quarter
Second Quarter

Dollars (in thousands)

This graph is called a **double bar graph** because there are two bars for each state.

From the key, you know that the colored bar shows the first-quarter sales, and the gray bar shows the second-quarter sales.

Jorge Ruiz needs to find the percent of change in sales from the first quarter to the second quarter for Virginia. He reads the graph for Virginia. The colored bar reaches the mark labeled 50. The gray bar reaches the mark labeled 150. What is the percent of increase from 50 to 150?

(1) $66\frac{2}{3}\%$ (4)   200%

(2)   75%   (5)   250%

(3)   100%

Answer **(4)** is correct. Jorge subtracts to find the part: $150 - 50 = 100$
To find the rate, he divides the part (100) by the base (50).
$100 \div 50 = 2 = 200\%$, so there was a 200% increase in sales in Virginia from the first quarter to the second quarter.

EXAMPLE   Next, Jorge finds the percent of change in sales for New York. He reads the graph for New York. The colored bar represents 175. The gray bar represents 225. To the nearest whole percent, what is the percent of increase from 175 to 225?

(1)   22%   (4)   50%

(2)   29%   (5)   72%

(3)   33%

Answer **(2)** is correct. Jorge subtracts to find the part: $225 - 175 = 50$
Then he divides the part (50) by the base (175) to find the rate.
$\frac{50}{175} = 0.285$, **which rounds to 0.29 = 29%**, so there was a 29% increase in sales in New York from the first quarter to the second quarter.

Circle the best answer for each item.

Items 1–7 refer to the following graph.

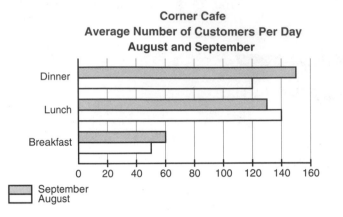

**Corner Cafe
Average Number of Customers Per Day
August and September**

1 How many more customers did the Corner Cafe have each day for dinner in September than in August?

(1) 15
(2) 20
(3) 30
(4) 40
(5) 270

2 To the nearest whole percent, what is the percent of increase in the number of customers each day for dinner from August to September?

(1) 13%
(2) 20%
(3) 25%
(4) 44%
(5) 56%

3 To the nearest whole percent, what percent of the total customers each day during September were breakfast customers?

(1) 11%
(2) 18%
(3) 25%
(4) 40%
(5) 60%

4 To the nearest whole percent, what is the percent of increase in the number of customers each day for breakfast from August to September?

(1) 12%
(2) 17%
(3) 20%
(4) 25%
(5) 83%

5 To the nearest whole percent, what is the percent of decrease in the number of customers each day for lunch from August to September?

(1) 7%
(2) 8%
(3) 14%
(4) 93%
(5) Not enough information is given.

6 How many customers did the Corner Cafe have during the months of August and September?

(1) 270
(2) 310
(3) 340
(4) 650
(5) Not enough information is given.

7 **Communicate** On a separate piece of paper, explain how to find the percent of increase in the total number of customers each day from August to September. Include in your explanation what numbers you would use for the part, base, and rate. Do not do any calculations.

Section 15 Review

Solve. Show your work.

1 27 is 30% of what number?

2 60 is 3% of what number?

3 $40.80 is 68% of what number?

4 4 is 20% of what number?

5 $35 is 125% of what number?

6 36 is 75% of what number?

Circle the best answer for each item.

7 The Riley's health insurance pays 80% of all medical expenses. The insurance paid $120 for Mona Riley's x-rays. Which is the correct expression to find the charge for the x-rays?
(1) 20% × $120
(2) 80% ÷ 120
(3) $120 ÷ 20%
(4) $120 × 80%
(5) $120 ÷ 80%

8 Consumer's Summary reported that 9,610 people surveyed prefer to drive a sedan. This number represents 62% of those surveyed. Which is the correct expression to find the number of people surveyed?
(1) 9,610 ÷ 38%
(2) 9,610 × 62%
(3) 9,610 ÷ 62%
(4) 38% × 9,610
(5) 62% ÷ 9,610

9 Last year the Hassans paid $216.00 each month for health insurance. This year, their monthly premium is $233.28. What is the percent of increase in their monthly premium?
(1) 7%
(2) 8%
(3) 17%
(4) 21%
(5) 93%

10 Stan works as a data entry operator. His current wage is $6.50 per hour. Because of Stan's excellent performance, his boss gave him a raise. His new hourly wage will be $7.15. What is the percent of increase in Stan's wages?
(1) 8%
(2) 9%
(3) 10%
(4) 65%
(5) 91%

Check your answers on page 310.

Solve. Show your work.

Items 11–12 refer to the following part of a price tag.

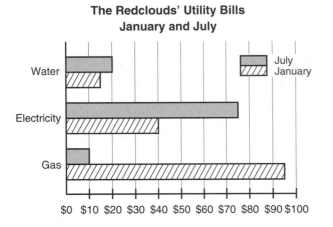

Discount Fashions
Shirt

Discount	25%
You save	$3.60

11 What is the regular price of the shirt?

12 What is the sale price of the shirt?

Items 13–14 refer to the following graph.

The Redclouds' Utility Bills
January and July

July
January

Water
Electricity
Gas

$0 $10 $20 $30 $40 $50 $60 $70 $80 $90 $100

13 To the nearest whole percent, what is the percent of decrease in the gas bill from January to July?

14 To the nearest whole percent, what is the percent of increase in the electric bill from January to July?

Items 15–17 refer to the following chart.

Salesperson	Base (Profit)	Rate	Part (Commission)
V. Murphy	$6,352.00	22%	
S. Santino		21%	$874.23
D. Hill	$5,240.00		$1,310.00

15 What is the amount of commission Val Murphy earned?

16 What amount of profit did Sal Santino make for his company?

17 What is Darwin Hill's rate of commission?

18 If Trudy earns 18 percent commission on her sales, what will she earn on the following amounts?

a. $4,280.00 _____

b. $7,342.00 _____

c. $6,015.00 _____

Special Topics Statistics and Probability/Circles

DATA ANALYSIS ## Basing Probability on Statistics

Many businesses gather information, or **data,** in order to make decisions. **Statistics** involves organizing and analyzing that data. Data can be gathered by taking a **survey.** When you gather information from an entire group, this group is called the **population.** Sometimes it is impractical to survey an entire population. In this case, gather information from a smaller group, or **sample,** that represents the population.

One uses statistics to make predictions based on the **probability** of something happening. Probability is the study of **chance** or **outcome.** Probability can be expressed as a fraction or as a percent.

EXAMPLE 1 Glenn works at Employment Op. He took a sample survey to find out how many people earn a living working from home. The circle graph at the right shows the results of his survey.

Preferred Place of Work

Work at Home

Work at Office 60%

What is the probability that someone surveyed works at an office?

(1) 0 or 0%   (4) $\frac{4}{5}$ or 80%

(2) $\frac{2}{5}$ or 40%   (5) 1 or 100%

(3) $\frac{3}{5}$ or 60%

Answer **(3)** is correct. Based on the percents shown on the circle graph, the probability that someone surveyed works at an office is $\frac{3}{5}$ **or 60%.**

EXAMPLE 2 Of 800 people surveyed, 52% named baseball as their favorite sport. What is the probability that the next person surveyed would *not* choose baseball as a favorite sport?

(1)   0%
(2)   48%
(3)   50%
(4)   52%
(5) 100%

Answer **(2)** is correct. To find the probability, subtract the percent of people who named baseball as their favorite sport (52%) from the total percent of people surveyed (100%). **100% − 52% = 48%.**

Circle the best answer for each item.

Items 1 and 2 refer to the following information and circle graph.

Neal is in charge of ordering inventory at Sporting Goods. He surveyed the customers in order to make decisions on what inventory to order. The results are shown below.

Popular Exercise

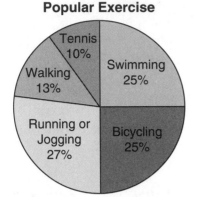

1. Of the 500 adults surveyed, 27% enjoyed running or jogging. How many people enjoy running or jogging as their favorite sport?

    (1)   5
    (2)   45
    (3)   90
    (4)   135
    (5)   450

2. According to the graph, what is the probability that the next customer Neal surveys would be interested in running or walking shoes?

    (1) $\frac{1}{4}$ or 25%
    (2) $\frac{2}{5}$ or 40%
    (3) $\frac{1}{2}$ or 50%
    (4) $\frac{3}{4}$ or 75%
    (5) 1 or 100%

Items 3 and 4 refer to the following information and circle graph.

Frank works at Auto Dealer. He took a survey of 60 people and asked them to name the kind of car they drive. The results are shown below.

Preferred Car to Drive

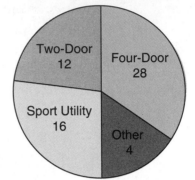

3. What is the probability that the next person surveyed drives a two-door?

    (1) 10%
    (2) 20%
    (3) 30%
    (4) 40%
    (5) 50%

4. What is the probability that the next person surveyed drives a four-door or a two-door?

    (1) 0%
    (2) 20%
    (3) 46%
    (4) 67%
    (5) Not enough information is given.

5. **Reason** Frank's survey is based on a sample of the population. The population is about 30,000 people who are potential customers at Auto Dealer. What percent of the population does the sample of 60 people surveyed represent? What do you think would make the survey a stronger predictor of probability?

Facts about Probability

Probability uses numbers to show how likely it is that an event will happen. When you cannot control how an event will turn out, the **outcome** or result is left to **chance.**

You can express probability by using the numbers from 0 to 1 or by using percents from 0% to 100%. A probability of 1 or 100% means that you are sure that a certain outcome will occur. A probability of 0 or 0% means that a certain outcome cannot possibly happen.

TIP

The name that Wanda picks is left to **chance.** In other words, Scott and Wanda cannot control how the drawing will turn out, and they cannot predict with certainty what will happen.

EXAMPLE   Scott is organizing the holiday office party. He writes the names of all ten employees on separate cards. Then he puts the ten cards in a box. Each person picks a name and buys that co-worker a gift. Wanda chooses first. What is the probability that Wanda will pick the name of one of the ten office employees?

Step 1.   Decide if the possible outcomes are equally likely to occur. *Yes, each of the ten outcomes is equally likely to occur.* This means that each card has an equal chance of being chosen.

Step 2.   Decide if there are any other possibilities. *There is no other possibility.*

The probability is **1 or 100%** that Wanda will pick a card with the name of an employee on it. There is no other possibility.

PRACTICE

Write each probability using the numbers 0 or 1 and using percents 0% or 100%. Explain your answer. The first one is done for you.

1 Read the example problem again. What is the probability that the card Wanda picks will be blank?

0 or 0%
None of the cards is blank, so there is no possibility that the card she chooses will be blank.

2 A box contains 15 blue marbles. What is the probability of picking a blue marble from this box?

3 Osami's sock drawer contains 11 pairs of socks: 3 pairs of blue socks, 4 pairs of black socks, and 4 pairs of white socks. What is the probability that she will pick a pair of brown socks?

4 Mr. Hernandez is a meteorologist. He told his television viewing audience that there is a 100% chance of rain tomorrow. According to Mr. Hernandez, what is the probability that it will *not* rain tomorrow?

Finding the Probability of an Outcome

You can write probabilities that are greater than 0 but less than 1 either as a fraction or as a decimal.

● **Probability of an outcome** $= \dfrac{\text{number of ways an outcome can occur}}{\text{number of possible outcomes}}$

EXAMPLE Ten co-workers decide to exchange gifts for the holidays. They each write their name on a card and place the cards in a box. Then each person will choose a name. Jesse picks first. What is the probability that Jesse will pick his own name?

Step 1. Find the number of possible outcomes. Because there are 10 cards in the box, there are 10 equally likely outcomes.

There are 10 possible outcomes.

Step 2: Find the number of ways the outcome (Jesse picking his own name) can occur. Since Jesse's name is written on only one of the cards, there is only one way the outcome can occur.

There is one way the outcome can occur.

Step 3: Write the probability as a fraction in lowest terms.

$\dfrac{1}{10}$

Step 4: To write the probability as a percent, change the fraction ($\frac{1}{10}$) to a decimal (0.1). Then change the decimal (0.1) to a percent (10%).

$1 \div 10 = 0.1 = 10\%$

The probability Jesse will pick his own name is 10%.

> **TIP**
>
> Remember to reduce the fraction to lowest terms.
>
> $\dfrac{2}{10} = \dfrac{1}{5}$

PRACTICE

Write each probability as a fraction and as a percent. The first one is done for you.

Items 1–4 refer to the following information. A number cube is a cube with six sides. Each side is shown below.

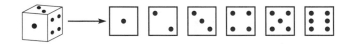

① If you roll a number cube, what is the probability that you will roll a 4?

Fraction: $\dfrac{1}{6}$ Percent: $16\frac{2}{3}\%$

② If you roll a number cube, what is the probability that you will roll a 9?

Fraction: _____ Percent: _____

③ If you roll a number cube, what is the probability that you will roll a number less than 7?

Fraction: _____ Percent: _____

④ If you roll a number cube, what is the probability that you will roll an even number?

Fraction: _____ Percent: _____

More than One Chance for the Same Outcome

Sometimes there is more than one way for an outcome to occur.

EXAMPLE   On a TV game show, a player spins a wheel to win a prize. The wheel has eight equal sections. Since the sections are the same size, there is an equal chance that the wheel will stop on any one section.

TIP

Four sections are marked *$100*.

Only one section is marked *New Car*. A player has a 1 in 8 chance of winning a new car. The probability that the wheel will stop on the section marked *New Car* is $\frac{1}{8}$ or 12.5%. What is the probability that the wheel will stop on $100?

Step 1.	Find the number of possible outcomes. There are eight sections on the wheel.	8
Step 2:	Find the number of ways an outcome can occur. There are four ways to win $100.	4
Step 3.	Write the probability of the outcome.	$\frac{\text{number of ways an outcome can occur}}{\text{number of possible outcomes}} =$ $\frac{4}{8} = \frac{1}{2}$
Step 4:	Change the fraction to a percent.	$\frac{1}{2} = 50\%$

The probability of a player winning $100 is $\frac{1}{2}$ or 50%.

EXAMPLE   A box contains 12 marbles. Nine marbles are black. Three marbles are white. If you take one marble from the box without looking, what is the probability that you will choose a white marble?

Step 1.	Find the number of possible outcomes. There are 12 marbles in the box.	12
Step 2.	Find the number of ways an outcome can occur. There are 3 chances of choosing a white marble.	3
Step 3.	Write the probability of the outcome.	$\frac{3}{12} = \frac{1}{4}$
Step 4:	Change the fraction to a percent.	$\frac{1}{4} = 25\%$

The probability of choosing a white marble is $\frac{1}{4}$ or 25%.

A. Write the probability for each problem. Show your work. The first one is done for you.

Items 1–4 refer to the following information. The carnival game spinner shown contains 10 equal sections.

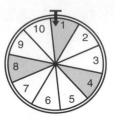

1 What is the probability that the wheel will stop on a shaded section?

$$\frac{number\ of\ shaded\ sections}{total\ number\ of\ sections} = \frac{3}{10} = 30\%$$

2 What is the probability that the wheel will stop on an even number (2, 4, 6, 8, 10)?

3 What is the probability that the wheel will stop on a number greater than 4?

4 What is the probability that the wheel will stop on a number less than 12?

B. Write the probability. Show your work. The first one is started for you.

Items 5–7 refer to the following information. A box contains 8 marbles: 3 black, 3 gray, and 2 white. You take one marble out of the box without looking.

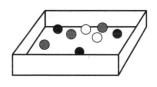

5 What is the probability you will choose a gray marble?

$$\frac{number\ of\ gray\ marbles}{number\ of\ marbles} =$$

6 What is the probability that you will choose a black or a gray marble?

7 What is the probability that you will not choose a black marble?

C. Find the probability. Show your work.

8 A box has 20 marbles: 6 black, 4 white, 5 blue, and 5 gray. What is the probability you will choose a white marble?

9 Out of every 500 pairs of jeans a company manufactures, 35 are defective. Out of the next batch of 500 pairs of jeans, what is the probability that the next pair of jeans is defective?

Finding a Dependent Probability

Sometimes the chance of something happening depends on another outcome. To find a **dependent probability,** you need to consider the other outcomes.

EXAMPLE M & M Furniture is having a drawing for its employees. The names of the six employees with the best attendance records will be written on cards. Two names will be drawn, and the employees whose names are drawn will win a trip. Of the six employees in the drawing, two work in shipping and four work in sales. What is the probability that the first name drawn will be an employee from the shipping department?

Step 1. Find the number of possible outcomes. 6
There are six employees in the drawing.

Step 2: Find the number of ways an outcome can occur. 2
Two of the employees work in shipping.

Step 3. Write the probability of the outcome. $\dfrac{\text{ways outcome can occur}}{\text{possible outcomes}} = \dfrac{2}{6} = \dfrac{1}{3}$

Step 4. Change the fraction to a percent. $\dfrac{1}{3} = 33\dfrac{1}{3}\%$

The probability that the first name drawn will be an employee from the shipping department is $33\dfrac{1}{3}\%$.

EXAMPLE After the first name was drawn, that card was *not* placed back in the drawing. Then a second name was drawn. What is the probability that the second name drawn will be an employee from sales?

Step 1. Find the number of possible outcomes. There are five names (cards) <u>left</u> in the drawing. 5 cards left

Step 2: Find the number of ways an outcome can occur. 4 people in sales
If the first name drawn was from shipping, then there are 4 names from sales and 1 name from shipping left.

Step 3. Write the probability of the outcome: $\dfrac{\text{4 people in sales}}{\text{5 cards left}}$

Step 4. Change the fraction to a percent. $\dfrac{4}{5} = 80\%$

The probability that the second name drawn will be an employee from sales is **80%**.

If the first name drawn was from sales, then there are 3 names from sales and 2 names from shipping left. That means the probability the second name drawn would be an employee from sales is $\dfrac{3}{5}$ **or 60%.**

A.   **Write the probability for each problem. Show your work. The first one is done for you.**

<u>Items 1–3</u> refer to the following information.
Five friends are planning a fishing trip. To decide who will drive, they place five cards the same size in a bag. *Drive* is written on one card. The other cards are blank. Each person will draw a card until someone draws the card with *Drive* on it.

1   What is the probability that the first card drawn will have *Drive* written on it?

$$\frac{number\ of\ cards\ with\ Drive}{number\ of\ cards} = \frac{1}{5} = 20\%$$

2   The first card drawn is blank. What is the probability that the second card drawn will have *Drive* written on it?

3   The first card drawn is blank. The second card drawn is blank. What is the probability that the third card drawn will also be blank?

B.   **Write the probability. Show your work. The first one is started for you.**

<u>Items 4–6</u> refer to the following information.
As part of Food Basket's grand opening, nine currency bills are placed in a box; six $20 bills and three $100 bills. Each hour, a customer will be selected to draw one bill from the box without looking.

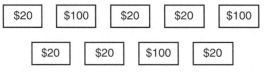

4   What is the probability that the first bill drawn will be a $100 bill?

$$\frac{number\ of\ \$100\ bills}{total\ number\ of\ bills} =$$

5   The first bill drawn is a $100 bill. What is the probability that the second bill drawn will also be a $100 bill?

6   The first bill drawn is a $20 bill. The second bill drawn is a $100 bill. What is the probability that the third bill drawn will also be a $100 bill?

C.   **Solve the following problems using a calculator.**

<u>Items 7 and 8</u> refer to the following information.
Amy's department is having a picnic. At the picnic there is going to be a raffle. There is one grand prize and five first-place prizes. Of the 25 people who bought a ticket, 8 are men and 17 are women. Tickets are drawn and *not* replaced.

7   What is the probability that the first ticket drawn will be Amy's ticket?

8   If the first three tickets drawn belonged to women, what is the probability that the fourth ticket drawn will belong to a man?

Finding the Circumference of a Circle

A **circle** is a closed figure with a curved edge. All points on the edge of a circle are the same distance from the circle's center.

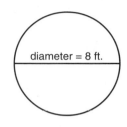

● The distance around the edge of a circle is called the circumference.

diameter = 8 ft.

EXAMPLE   Roger is building a low wall around a circular fountain. If the diameter of the fountain is 8 feet, what is the circumference of the fountain?

Because the edge of a circle is curved, you cannot measure the length of its edge in the same way that you measure the perimeter of squares or rectangles. Instead, you can use information about a circle to find its circumference.

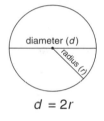

● The diameter is a straight line drawn from the edge of the circle, through the center, to the other edge.

diameter (*d*)
radius (*r*)

● The radius is a straight line connecting the center of the circle to any point on the edge. The radius is half the length of the diameter.

$d = 2r$

For any circle, the ratio of its circumference to its diameter is the same. The ratio is represented by the Greek letter π (pi). The value of π is about $\frac{22}{7}$, or 3.14. Using the ratio, you can write a formula for finding the circumference, *C*, of a circle.

● $C = \pi \times d$, where π is approximately 3.14 and *d* = diameter

Which is the correct expression to find the circumference of the fountain in the example?
(1) $3.14 + 8$
(2) 3.14×8
(3) 3.14×3.14
(4) 8×8
(5) $8 \div 3.14$

Answer **(2)** is correct. To find the circumference of the fountain, multiply π (3.14) by the diameter (8 feet). $C = \pi \times d = 3.14 \times 8 = 25.12$, so the circumference of the fountain is **25.12 feet**.

Circle the best answer for each item.

1 What is the circumference of this circle to the nearest foot?
(Hint: diameter = 2 × radius)

(1)   7
(2)   13
(3)   25
(4)   36
(5)   50

4 ft.

2 Raul is building a deck around a circular pool. The diameter of the pool is 24 feet. What is the circumference of the pool to the nearest foot?

(1)   10
(2)   27
(3)   48
(4)   75
(5)   576

3 What is the circumference of this circle to the nearest meter?

(1)   19
(2)   38
(3)   48
(4)   113
(5)   452

12 m

4 Kevin is a mechanic at Bill's Bike Shop. He needs to replace a bicycle tire. The radius of the wheel is 10 inches. What is the circumference of the wheel to the nearest inch? (Hint: diameter = 2 × radius)

(1)   31
(2)   63
(3)   100
(4)   314
(5)   396

5 Sylvia works at Frame It. This week a customer brought in a circular needlepoint craft to be framed. Which is the correct expression to find the circumference of the craft to be framed?

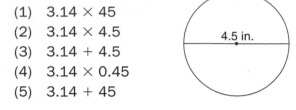

(1)   3.14 × 45
(2)   3.14 × 4.5
(3)   3.14 + 4.5
(4)   3.14 × 0.45
(5)   3.14 + 45

4.5 in.

6 Keesha is buying lace trim to sew around the edge of a round tablecloth. The radius of the tablecloth is 32 inches. What is the circumference of the tablecloth to the nearest inch?

(1)   64
(2)   100
(3)   128
(4)   201
(5)   Not enough information is given.

7 Yu-lan is jogging on a circular track. The diameter of the track is 25 feet. What is the circumference of the track?

(1)   39.25 ft.
(2)   50 ft.
(3)   78.5 ft.
(4)   157.0 ft.
(5)   Not enough information is given.

8 **Connect** On a separate sheet of paper, draw several circles using different-sized items such as soda cans, jars, lids, and so on. For each circle, use a ruler to measure the diameter and string to help you measure the circumference. For each circle, calculate the ratio of the circumference to the diameter. What pattern do you see? What does this pattern tell you?

Finding the Area of a Circle

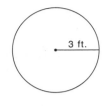

EXAMPLE Lin is putting tiles on the bottom of a circular pond. To buy tiles, he needs to know the area of the pond. Lin uses a formula to find the area, *A*, of a circle.

$A = \pi \times r^2 = \pi r^2$, where $\pi = 3.14$ and $r = $ radius

If the radius of the pond is 3 feet, which is the correct expression to find the area of the pond?

(1) 3.14×3
(2) $3.14 \times 3.14 \times 3.14$
(3) $3.14 \times 3.14 \times 3$
(4) $3.14 \times 3 \times 3$
(5) $3 \times 3 \times 3$

Answer **(4)** is correct. Lin multiplies 3.14 by the square of the radius (3×3). **$3.14 \times 3 \times 3 = 3.14 \times 9 = 28.26$ square feet**, so the area of the pond is **28.26 square feet.**

EXAMPLE Artemis is planting a circular rose garden that has a radius of 7 yards. What is the area of the garden to the nearest square yard?

(1)   22
(2)   44
(3)   49
(4) 154
(5) 343

Answer **(4)** is correct. To find the area of a circle, multiply 3.14 by the square of the radius (7×7). **$3.14 \times 7 \times 7 = 3.14 \times 49 = 153.86$ square yards,** so the area of the rose garden to the nearest square yard is **154 square yards.**

Circle the best answer for each item.

1 A circular window on a side of a building has a radius of 8 feet. What is the area of the window to the nearest square foot?

(1)   24
(2)   50
(3)   79
(4)   201
(5)   330

2 Rita has a circular pond in her backyard. The diameter of the pool is 18 feet. To the nearest square foot, what is the minimum area her pool cover will need to cover? (Hint: To find the radius, divide the diameter by 2.)

(1)   50
(2)   55
(3)   60
(4)   110
(5)   254

3 Colin works at Dino's Pizza. An extra large pizza has a diameter of 16 inches. What is the area of Dinos's extra large pizza to the nearest inch?

(1)   25
(2)   50
(3)   100
(4)   201
(5)   402

4 Kafia is a landscaper. She is building a fence around a circular garden. The garden has a radius of 12 feet. Which is the correct expression to find the area of the garden?

(1)   3.14 × 12 + 12
(2)   3.14 × 3.14 × 12
(3)   3.14 × 12 × 12
(4)   3.14 + 12 × 12
(5)   12 × 12 × 12

5 Marge built a circular fish pond in her backyard. The radius of the pond is 7 feet. What is the area of the bottom of the pond to the nearest square foot?

(1)   44
(2)   69
(3)   154
(4)   343
(5)   Not enough information is given.

6 Hector needs mulch for a circular flower bed. One bag of mulch covers 25 square feet. How many bags of mulch will Hector need to cover the flower bed?

(1)   38
(2)   49
(3)   79
(4)   113
(5)   Not enough information is given.

7 **Reason** On a separate piece of paper, explain how you would find the area of the colored section in the figure shown. Do not do any calculations. (Hint: How many circles are there?)

Section 16 Review

Solve each problem. Show your work.

Items 1–3 refer to the graph at the right.

1 What is the total number surveyed?

2 What percent of those surveyed preferred social studies?

3 According to the graph, what is the probability that someone will name English as his or her preferred subject?

Preferred Subject

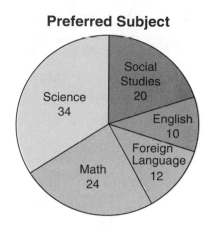

Write each probability using the numbers 0 or 1 and using percents 0% or 100%.

4 A drawer contains 12 multicolored scarves. No other items are in the drawer. What is the probability of choosing a multicolored scarf from this drawer?

5 The forecast is for 100% chance of snow today. According to the forecast, what is the probability that it will *not* snow today?

Write each probability as a fraction and as a percent.

Items 6–9 refer to the following information.
A box contains 5 marbles: a red, a green, a blue, an orange, and a yellow marble. You take one marble out of the box without looking. After a marble is taken, it is put back in the box.

6 What is the probability you will choose a green marble?

Fraction: _____ Percent: _____

8 What is the probability you will choose an orange or a yellow marble?

Fraction: _____ Percent: _____

7 What is the probability you will choose a marble that is *not* blue?

Fraction: _____ Percent: _____

9 What is the probability you will choose a black marble?

Fraction: _____ Percent: _____

Check your answers on page 313.

Items 10–17 refer to the following information.
A box contains 10 marbles: 6 black, 1 gray, and 3 white.
You take one marble out of the box without looking.
After a marble is taken, it is put back in the box.

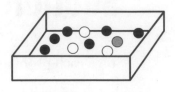

Circle the best answer for each item.

10 What is the probability that you will choose a black marble?

(1) $\frac{2}{5}$ or 40%

(2) $\frac{3}{5}$ or 60%

(3) $\frac{2}{3}$ or $66\frac{2}{3}$%

(4) 1 or 100%

(5) $\frac{3}{2}$ or 150%

12 What is the probability that you will choose a gray marble?

(1) 0 or 0%

(2) $\frac{1}{10}$ or 10%

(3) $\frac{1}{9}$ or $11\frac{1}{9}$%

(4) $\frac{9}{10}$ or 90%

(5) 1 or 100%

11 What is the probability that you will choose a green marble?

(1) 0 or 0%

(2) $\frac{1}{10}$ or 10%

(3) $\frac{1}{9}$ or $11\frac{1}{9}$%

(4) $\frac{9}{10}$ or 90%

(5) 1 or 100%

13 What is the probability that you will *not* choose a white marble?

(1) 0 or 0%

(2) $\frac{3}{10}$ or 30%

(3) $\frac{2}{5}$ or 40%

(4) $\frac{3}{5}$ or 60%

(5) $\frac{7}{10}$ or 70%

Solve. Show your work. Assume after each marble is drawn, it is *not* put back in the box.

14 The first marble drawn is gray. A second marble is drawn. What is the probability that it is white?

16 The third marble drawn is black. A fourth marble is drawn. What is the probability that it is also black?

15 The second marble drawn is white. A third marble is drawn. What is the probability that it is also white?

17 The fourth marble drawn is black. A fifth marble is drawn. What is the probability that it is white?

Find the circumference and area of each circle. Round your answer to the nearest whole number.

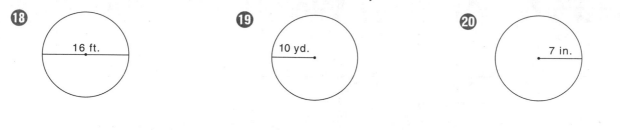

⑱ 16 ft.

⑲ 10 yd.

⑳ 7 in.

Circle the best answer for each item.

㉑ Gloria has a circular rug in her living room. The radius of the rug is 16 inches. Which is the correct expression to find the circumference of the rug?
(1) 3.14 × 16
(2) 3.14 × 32
(3) 3.14 × 8
(4) 3.14 ÷ 16
(5) 16 ÷ 3.14

㉒ The Mendez family has a circular flower bed. The diameter of the circle is 6 feet. Which is the correct expression to find the area of the flower bed?
(1) 3.14 × 6
(2) 3.14 × 6 × 6
(3) 3.14 × 3 × 3
(4) 3.14 × 3
(5) 6 × 6 × 6

Solve. Show your work.

㉓ Marcus is planning to lay a sidewalk around the circular fountain in City Park. The radius of the fountain is 13 feet. What is the circumference of the fountain to the nearest foot?

㉔ Andrea is placing mulch in circles around the young trees she has just planted. The radius of each mulched area is 9 inches. What is the area covered by each tree and its surrounding mulch to the nearest square inch?

Math at Work

Hospitality: Cook

Some Careers in Hospitality

Assistant Cook prepares recipe ingredients for chief cook by measuring, chopping, peeling, and cleaning

Hotel Desk Clerk checks in and checks out hotel guests, calculates bills, answers questions, and assists guests

Restaurant Host welcomes and seats diners, makes and checks reservations, and monitors number of diners in each section

Waitperson takes food orders, serves customers, calculates bills, collects money, and makes change

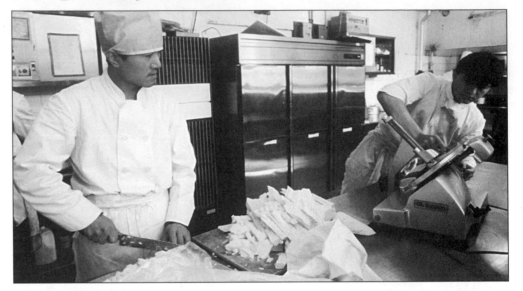

Cooks use math skills to prepare recipes.

Almost everybody enjoys a tasty meal. Many people like having that meal prepared by a professional cook or chef and served in a restaurant outside their home.

Cooks have the ability to combine a variety of ingredients and turn them into sweet or savory dishes to be enjoyed by diners in a wide range of settings. Cooks work in casual and elegant restaurants, commercial kitchens, bakeries, grocery stores, and school cafeterias.

Many cooks are responsible for planning menus; creating new recipes; buying meats, spices, produce and other foodstuffs; and preparing a variety of foods to be served. Often restaurant cooks help set the prices of the dishes on the menu. Other cooks follow the directions of the chief chef. They follow recipes, cook the foods, and present them in an appetizing display.

Measuring ingredients, following recipes, and **adjusting recipes** to serve a greater number or smaller number of people are just some of the ways in which cooks apply math skills. In these types of situations, cooks need to be able to use their knowledge of **percents, ratio,** and **proportion.**

Look at the Some Careers in Hospitality chart.

- Do any of the careers interest you? If so, which ones?
- What information would you need to find out more about those careers? On a separate piece of paper, write some questions that you would like answered. You can find more information about those careers in the *Occupational Outlook Handbook* at your local library.

Use the recipe below to answer the questions that follow.

Roasted Red Pepper Pesto
Serves 6 people

Ingredients

4 red peppers, halved, seeded, and roasted

3 garlic cloves

1 cup Parmesan cheese, shredded

$\frac{1}{3}$ cup pine nuts, toasted

$\frac{1}{2}$ cup olive oil

$\frac{1}{4}$ teaspoon salt

$\frac{1}{4}$ teaspoon freshly ground black pepper

Grind garlic and pine nuts in food processor until finely minced. Add all remaining ingredients except olive oil. Grind until mixture becomes a coarse puree. Keep processor running and drizzle in olive oil. Process until all ingredients are combined in a fine puree.

---

1 In the recipe, what is the ratio of the number of garlic cloves to the number of cups of cheese?

(1) 4:3

(2) 3:4

(3) 4:1

(4) 3:1

(5) 1:3

2 Danielle needs to prepare the recipe above for 8 people. Which of the following expressions should she use to find the number of red peppers she should use?

(1) $\frac{6}{8} = \frac{4}{x}$

(2) $\frac{8}{6} = \frac{4}{x}$

(3) $\frac{4+6}{4+8} = \frac{6+8}{x+8}$

(4) $6 \times 8 = 4x$

(5) Not enough information is given.

3 Based on the recipe, which percentage represents the approximate amount of a garlic clove that each person would get in an individual serving?

(1) 10%

(2) 25%

(3) 33%

(4) 50%

(5) 300%

Solve each problem. Show your work.

1 $\frac{8}{20} = \frac{?}{35}$

2 What is 450% of 72?

3 What percent of 320 is 8?

4 $4.05 is 9% of what amount?

Find the percent of increase or decrease. Round your answer to the nearest percent. Show your work.

5 Original amount   New Amount   Percent of Change

     $12.50 $17.50   _____

6 Original amount   New Amount   Percent of Change

     45 pounds   15 pounds   _____

7 Original amount   New Amount   Percent of Change

     60 gallons   50 gallons   _____

Solve. Show your work.

Donita bought a washing machine that cost $520. She paid $100 down. She will pay the rest in monthly payments over 2 years at 18% interest.

8 Estimate how much interest Donita will pay.

9 Estimate what Donita's monthly payment will be.

10 Auto Needs has oil on sale at 3 quarts for $2.85. Minh wants to buy 8 quarts. How much will he pay for the oil?

11 Carlos brought in profits of $4,527. He earns a 22% commission rate. What is the amount of commission Carlos earned?

Circle the best answer for each item.

Items 12–13 refer to the following information.

A bag contains 3 nickels and 2 dimes. You choose one coin without looking.

12 What is the probability that you will choose a nickel?

  (1) $\frac{1}{5}$ or 20%

  (2) $\frac{2}{5}$ or 40%

  (3) $\frac{3}{5}$ or 60%

  (4) $\frac{4}{5}$ or 80%

  (5) 1 or 100%

13 The first coin you choose is a nickel. You did not put it back in the bag. You choose another coin without looking. What is the probability that the second coin you choose will also be a nickel?

  (1) 0 or 0%

  (2) $\frac{2}{5}$ or 40%

  (3) $\frac{1}{2}$ or 50%

  (4) $\frac{3}{4}$ or 75%

  (5) 1 or 100%

14 The Wilsons want to make a circle graph of their family budget. They budget 30% of their take-home pay each month for rent. How many degrees should be in the section representing rent?

  (1) 30°

  (2) 70°

  (3) 108°

  (4) 120°

  (5) 130°

15 A round dining room table has a radius of 30 inches. What is the circumference of the table to the nearest whole inch?

  (1) 60

  (2) 188

  (3) 900

  (4) 942

  (5) 2,826

16 Hector needs mulch for a circular flower bed that is 6 feet in diameter. One bag of mulch covers 25 square feet. What is the area of the garden to the nearest square foot?

Items 17 and 18 refer to the following graph.

Preferred Reading Survey
(in hundreds of adults)

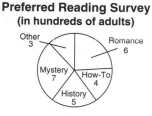

17 What is the probability that the next person surveyed prefers romance or history?

18 What is the probability that the next person surveyed prefers mysteries?

Math Extension　For the next week, keep track of how much money you spend on a hobby or favorite pastime. What percent of your weekly pay or allowance does this amount represent?

Math Connection: Ratios, Proportions, Percents, and Earth Science

Cold, Old Water

They're old. They're cold. They're gigantic frozen masses. They are formed from the lightest of snowflakes transformed into some of the densest ice found on Earth. Given enough time, they're powerful enough to sculpt mountains and carve enormous lakes. When we look at the Matterhorn mountain in Switzerland or the Great Lakes in the United States, we see evidence of their mighty force. What are they? They are glaciers.

Scientists believe that during the Ice Ages, glacial ice covered almost 32% of Earth's land masses and about 30% of its oceans. Today, glacial ice covers only 10% of Earth's surface and holds about 75% of Earth's fresh water. The eight Ice Ages and the warmer periods that occurred between them contributed to the formation of Earth's giant ice masses.

One scientist who studies glaciers is Mark Meier. He estimates that globally, glaciers lost 11% of their volume between the 1850s and 1990s. For some glaciers, the loss has been even greater; they now measure only half of what they did in the mid-1800s. Scientists such as Meier are concerned because glacier reduction is occurring at an increasing rate.

A mountain glacier.

Earth Science: Thick Blue Ice

Glaciers are found in cold climates that stay cold all year round. We find them at Earth's poles and in mountains at high altitudes—even near the equator. As layers of snow fall and become buried by new layers, the snow crystals melt and change into ice crystals called *firn*. The weight from the accumulated masses compresses the firn into even larger, denser ice crystals. Some grow to become the size of baseballs.

Some of the largest glaciers are found in the polar areas of Greenland and Antarctica. These continental ice sheets can be more than 1,000 meters (3,280 feet) thick. About three meters below the surface, the dense glacial ice begins to look blue. The deeper you look, the bluer the ice appears.

Use the information on the previous page to answer the questions that follow.

 1 Which of the following percents is the approximate surface area currently covered by Earth's glaciers?
(1) 10%
(2) 30%
(3) 32%
(4) 62%
(5) 75%

2 Glaciers are sometimes referred to as blue ice because
(1) the ice becomes so cold and hard that it turns blue.
(2) people who go to see them become cold and turn blue.
(3) at a certain depth, the ice in the glacier appears blue.
(4) touching a glacier makes your teeth chatter and skin turn blue.
(5) the sun and the blue sky melts the snow and turns the glacial ice blue.

3 Which ratio best represents the amount of glacial ice covering Earth's surface (land and water) in the Ice Ages compared to the amount of glacial ice covering Earth's surface today?
(1) 1:3
(2) 1:6
(3) 3:1
(4) 3:7
(5) 6:1

4 At which depth does glacial ice begin to look blue?
(1)   1 meter
(2)   3 meters
(3)   6 meters
(4)   9 meters
(5) 3,280 feet

5 Which of the following is not a true statement about glaciers?
(1) Glaciers are found at the north and south poles.
(2) Glaciers are found in high altitudes.
(3) Glaciers began forming thousands of years ago.
(4) Glaciers are not found near the equator.
(5) Glaciers are receding at an accelerating rate.

6 Scientists are concerned about the rate at which glaciers are receding. Use a separate piece of paper to write a sentence or two to explain why you think they are concerned.

Cumulative Review

Write each number in words.

1 6,030,017 _____

2 7.096 _____

Compare each pair of numbers. Write >, <, or =.

3 54,277 ____ 54,227   **4** 0.083 ____ 0.83   **5** 5.10 ____ 5.1

Solve. Show your work.

6 $\begin{array}{r} 5\frac{2}{3} \\ +3\frac{3}{4} \\ \hline \end{array}$ **7** $\begin{array}{r} 7\frac{1}{6} \\ -4\frac{4}{5} \\ \hline \end{array}$

8 $5 \times 4 - 2 + 9 \div 3 =$ **9** $\sqrt{25} =$ **10** $9^2 =$

11 $\$27.36 \div 9 =$ **12** $76 \times 60 =$ **13** $0.097 \times 0.05 =$

14 $457 + 32 + 619 =$ **15** $\$9.00 - \$3.48 =$

16 $3\frac{3}{8} \times 3\frac{1}{3} =$ **17** $2\frac{1}{2} \div 3\frac{1}{4} =$

18 $2.54 + 3.408 + 53.1 =$ **19** $8.4 - 3.726 =$

Check your answers on page 316.

Circle the best answer for each item.

20 Ben is building a rectangular dog pen that is 4.3 meters long and 2.5 meters wide. Which is the correct expression to find the perimeter of the dog pen?

(1) 2.5^2
(2) 4.3^2
(3) $4.3 + 2.5$
(4) 4.3×2.5
(5) $4.3 + 4.3 + 2.5 + 2.5$

21 A one-ounce serving of cereal has 120 calories and 2 grams of fat. What percent of the calories comes from fat? (Reminder: 1 gram of fat contains 9 calories.)

(1) 8%   (4) 15%
(2) 10%   (5) 25%
(3) 12%

22 The early morning low temperature was −5 degrees. By noon, the temperature had increased 8 degrees. What was the temperature at noon?

(1) −40°   (4)   3°
(2) −13°   (5) 13°
(3)   −3°

Item 23 refers to the following table.

Funship Park Admission

	1-Day Pass	2-Day Pass
Children (18 and under)	$ 9.95	$11.95
Adults	$19.95	$26.95
Senior Citizens (over 65)	$14.95	$21.95

23 Marti, who is 32, wants to buy a 2-day pass for herself and each of her three school-age children. How much will the passes cost?

(1) $35.85
(2) $47.80
(3) $49.80
(4) $62.80
(5) $99.60

Answer the following questions. Show your work.

Items 24 and 25 refer to the following information.

During one week, the employees at RV Sales worked the following hours: 46, 39.5, 41.5, 38.5, 39, and 41.5.

24 What is the mean number of hours worked?

25 What is the median of the data?

Items 26 and 27 refer to the following graph.

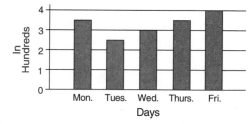

Crosstown Courier Service Deliveries, 3/24–3/28

26 What is the difference between the number of deliveries made on the busiest day and the number of deliveries made on the slowest day?

27 How many deliveries in all did Crosstown Courier Service have during the days from 3/24 to 3/28?

Posttest

Write the value of the underlined digit in words.

1. 2,043,109 _____

2. 4.0<u>5</u>6 _____

Compare each pair of numbers. Write >, <, or = between the two numbers.

3. 25,744 _____ 25,447   4. 0.83 _____ 0.083

Write your answers in the blanks.

5. Round 420,536 to the nearest thousand. _____

6. Round 7.839 to the nearest tenth. _____

7. Write the mixed number that names the shaded portion. _____

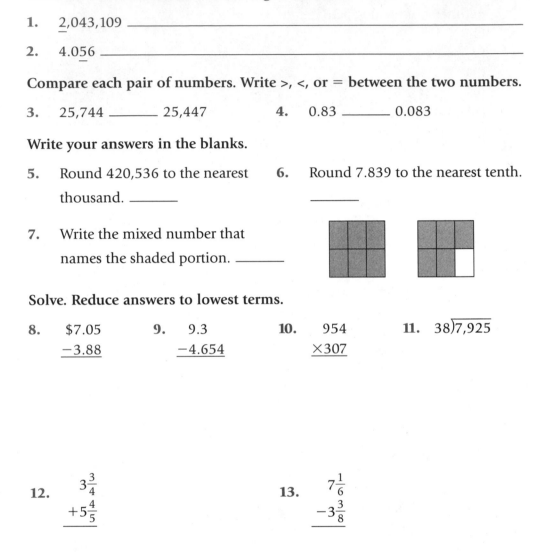

Solve. Reduce answers to lowest terms.

8. $7.05
    −3.88

9. 9.3
    −4.654

10. 954
    ×307

11. 38)7,925

12. $3\frac{3}{4}$
    $+5\frac{4}{5}$

13. $7\frac{1}{6}$
    $-3\frac{3}{8}$

14. $8 + 2 \times 9 \div (7 - 4) =$

15. $76 + 58 + 429 =$

16. $7.29 + 28.306 + 4.6 =$

17. $800 - 416 =$

18. $0.067 \times 0.08 =$

19. $2\frac{2}{3} \times 2\frac{1}{10} =$

20. $\$27.68 \div 8 =$

21. $2.5 \div 0.04 =$

22. $6\frac{2}{5} \div 5\frac{1}{3} =$

23. $6^3 =$

24. Change $3\frac{1}{2}\%$ to a decimal.

25. Change 6.5 to a percent.

26. Change 75% to a fraction.

27. Change $\frac{3}{5}$ to a percent.

28. What is 40% of $36?

29. What percent of 16 is 20?

30. 8 is 2% of what number?

31. $\frac{12}{40} = \frac{9}{?}$

Circle the best answer for each item.

32. Rita wants to tile a rectangular floor that is 8 feet wide and 18 feet long. What is the area of Rita's floor in square feet?

    (1) 26
    (2) 52
    (3) 64
    (4) 144
    (5) Not enough information is given.

33. Of 121 customers who made a purchase at Sports World on Thursday, 63 paid by check or with cash. Estimate the number of charge customers.

    (1) 40
    (2) 50
    (3) 60
    (4) 175
    (5) 185

34. A box has 10 marbles in it: 5 black, 4 white, and 1 blue. What is the probability of choosing a white marble?

    (1) $\frac{1}{10}$ or 10%
    (2) $\frac{2}{5}$ or 40%
    (3) $\frac{1}{2}$ or 50%
    (4) $\frac{3}{5}$ or 60%
    (5) 1 or 100%

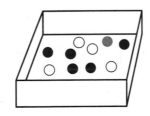

35. Karen's recipe for pudding uses $2\frac{1}{2}$ cups of milk. Karen is doubling the recipe. Which is the correct expression to find the number of cups of milk she needs?

    (1) $\frac{1}{2} \div 2\frac{1}{2}$
    (2) $2 \div 2\frac{1}{2}$
    (3) $2\frac{1}{2} \times \frac{1}{2}$
    (4) $2\frac{1}{2} \times 2$
    (5) $2\frac{1}{2} \div 2$

36. Which expression best describes how to find the length of the rectangle?

    (1) 10 + 200
    (2) 200 − 10
    (3) 10 × 200
    (4) 200 ÷ 10
    (5) Not enough information is given.

    | Area = 200 sq. ft. | 10 ft. |

37. Charlene works 8 hours per day as a salesperson. She spends 25% of her time preparing sales reports. How many hours each day does she spend preparing sales reports?

    (1) 0.25
    (2) 2
    (3) 4
    (4) 8
    (5) 20

38. A department store is having a summer clearance sale. Men's sport shirts are on sale at 3 for $25.98. Rudy wants to buy one shirt. Which is the correct expression to find the cost of one shirt?

    (1) 3 × $25.98
    (2) 3 ÷ $25.98
    (3) $25.98 × 3
    (4) $25.98 ÷ 0.3
    (5) $25.98 ÷ 3

39. Lisa's gross pay is $1,200.00 each month. She has $75.40 deducted for health insurance. Which is the correct expression to find the percent of Lisa's gross pay that is deducted for health insurance?

    (1) $75.40 × $1,200.00
    (2) $75.40 ÷ $1,200.00
    (3) $75.40 − $1,200.00
    (4) $1,200.00 − $75.40
    (5) $1,200.00 ÷ $75.40

40. Ray bought a sweater for his son at Discount Fashions. The price tag is shown below. What is the original price of the sweater?

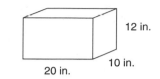

Discount Fashions

Discount	40%
You save	$6.80
Original price	???

(1) $2.72
(2) $9.52
(3) $10.20
(4) $17.00
(5) $23.80

41. Best Value has tube socks on sale at 6 pairs for $4.98. Darrell wants to buy 10 pairs. Which is the correct expression to find the cost of 10 pairs?

(1) $\frac{6}{?} = \frac{\$4.98}{10}$

(2) $\frac{6}{10} = \frac{?}{\$4.98}$

(3) $\frac{6}{\$4.98} = \frac{?}{10}$

(4) $\frac{6}{\$4.98} = \frac{10}{?}$

(5) $\frac{10}{\$4.98} = \frac{6}{?}$

42. Marge works the third shift at Capital Products. She earns $8.60 per hour. If she were to switch to first shift, her wages would be $8.30 per hour. Which is the correct expression to find the percent of decrease in her wages?

(1)   ($8.30 − $8.60) ÷ $8.30
(2)   ($8.30 + $8.60) ÷ $8.60
(3)   ($8.60 − $8.30) ÷ $8.30
(4)   ($8.60 − $8.30) ÷ $8.60
(5)   ($8.60 ÷ $8.30) − $8.60

Write your answers and show your work.

43. What is the volume in cubic inches of the shipping carton shown?

12 in.

10 in.

20 in.

44. Gloria delivers newspapers. She drives $2\frac{1}{5}$ miles on Broad Street, $3\frac{9}{10}$ miles through a housing subdivision, and $1\frac{3}{10}$ miles on Hilltop Road. Estimate how many miles Gloria drives on her route.

Item 45 refers to the following table.

Northbrook Hotel
Banquet Reservations, Week of 4/15

	Ballroom	Salon	Terrace
Friday	416	184	112
Saturday	504	224	136

45. Banquet guests are seated 8 guests to a table. How many tables must be set up for a banquet on Saturday in the ballroom?

46. The Chans bought a circular pool for their children. The radius of the pool is 7 feet. What is the area to the nearest square foot covered by this pool?

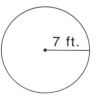

7 ft.

47. George works at a hardware store. His customer bought a gallon of paint for $11.37 and a paintbrush for $2.89. Sales tax on this purchase was $0.72. The customer gave George $20.00 in cash. How much change should George have given his customer?

Item 50 refers to the following data.

Number of vehicles serviced by a mechanic during the 4 weeks in September: 42, 36, 19, 31

50. What is the mean of the data?

51. At Good's Supply, 18 employees work in the warehouse and 12 drive delivery trucks. What is the ratio in lowest terms of warehouse employees to drivers?

Item 48 refers to the following graph.

Plainview Rainfall

48. What was the total amount of rainfall for August and September?

Item 52 refers to the following graph.

Swansons' Monthly Budget (in hundreds of dollars)

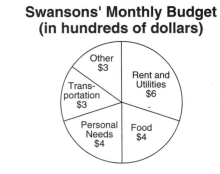

52. What percent of the Swansons' budget is for transportation?

Item 53 refers to the following number line.

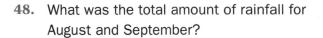

53. In the morning, the temperature was ⁻3 degrees. By noon the temperature had increased 6 degrees. What was the temperature at noon?

49. Find the circumference of this circle to the nearest inch.

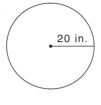

20 in.

Check your answers on pages 317–318.

Posttest Correlation Chart

Section/Skill	Item Numbers	Total Correct	Pages
1. Number Sense	1, 3, 5, 14, 53	_____ out of 5	10–21
2. Adding and Subtracting Whole Numbers	15, 17, 33	_____ out of 3	22–33
3. Multiplying and Dividing Whole Numbers	10, 11, 32, 36, 45, 50	_____ out of 6	34–51
4. Squares, Cubes, and Square Roots	23, 43	_____ out of 2	52–67
5. Fraction Basics	7	_____ out of 1	68–81
6. Adding and Subtracting Fractions	12, 13, 44	_____ out of 3	82–95
7. Multiplying and Dividing Fractions	19, 22, 35	_____ out of 3	96–113
8. Decimal Basics	2, 4, 6	_____ out of 3	114–125
9. Adding and Subtracting Decimals	8, 9, 16, 47, 48	_____ out of 5	126–135
10. Multiplying and Dividing Decimals	18, 20, 21, 38	_____ out of 4	136–155
11. Ratio and Proportions	31, 41, 51	_____ out of 3	156–169
12. Percent Basics	24, 25, 26, 27	_____ out of 4	170–175
13. Solving for the Part (p)	28, 37	_____ out of 2	176–186
14. Solving for the Rate (r)	29, 39, 52	_____ out of 3	187–198
15. Solving for the Base (b) and Percent of Change	30, 40, 42	_____ out of 3	199–212
16. Special Topics	34, 46, 49	_____ out of 3	213–234

TOTAL CORRECT FOR POSTTEST _____ **out of 53**

If you answered fewer than 48 items correctly, determine which of the areas you need to study further. Go back and review the content in those pages.

Calculator Handbook

Calculator Features

A calculator is an useful math tool. It is designed to help us find amounts. However, there are decisions you need to make *before* entering numbers in a calculator. They are:

● What do I need to find?

● What numbers do I need?

● What operations do I need to solve the problem?

● Approximately what amount should I see in the final calculator display?

There are many different types of calculators, but all of them have these common features.

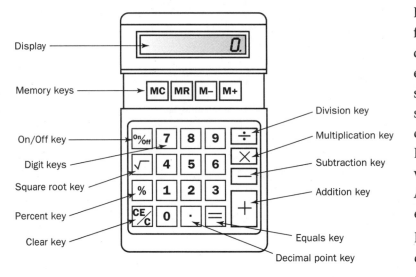

However, even these common features may differ slightly from calculator to calculator. For example, some calculators have a separate CE key (Clear Entry) and a separate C key (Clear). On some calculators, you can press CE (Clear Entry) to erase your last entry without erasing the whole problem. A separate C key will usually erase everything you entered.

If your calculator features a combined CE\C key, press once to clear your last entry or twice to clear everything. Develop the good habit of completely clearing the display before calculating a new problem.

PRACTICE

 Get to know your calculator. Practice entering and clearing the following numbers and symbols on your calculator.

CE/C 4 **+** 8 **+** 2 **=**   Did you get an answer of 14 on your calculator display?

CE/C 4 **+** 8 **CE/C** **+** 2 **=**   Did you get an answer of 6 on your calculator display? Which digit did your calculator erase?

Adding Whole Numbers

Practice this example of adding using your calculator.

EXAMPLE 279 + 36

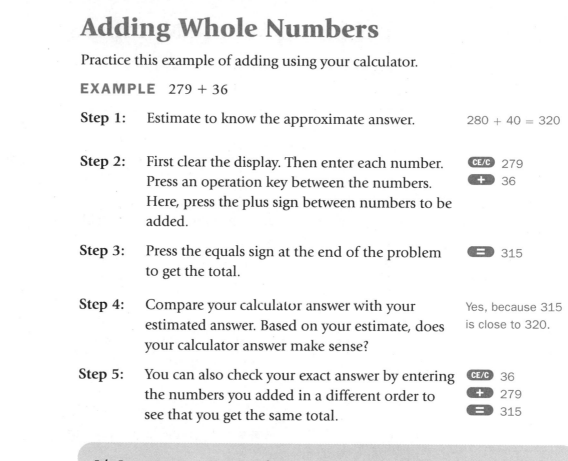

Step 1: Estimate to know the approximate answer.

280 + 40 = 320

Step 2: First clear the display. Then enter each number. Press an operation key between the numbers. Here, press the plus sign between numbers to be added.

CE/C 279
+ 36

Step 3: Press the equals sign at the end of the problem to get the total.

= 315

Step 4: Compare your calculator answer with your estimated answer. Based on your estimate, does your calculator answer make sense?

Yes, because 315 is close to 320.

Step 5: You can also check your exact answer by entering the numbers you added in a different order to see that you get the same total.

CE/C 36
+ 279
= 315

It's Important to Know: When you *add* numbers on a calculator, you can enter the numbers *in any order*.

Example The numbers in the problem 20 + 8 + 43 could also be entered as 8 + 20 + 43 **or** 8 + 43 + 20 **or** 43 + 20 + 8 and so on.

PRACTICE

Follow the process shown above to solve these problems using a calculator.

❶

SOSA'S APPLIANCES INVOICE	
Oven	$499
Tax	$ 29
Delivery	$ 25
Installation	$ 75
TOTAL	

❷

McHENRY PLUMBING INVOICE	
Fixtures	$780
Pipes	$345
Faucets	$290
Connectors	$ 85
Labor	$320
TOTAL	

Estimated answer: _____

Exact answer: _____

Checked answer: _____

Estimated answer: _____

Exact answer: _____

Checked answer: _____

Check your answers on page 318.

Subtracting Whole Numbers

Practice this example of subtracting using your calculator.

EXAMPLE 346 − 88

Step 1:	Estimate to know the approximate answer.	350 − 90 = 260
Step 2:	First clear the display. Then enter the number you are subtracting from. Use the minus sign between numbers. Then enter the number you are subtracting.	**CE/C** 346 **−** 88
Step 3:	Press the equals sign at the end of the problem to find the answer.	**=** 258
Step 4:	Compare your calculator answer with your estimated answer. Based on your estimate, does your calculator answer make sense?	Yes, because 258 is close to 260.
Step 5:	To check your exact answer add your calculator answer to the number you subtracted. You should get the amount you started with.	**CE/C** 258 **+** 88 **=** 346

It's Important to Know: When you *subtract* numbers on a calculator, you must enter the numbers *in order* of the whole amount minus the number being subtracted.

Example The numbers in the problem 547 − 95 must be entered in that order: first 547, then the minus sign, then 95.

PRACTICE

Follow the process shown above to solve these problems using a calculator.

	1st Half	Final
Bluebirds	39	89
Redwings	47	116

1 By how many points were the Redwings leading at the end of the first half?

Estimated answer: _____

Exact answer: _____

Checked answer: _____

2 By how many points did the Redwings beat the Bluebirds?

Estimated answer: _____

Exact answer: _____

Checked answer: _____

3 14,899 people attended this game. Paid attendance was 13,968. How many people didn't have to pay for their tickets?

Estimated answer: _____

Exact answer: _____

Checked answer: _____

Check your answers on page 318.

Multiplying Whole Numbers

Practice this example of multiplying using your calculator.

EXAMPLE 83×17

Step 1:	Estimate to know the approximate answer.	$80 \times 20 = 1{,}600$
Step 2:	First clear the display. Then enter each number. Use the multiplication sign between numbers to be multiplied.	**CE/C** 83 **✕** 17
Step 3:	Press the equals sign at the end of the problem to get the product.	**=** 1411
Step 4:	Compare your calculator answer with your estimated answer. Based on your estimate, does your calculator answer make sense?	Yes, because 1,411 is close to 1,600.
Step 5:	You can also check your exact answer by entering the numbers you multiplied in a different order to see that you get the same product.	**CE/C** 17 **✕** 83 **=** 1411

It's Important to Know: When you *multiply* numbers on a calculator, you can enter the numbers *in any order*.

Example The numbers in the problem $2 \times 8 \times 4$ could also be entered as $8 \times 2 \times 4$ **or** $8 \times 4 \times 2$ **or** $4 \times 2 \times 8$ and so on.

PRACTICE

Follow the process shown above to solve these problems using a calculator.

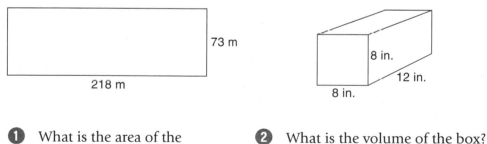

73 m

218 m

8 in.

8 in.

12 in.

❶ What is the area of the rectangle?

Estimated answer: _____

Exact answer: _____

Checked answer: _____

❷ What is the volume of the box?

Estimated answer: _____

Exact answer: _____

Checked answer: _____

Dividing Whole Numbers

Practice this example of dividing using your calculator.

EXAMPLE 256 ÷ 4

Step 1: Estimate to know the approximate answer.

250 ÷ 5 = 50

Step 2: First clear the display. Then enter the number to be divided. Use the division sign between numbers. Then enter the number you are dividing by.

CE/C 256 **÷** 4

Step 3: Press the equals sign at the end of the problem to get the answer.

= 64

Step 4: Compare your calculator answer with your estimated answer. Based on your estimate, does your calculator answer make sense?

Yes, because 64 is close to 50.

Step 5: You can also check your exact answer by multiplying your calculator answer by the number you divided by. You should get the same amount you started with.

CE/C 64 **×** 4
= 256

> **It's Important to Know:** When you *divide* numbers on a calculator, you must enter the numbers *in order* of the whole amount then the number you are dividing by.
>
> **Example** The numbers in the problem 96 ÷ 6 must be entered in that order: first 96, then the division sign, then 6.

PRACTICE

Follow the process shown above to solve these problems using a calculator.

13 FACTORY WORKERS ★ WIN LOTTERY! ★
★★★★★★★★★★★
After Taxes, Split $32,500,000

$93,000 STATE GRANT AWARDED
6 Counties to Divide Evenly for Joint Project

❶ How much will each lottery winner receive?

Estimated answer: _____

Exact answer: _____

Checked answer: _____

❷ How much grant money will each of the 6 counties receive?

Estimated answer: _____

Exact answer: _____

Checked answer: _____

Check your answers on page 319.

Exponents and Square Roots

Calculators are very useful when solving problems with exponents. For example, to calculate 26^2 you would need to multiply 26×26. To calculate 8^3 you would need to multiply $8 \times 8 \times 8$. Using a calculator can make both problems much easier. Practice these examples using your calculator.

EXAMPLE   base→ 11^3 ← exponent

Step 1:	Estimate to know the approximate answer.	$10 \times 10 \times 10$ = 1,000

Step 2: First clear the display. Then enter the base number (11). Multiply that number by itself the number of times shown by the exponent (3). Use the multiplication sign between numbers to be multiplied.

 CE/C 11 ✖ 11 ✖ 11

Step 3: Press the equals sign at the end of the problem to get the product.

 = 1331

Step 4: Compare your calculator answer with your estimated answer. Based on your estimate, does your calculator answer make sense?

Yes, because 1,331 is close to 1,000.

Step 5: You can also check your exact answer by entering the numbers again to see that you get the same product.

CE/C 11 ✖ 11
✖ 11
= 1331

EXAMPLE $\sqrt{75}$

Step 1: Estimate to know the approximate answer.

$\sqrt{64}$ = 8 and $\sqrt{81}$ = 9, so the answer should be between 8 and 9.

Step 2: First clear the display. Enter the number to find the square root of. Then press the square root key. Your answer will appear in the display.

 CE/C 75 √
8.660254

Step 3: Compare your calculator answer with your estimated answer. Based on your estimate, does your calculator answer make sense?

Yes, because 8.66 is between 8 and 9.

Step 4: You can check your exact answer by repeating Step 2. You can also check your answer by multiplying the calculator answer by itself.

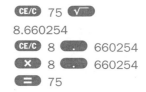 CE/C 75 √
8.660254
CE/C 8 . 660254
✖ 8 . 660254
= 75

PRACTICE

 Follow the steps shown above to solve these problems using a calculator.

1 5^5 **2** 2.5^2 **3** $\sqrt{625}$ **4** $\sqrt{289}$

Adding Decimals

Practice this example of adding using your calculator.

EXAMPLE 45.67 + 24.069

Step 1:	Estimate to know the approximate answer.	45 + 25 = 70
Step 2:	First clear the display. Then enter each number. Use the decimal point key to enter the decimal point at the correct place in each decimal number. Use the plus sign between numbers to be added.	[CE/C] 45 [.] 67 [+] 24 [.] 069
Step 3:	Press the equals sign at the end of the problem to get the total.	[=] 69.739
Step 4:	Compare your calculator answer with your estimated answer. Based on your estimate, does your calculator answer make sense?	Yes, because 69.739 is close to 70.
Step 5:	You can also check your exact answer by entering the numbers you added in a different order to see that you get the same total.	[CE/C] 24 [.] 069 [+] 45 [.] 67 [=] 69.739

It's Important to Know: Whether you're adding whole numbers or decimals on a calculator, you can enter the numbers *in any order*.

Example The problem 20.4 + 8.02 + 43.9 could also be entered as 8.02 + 20.4 + 43.9 **or** 8.02 + 43.9 + 20.4 **or** 43.9 + 20.4 + 8.02 and so on.

PRACTICE

Follow the process shown above to solve these problems using a calculator.

1

Oil Change — Every 3,500 Miles

Oil Changed at: | 16,632.8 | Miles

Next Oil Change: [   ] Miles

Estimated answer: _____

Exact answer: _____

Checked answer: _____

2

INVOICE

Filter	$5.88
4 qt. oil	4.25
O-ring	0.60
Parts Total	

Estimated answer: _____

Exact answer: _____

Checked answer: _____

Check your answers on page 319.

Subtracting Decimals

Practice this example of subtracting using your calculator.

EXAMPLE 36.78 − 18.9

Step 1: Estimate to know the approximate answer.

40 − 20 = 20

Step 2: First clear the display. Then enter the number you are subtracting from. Use the decimal point key to enter the decimal point at the correct place in each decimal number. Use the minus sign between numbers to be subtracted. Now enter the number you are subtracting.

CE/C 3 6 • 7 8
─ 1 8 • 9

Step 3: Press the equals sign at the end of the problem to find the answer.

= 17.88

Step 4: Compare your calculator answer with your estimated answer. Based on your estimate, does your calculator answer make sense?

Yes, because 17.88 is close to 20.

Step 5: You can also check your exact answer by adding your calculator answer to the number you subtracted. You should get the total amount you started with.

CE/C 1 7 • 8 8
+ 1 8 • 9
= 36.78

It's Important to Know: Whether you're subtracting whole numbers or decimals on a calculator, you must enter the numbers *in order* of the whole amount minus the number being subtracted.

Example The numbers in the problem 54.7−9.5 must be entered in that order: first 54.7, then the minus sign, then 9.5.

PRACTICE

Follow the process shown above to solve these problems using a calculator.

May 10 — Odometer Reading	21,383.7
May 15 — Odometer Reading	21,734.6
May 20 — Odometer Reading	21,862.2

Start With:	$20.00
Spend:	$11.75 (Gasoline)
	$1.39 (Large Fries)
	$1.01 (Soda)

1 How many miles were driven between May 10 and May 15?

Estimated answer: _____

Exact answer: _____

Checked answer: _____

2 How much change is left?

Estimated answer: _____

Exact answer: _____

Checked answer: _____

Check your answers on page 319.

Multiplying Decimals

Practice this example of multiplying using your calculator.

EXAMPLE 7.3 × 8.17

Step 1: Estimate to know the approximate answer.

7 × 8 = 56

Step 2: First clear the display. Then enter each number. Use the decimal point key to enter the decimal point at the correct place in each decimal number. Use the multiplication sign between numbers to be multiplied.

 CE/C 7 ● . 3
× 8 ● . 17

Step 3: Press the equals sign at the end of the problem to get the product.

= 59.641

Step 4: Compare your calculator answer with your estimated answer. Based on your estimate, does your calculator answer make sense?

Yes, because 59.641 is close to 56.

Step 5: You can also check your exact answer by entering the numbers you multiplied in a different order to see that you get the same product.

 **CE/C** 8 ● . 17
× 7 ● . 3
= 59.641

It's Important to Know: Whether you're multiplying whole numbers or decimals on a calculator, you can enter the numbers *in any order*.

Example The numbers in the problem 2.67 × 8.1 × .04 could also be entered as 8.1 × 2.67 × .04 **or** 8.1 × .04 × 2.67 **or** .04 × 2.67 × 8.1 and so on.

PRACTICE

 Follow the process shown above to solve these problems using a calculator.

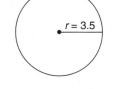
12.4 m

20.7 m

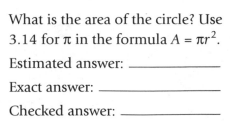
r = 3.5

1 What is the area of the space shown here?

Estimated answer: _____

Exact answer: _____

Checked answer: _____

2 What is the area of the circle? Use 3.14 for π in the formula $A = \pi r^2$.

Estimated answer: _____

Exact answer: _____

Checked answer: _____

Check your answers on page 319.

Dividing Decimals

Practice this example of dividing using your calculator.

EXAMPLE 25.6 ÷ .8

Step 1: Estimate to know the approximate answer.

26 ÷ 1 = 26

Step 2: First, clear the display. Then enter each number. Use the division sign between numbers to be divided.

 **CE/C** 25 **.** 6
÷ **.** 8

Step 3: Press the equals sign at the end of the problem to get the answer.

= 32

Step 4: Compare your calculator answer with your estimated answer. Based on your estimate, does your calculator answer make sense?

Yes, because 32 is close to 26.

Step 5: You can also check your exact answer by multiplying your calculator answer by the number you divided by. You should get the same amount you started with.

CE/C 32 **X**
. 8 **=** 25.6

It's Important to Know: Whether you're dividing whole numbers or decimals on a calculator, you must enter the numbers *in order* of the whole amount then the number you are dividing by.

Example The numbers in the problem 43.28 ÷ .4 must be entered in that order: first 43.28, then the division sign, then .4.

PRACTICE

 Follow the process shown above to solve these problems using a calculator.

|←————— 10.5 ft. —————→|

RECEIPT:	
Bakery	$15.83
Plates	$ 3.25
Napkins	$ 2.15
Subtotal	$21.23
Tax	$ 1.27
Total	$22.50

1 If the board were cut into 4 equal pieces, how long would each length be?

Estimated answer: _____

Exact answer: _____

Checked answer: _____

2 If three people split the cost of these birthday items, how much would each person pay?

Estimated answer: _____

Exact answer: _____

Checked answer: _____

Converting Fractions, Decimals, and Percents

Calculators are tools that help simplify converting between fractions, decimals, and percents.

Practice converting the following fraction using your calculator.

EXAMPLE Change $\frac{3}{4}$ to a decimal.

Step 1: First, clear the display. Then enter the numerator (3). Divide the numerator by the denominator (4). Use the division sign between numbers to be divided. **CE/C** 3 **÷** 4

Step 2: Press the equals sign at the end of the problem to get the answer. **=** .75

You can then change .75 to a percent. Multiply by 100 and add a percent sign: .75 = 75%.

PRACTICE

Complete the chart below to show equivalent fractions, decimals, and percents. Use your calculator and the process shown above when needed.

Fraction	Decimal	Percent
$\frac{1}{10}$		10%
$\frac{1}{8}$		
	.2	
		25%
$\frac{1}{3}$	.33	$33\frac{1}{3}\%$
$\frac{2}{5}$		
		50%
	.6	
$\frac{2}{3}$	.67	$66\frac{2}{3}\%$
		75%
$\frac{4}{5}$		
	.9	
1	1	100%

Check your answers on page 319.

Solving Percent Problems

Use your calculator to practice finding the part in this percent example.

EXAMPLE What is 40% of 60?

Step 1: First, clear the display. Then enter the whole (60). Press the multiplication sign.
[CE/C] 60 [×]

Step 2: Enter the number that represents the percent (40). Then press the percent key. Your answer should appear in the display. (Some calculators may require you to press the equals sign to display the answer.)
40 [%] 24

Use your calculator to practice finding the percent in this percent example.

EXAMPLE What percent is 50 of 200?

Step 1: First, clear the display. Then enter the part (50). Press the division sign.
[CE/C] 50 [÷]

Step 2: Enter the number that represents the whole (200). Then press the equals sign at the end of the problem to get the decimal equivalent of the percent.
200 [=] .25

Step 3: Write the decimal answer as a percent.
.25 = 25%

Use your calculator to practice finding the whole in this percent example.

EXAMPLE $30 is 25% of what amount?

Step 1: First, clear the display. Then enter the part (30). Press the division sign.
[CE/C] 30 [÷]

Step 2: Manually convert the percent to a decimal.
25% = .25

Step 3: Enter the decimal equivalent (.25). Then press the equals sign at the end of the problem to get the whole amount. Write the answer as $120.
[.] 25 [=] 120
120. = $120

PRACTICE

Follow the examples shown above to solve these problems using a calculator.

1 Purchase $35.00
Tax at 6.25% _____

What is the dollar amount of the tax on this purchase? (Hint: Round the amount of the tax to the nearest cent.)

2 5 defective parts
500 parts tested for defects

Based on the number of defective parts and the number of parts tested, what percent of the parts are defective?

Solving Multi-step Problems

Apply what you have learned in this handbook to problems that require more than one step.

Practice this example using your calculator.

EXAMPLE A square plot of land measuring 24 feet on each side is to be fenced in. Two gates are going to be installed: one 3-foot gate and one 6-foot gate. How much fencing is needed?

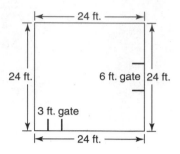

Step 1: Estimate to know the approximate answer.

$25 \times 4 = 100$
$100 - 10 = 90$

Step 2: First, clear the display. Then enter the length of one side. Multiply that number by the number of sides to be fenced in (4). This will find the length of fencing needed *without* the two gates. Press the equals sign at the end of the problem to get the product.

CE/C 24 **×** 4
= 96

Step 3: Subtract the measurements of the gates to find the amount of fencing needed. (Use mental math to get this amount: 3 + 6 = 9.)

− 9 **=** 87

Step 4: Compare your calculator answer with your estimated answer. Based on your estimate, does your calculator answer make sense?

Yes, because 87 is close to 90.

PRACTICE

Find the subtotal of the checks. Then find the amount of the deposit after subtracting the amount for *cash received*.

1

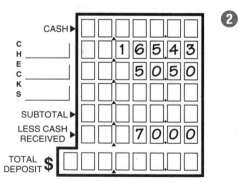

2

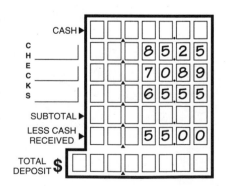

Check your answers on page 319.

Answers and Explanations

PAGES 1–6

1. **three hundred thousand**
2. **seven thousandths**
3. $46,\underline{0}23 < 46,\underline{2}03$ Zero is less than 2.
4. $0.76 > 0.456$ Add a zero: $0.7\underline{6}0 \quad 0.\underline{4}56$
   Then compare. 0.760 is greater than 0.456.
5. **530,000**
   $5\underline{\textcircled{3}4},103$ The number to the right of the 3 is less than 5. Do not change the circled digit.
6. **3.73**
   $3.7\textcircled{2}5$ The number to the right of the 2 is 5; add 1 to the circled digit.
7. $2\frac{1}{4}$ Each figure is divided into 4 equal parts. Two figures are completely shaded, and 1 of the 4 parts of the last figure is shaded.
8. **272**

   $$\begin{array}{r} \overset{5\ 9\ 10}{\cancel{6}\cancel{0}\cancel{0}} \\ -328 \\ \hline 272 \end{array}$$

9. **2.854**

   $$\begin{array}{r} \overset{5\ 16\ 9\ 10}{\cancel{6}.\cancel{7}\cancel{0}\cancel{0}} \\ -3.846 \\ \hline 2.854 \end{array}$$

10. **451,744**

    $$\begin{array}{r} 743 \\ \times 608 \\ \hline 5\ 944 \\ +445\ 80 \\ \hline 451,744 \end{array}$$

11. **$3.07**

    $$\begin{array}{r} \$3.07 \\ 4\overline{)\$12.28} \\ -12 \\ \hline 0\ 28 \\ -\ 28 \\ \hline 0 \end{array}$$

12. $7\frac{5}{24}$

    $$\begin{array}{r} 4\frac{7}{8} = 4\frac{21}{24} \\ +2\frac{1}{3} = 2\frac{8}{24} \\ \hline 6\frac{29}{24} = 6 + 1\frac{5}{24} = 7\frac{5}{24} \end{array}$$

13. $1\frac{11}{15}$

    $$\begin{array}{r} 6\frac{2}{5} = 6\frac{6}{15} = \overset{5\ 21}{\cancel{6}\frac{\cancel{6}}{15}} \\ -4\frac{2}{3} = 4\frac{10}{15} = 4\frac{10}{15} \\ \hline 1\frac{11}{15} \end{array}$$

14. **20**

    $$\begin{array}{l} 12 \div (2 + 1) \times 6 - 4 \\ 12 \div \quad 3 \quad \times 6 - 4 \\ \qquad 4 \qquad \times 6 - 4 \\ \qquad\qquad 24 \quad - 4 \\ \qquad\qquad\qquad 20 \end{array}$$

15. **475**

    $$\begin{array}{r} \overset{1\ 1}{42} \\ 376 \\ +\ 57 \\ \hline 475 \end{array}$$

16. **38.201**

    $$\begin{array}{r} \overset{1\ 1\ 1}{3.270} \\ 26.400 \\ +\ 8.531 \\ \hline 38.201 \end{array}$$

17. **$2.09**

    $$\begin{array}{r} \overset{4\ 9\ 18}{\$\cancel{5}.\cancel{0}\cancel{8}} \\ -\ 2.99 \\ \hline \$2.09 \end{array}$$

18. **0.0162**

    $$\begin{array}{r} 0.054 \\ \times\ 0.3 \\ \hline 0.0162 \end{array}$$

19. $16\frac{1}{2}$ $\quad 4\frac{2}{5} \times 3\frac{3}{4} = \dfrac{\overset{11}{\cancel{22}}}{\underset{1}{\cancel{5}}} \times \dfrac{\overset{3}{\cancel{15}}}{\underset{2}{\cancel{4}}} = \dfrac{33}{2} = 16\frac{1}{2}$

20. **182 r14**

    $$\begin{array}{r} 182\ \text{r}14 \\ 42\overline{)7,658} \\ -4\ 2 \\ \hline 3\ 45 \\ -3\ 36 \\ \hline 98 \\ -84 \\ \hline 14 \end{array}$$

21. **520**

    $$\begin{array}{r} 5\ 20 \\ 0.06.\overline{)31.20.} \\ -30 \\ \hline 1\ 2 \\ -1\ 2 \\ \hline 0 \end{array}$$

22. $\frac{4}{5}$ $\quad 3\frac{1}{2} \div 4\frac{3}{8} = \frac{7}{2} \div \frac{35}{8} = \dfrac{\overset{1}{\cancel{7}}}{\underset{1}{\cancel{7}}} \times \dfrac{\overset{4}{\cancel{8}}}{\underset{5}{\cancel{35}}} = \frac{4}{5}$

23. **49** $\quad 7^2 = 7 \times 7 = 49$
24. **3.75** $\quad 375\% \div 100 = 3.\underset{\smile}{75}. = 3.75$
25. **7%** $\quad 0.07 \times 100 = 0.\underset{\smile}{07}. = 7\%$
26. $\frac{4}{5}$ $\quad 80\% = \frac{80}{100} = \frac{80 \div 20}{100 \div 20} = \frac{4}{5}$

27. 75% $4\overline{)3.00}$ $0.75 = 75\%$
    $\underline{-2\,8}$
    20
    $\underline{-20}$
    0
    (0.75 shown above the division)

28. 12   part = base × rate
    8% = 0.08 150
    $\underline{\times 0.08}$
    12.00

29. 25%   rate = part ÷ base
    0.25
    $\$72\overline{)\$18.00}$ $0.25 = 25\%$
    $\underline{-14\,4}$
    $3\,60$
    $\underline{-3\,60}$
    0

30. 36   base = part ÷ rate
    $3\,6.$
    150% = 1.5 $1.5.\overline{)54.0.}$
    $\underline{-45}$
    $9\,0$
    $\underline{-9\,0}$
    0

31. 14 $8 \times 35 = 280; 280 \div 20 = 14$

32. **(3) 144 − 48** Subtract the number of gallons of interior paint sold (48) from the number of gallons of interior paint in stock (144). You do not need the number of gallons of exterior paint.

33. **(2) 36** Two sides measure 12 feet, and two sides measure 6 feet. Add all four measurements to find the perimeter.
    $12 + 12 + 6 + 6 = 36$

34. **(3) 20^2** The room is a square because all four sides have the same length. Use the formula $A = s^2$, and substitute 20 for s.

35. **(3) 170** Round the three to the nearest ten, and add to estimate the total.

79	rounds to	80
35	rounds to	40
+52	rounds to	+50
		170

36. **(4) 16 × 8** To find the total amount, multiply the number of packages (16) by the number of buns in each package (8).

37. **(3) $\frac{1}{2}$ or 50%**
    $$\frac{\text{number of sections with 2}}{\text{number of sections}} = \frac{3}{6} = \frac{1}{2} \text{ or } 50\%$$

38. **(2) $\frac{3}{4}$** To make one third of the recipe, divide the amount of broth ($2\frac{1}{4}$ cups) by 3.
    $$2\frac{1}{4} \div 3 = \frac{9}{4} \div \frac{3}{1} = \frac{\overset{3}{\cancel{9}}}{4} \times \frac{1}{\underset{1}{\cancel{3}}} = \frac{3}{4}$$

39. **(2)** $\frac{3}{\$1.56} = \frac{5}{?}$ Both ratios are in the same order. The first ratio means that 3 pounds cost $1.56. The second ratio means that 5 pounds cost an unknown amount.

40. **(4) 40 × 15%** Multiply the base (40) times the rate (15%) to find the part.

41. **(4) 200** Multiply by 1,000 to convert kilograms to grams. Move decimal point 3 places to the right to multiply by 1,000 (3 zeros).
    $0.2 \times 1,000 = 0.200. = 200$

42. **(4) $2.08 × 4** Find the total cost by multiplying the unit cost ($2.08) by the number of pounds (4).

43. **(2) 6%** Divide the part ($25.20) by the base ($420.00) to find the rate.

    0.06 $0.06 = 6\%$
    $\$420\overline{)\$25.20}$
    $\underline{-25\,20}$
    0

44. **(3) $18.60 ÷ 30%** Divide the part ($18.60) by the rate (30%) to find the base.

45. **(4) 60 ÷ 12** The area of the rectangle and the length of one side are given. Divide the area by the length to find the width.

46. **(2) 4%** Subtract Abdul's current wage ($5.25) from his new wage ($5.46) to find the amount of change. Then divide by the original amount, his current wage.

    0.04 $0.04 = 4\%$
    $\$5.46\$5.25.\overline{)\$0.21.00}$
    $\underline{-5.25}\underline{-21.00}$
    $\$0.210$

47. **1,920 cubic inches**
    $V = l \times w \times h$
    $ = 10 \times 12 \times 16$
    $ = 1,920$

48. **18 miles** Round each distance and subtract.

$36\frac{9}{10}$ rounds to   37

$-19\frac{2}{5}$ rounds to -19

                             18

49. **345 miles** Multiply the number of gallons of gas (15) by the miles car B can travel on the highway per gallon of gas (23).

$$\begin{array}{r} 23 \\ \times 15 \\ \hline 115 \\ +23 \\ \hline 345 \end{array}$$

50. **$4\frac{17}{24}$ feet** Since there are 12 inches in a foot and inches are smaller than feet, divide $56\frac{1}{2}$ by 12.

$$56\frac{1}{2} \div 12 = \frac{113}{2} \div \frac{12}{1} = \frac{113}{2} \times \frac{1}{12} = \frac{113}{24} = 4\frac{17}{24}$$

51. **$3\frac{3}{4}$ hours** Since 43 is close to 45, think of 8:43 as $8\frac{3}{4}$. Since 36 is close to 30, think of 12:36 as $12\frac{1}{2}$. Then subtract.

$$\begin{array}{r} 12\frac{1}{2} = 12\frac{2}{4} = \cancel{12}^{\,11}\cancel{\frac{2}{4}}^{\frac{6}{4}} \\ - \ 8\frac{3}{4} = \ 8\frac{3}{4} = \ 8\frac{3}{4} \\ \hline 3\frac{3}{4} \end{array}$$

52. **50 square feet** Divide the diameter (8 ft.) by 2 to find the radius (8 ÷ 2 = 4 ft.).

$A = \pi r^2$

$A = 3.14 \times 4 \times 4$

$A = 3.14 \times 16$

$A = 50.24$ square feet

Then 50.24 square feet rounds to 50 square feet.

53. **$4.04** Add the cost of the statue and the card to find the subtotal. Add the tax to find the total. Then subtract the total from the amount of cash the customer gives Tawanna to find the amount of change.

$$\begin{array}{ccc} \overset{1}{} & & \overset{3\ 9\ \ 910}{} \\ \$32.50 & \$34.25 & \$\cancel{40.00} \\ +\ \ 1.75 & +\ \ 1.71 & -\ 35.96 \\ \hline \$34.25 & \$35.96 & \$4.04 \end{array}$$

54. **1,500** The bar for Lincoln County ends between the 3 and 4 marks and represents 3,500 customers. The bar for Hamilton County reaches the 2 mark and represents 2,000 customers. Subtract

to find the   3,500

difference. $-2,000$

                      1,500

55. **28 feet**

$C = \pi d$

$C = 3.14 \times 9$

$C = 28.26$ feet

Then 28.26 feet rounds to 28 feet.

56. **25** Find the total of the numbers in the set.

$37 + 18 + 21 + 24 = 100$

Then divide by the number of items in the set.

$100 \div 4 = 25$

57. **22.5** Arrange the numbers in order. Then find the middle numbers.

18, <u>21</u>, <u>24</u>, 37

Then find the mean of the middle numbers.

$21 + 24 = 45; 45 \div 2 = 22.5$

58. **$\frac{5}{9}$**

$$\frac{20 \text{ in factory}}{36 \text{ employees}} = \frac{20 \div 4}{36 \div 4} = \frac{5}{9}$$

59. **$750**

$$\begin{aligned} interest &= principal \ \times \ rate \ \times \ time \\ &= \$3,000 \ \times \ 0.05 \ \times \ 5 \\ &= \$750 \end{aligned}$$

60. **$^-$5 degrees** Start at $^+$2 on the number line. Count 7 units to the left.

UNIT 1: WHOLE NUMBERS

SECTION 1

PAGE 11

1. **(5) laser printer** Look in the column labeled *Item Number* to find the item number listed in the question. Then look in the column labeled *Description* to find the matching item.

2. **(3) monitor, scanner, printer** Look under the column labeled *Item Number* to find the three item numbers that begin with the same three digits—in this case, 414. Look under the column labeled *Description* to find the matching items.

3. **(2) 5 × $6.99** Look in the column labeled *Description* for laser paper. Then look in the column labeled *Qty* to find the number of items ordered and the column labeled *Item Price* to find the price for each item ordered. Multiply the quantity and the price per item to find the total price for that item.

4. **(4) $55 + $35** Look in the column labeled *Total* to find how much the items ordered will cost.

5. **(4) 1 + 5 + 1 + 1 + 5** Kelly is placing an order for 1 monitor, 5 diskettes, 1 scanner, 1 laser printer, and 5 packages of laser paper.

6. **(5) $699.99 − $459.99** To find the amount of savings, compare prices by subtracting the cheaper price from the more expensive price.

7. **The total amount for diskettes is $54.95 because Kelly ordered five diskettes.** Multiply $10.99 by 5 to get the total amount.

PAGE 12
2. ones
3. thousands
4. hundreds
5. ten thousands
6. hundred thousands
7. thousands
8. millions
9. hundreds
10. hundred thousands
11. tens
12. ten thousands
13. ones
14. ten millions
15. thousands
16. hundreds

PAGE 13
2. forty-three thousand, eighteen
3. one hundred fifteen thousand, two hundred
4. five million, four hundred thousand, twelve
6. 250,911
7. 12,016
8. 9,014,560

PAGE 14
2. 38,<u>0</u>00 < 38,<u>5</u>00 0 is less than 5.

3. 179 = 179 The numbers are the same.
4. 210,<u>5</u>80 > 210,<u>4</u>80 5 is greater than 4.
5. 1,000,000 < 10,000,000 10,000,000 has 8 digits; 1,000,000 has only 7 digits.
6. 496 < 4,690 496 has fewer digits than 4,690.
7. 13,415 = 13,415 The numbers are the same.
8. 802,<u>1</u>65 < 803,<u>9</u>80 2 is less than 3.
9. 5,000 < 50,000 5,000 has fewer digits than 50,000.
10. 1,<u>3</u>45 < 1,<u>4</u>35 3 is less than 4.
11. 10,334 = 10,334 The numbers are the same.
12. 47<u>9</u> > 47<u>6</u> 9 is greater than 6.
13. 340,<u>6</u>35 < 340,<u>8</u>35 6 is less than 8.
14. 5,0<u>1</u>0 > 5,0<u>0</u>1 1 is greater than 0.
15. 682,489 = 682,489 The numbers are the same.
16. 3,8<u>0</u>0 < 3,8<u>5</u>0 0 is less than 5.

PAGE 15
2. **−6 yards** +4 + (−10) = −6
3. **+$8** 3 + 5 = 8
4. **−$11** −8 + (−3) = −11
5. **2 points** 12 − 10 = 2
6. **+3%** −7 + 10 = 3
7. **−1 degree** −6 + 5 = −1
8. **+2 yards** −4 + 6 = 2

PAGE 17
2. 1653.
3. 10241.
4. 18.92
6. 73. × 46. = 3358.
7. 2187. ÷ 3. = 729.
8. 1406.19 − 94.37 = 1311.82
9. $94.85
10. $6.30

PAGE 18
2. **1,700** 1,<u>7</u>23 Since 2 is less than 5, do not change the underlined digit.
3. **7,000** <u>6</u>,509 The number to the right of 6 is 5; add 1 to the underlined digit.
4. **900** <u>8</u>61 Since 6 is greater than 5, add 1 to the underlined digit.
5. **20,000** 1<u>9</u>,580 The number to the right of 9 is 5; add 1 to the underlined digit. The 9 becomes a 10. Write a zero and add 1 to the next place value to the left.

6. **210,000** 2<u>0</u>9,320 Since 9 is greater than 5, add 1 to the underlined digit.

7. **64,000** 6<u>4</u>,299 Since 2 is less than 5, do not change the underlined digit.

8. **5,300,000** 5,2<u>5</u>6,000 The number to the right of 2 is 5; add 1 to the underlined digit.

PAGE 19

1. **Estimate: $50** $10 + $15 + $25
   Exact: $49

2. **Estimate: 100 feet**
   60 feet + 10 feet + 30 feet
   Exact: 99 feet

3. **$155** $20 + $120 + $15

4. **$600** 18 + 22 = 40   40 × $15 = $600

PAGES 20–21

1. **hundreds**

2. **ten thousands**

3. **millions**

4. **hundred thousands**

5. **twenty-eight thousand, three hundred two**

6. **one million, seventy-six thousand, five hundred**

7. **42,057**

8. **3,400,590**

9. **5,680 > 856**
   5,680 has more digits than 856.

10. **32,457 = 32,457** The numbers are the same.

11. **82,<u>3</u>46 < 82,<u>5</u>46** 3 is less than 5.

12. **7<u>9</u>0,300 > 7<u>0</u>9,300** 9 is greater than 0.

13. **40** <u>4</u>3 Since 3 is less than 5, do not change the underlined digit.

14. **2,500** 2,<u>4</u>53 The number to the right of the underlined digit is 5; add 1 to the underlined digit.

15. **310,000** 3<u>0</u>7,216 Since 7 is greater than 5, add 1 to the underlined digit.

16. **4,000,000** <u>4</u>,293,785 Since 2 is less than 5, do not change the underlined digit.

17. **648**

18. **145,800**

19. **1,985**

20. **10,401**

21. **(2) Fewer parts were produced during week 1 than during week 3.**
    12,<u>4</u>35 (week 1) > 12,<u>3</u>45 (week 3) because 4 > 3. Therefore, <u>more</u> parts were produced during week 1 than during week 3.

22. **(4) week 4**
    1<u>4</u>,814 (week 4) > 1<u>2</u>,435 (week 1) because 4 > 2.
    14,<u>8</u>14 (week 4) > 14,<u>5</u>26 (week 2) because 8 > 5.
    1<u>4</u>,814 (week 4) > 1<u>2</u>,345 (week 3) because 4 > 2.
    1<u>4</u>,814 (week 4) > 1<u>3</u>,706 (week 5) because 4 > 3.

23. **(5) week 5** 13,<u>7</u>06 Since 7 is greater than 5, add 1 to the underlined digit. 13,706 (week 5) rounds to 14,000.

24. **(2) one million, thirty thousand, four hundred two**

25. **(4) $2,300**
    $2,<u>2</u>64 Since 6 is greater than 5, add 1 to the underlined digit.

26. **(5) −2**

SECTION 2

PAGE 23

1. **(4) 40 + 40 + 30 + 30** You know that two sides measure 40 feet and two sides measure 30 feet. Add all four sides to find the perimeter.

2. **(5) 20 + 20 + 24 + 24** Since it is a rectangle, two sides of the painting measure 20 inches and two sides measure 24 inches. Add all four measurements to find the perimeter.

3. **(2) 80 − 3** Find the difference in the width of the yard and the opening needed for the sidewalk. Subtract the smaller number from the larger one.

4. **(4) 30 + 30 + 6 + 6** Since the dog run is rectangular in shape, if you know one length and one width, you know the remaining length and width. Total all four measurements to find the perimeter.

5. **(1) 10 + 10 + 14 + 14** You know that two walls measure 10 feet across and two walls measure 14 feet across. Add all four wall measurements to find the perimeter.

6. **(1) 10 + 10 + 18** The deck is a rectangle, but the railing goes along only three sides. Two of the sides measure 10 feet, and the third side measures 18 feet. Add the three measurements to find the total railing length.

7. You can find the perimeter of a rectangular room by measuring only one length and one width because in a rectangle you know that the opposite sides are equal in length. Therefore, you really know the lengths of all the sides.

PAGE 25

2. 494
$$\begin{array}{r} 1 \\ 476 \\ +\ 18 \\ \hline 494 \end{array}$$
Check
$$\begin{array}{r} 1 \\ 18 \\ +476 \\ \hline 494 \end{array}$$

3. 651
$$\begin{array}{r} 1\ 1 \\ 387 \\ +264 \\ \hline 651 \end{array}$$
Check
$$\begin{array}{r} 1\ 1 \\ 264 \\ +387 \\ \hline 651 \end{array}$$

4. 655
$$\begin{array}{r} 1 \\ 473 \\ +182 \\ \hline 655 \end{array}$$
Check
$$\begin{array}{r} 1 \\ 182 \\ +473 \\ \hline 6.55 \end{array}$$

5. 707
$$\begin{array}{r} 1\ 1 \\ 148 \\ 327 \\ +232 \\ \hline 707 \end{array}$$
Check
$$\begin{array}{r} 1\ 1 \\ 232 \\ 327 \\ +148 \\ \hline 707 \end{array}$$

6. 1,375
$$\begin{array}{r} 1\ 1 \\ 135 \\ 456 \\ +784 \\ \hline 1,375 \end{array}$$
Check
$$\begin{array}{r} 1\ 1 \\ 784 \\ 456 \\ +135 \\ \hline 1,375 \end{array}$$

7. 735
$$\begin{array}{r} 1\ 1 \\ 542 \\ 125 \\ +\ 68 \\ \hline 735 \end{array}$$
Check
$$\begin{array}{r} 1\ 1 \\ 68 \\ 125 \\ +542 \\ \hline 735 \end{array}$$

8. 1,346
$$\begin{array}{r} 1\ 1 \\ 817 \\ 76 \\ +453 \\ \hline 1,346 \end{array}$$
Check
$$\begin{array}{r} 1\ 1 \\ 453 \\ 76 \\ +817 \\ \hline 1,346 \end{array}$$

9. 195
$$\begin{array}{r} 1 \\ 38 \\ +157 \\ \hline 195 \end{array}$$
Check
$$\begin{array}{r} 1 \\ 157 \\ +\ 38 \\ \hline 195 \end{array}$$

10. 964
$$\begin{array}{r} 1\ 1 \\ 450 \\ 326 \\ +188 \\ \hline 964 \end{array}$$
Check
$$\begin{array}{r} 1\ 1 \\ 188 \\ 326 \\ +450 \\ \hline 964 \end{array}$$

11. 23,068
$$\begin{array}{r} {}^{2}\ 1\ 1 \\ {}_1 3,947 \\ 18,889 \\ +\ \ \ 232 \\ \hline 23,068 \end{array}$$
Check
$$\begin{array}{r} 1\ 1 \\ {}_{1\ 2}\ 232 \\ 18,889 \\ +\ 3,947 \\ \hline 23,068 \end{array}$$

12. 954
$$\begin{array}{r} 1 \\ 142 \\ 7 \\ 802 \\ +\ \ 3 \\ \hline 954 \end{array}$$
Check
$$\begin{array}{r} 1 \\ 3 \\ 802 \\ 7 \\ +142 \\ \hline 954 \end{array}$$

13. **Calculator answer: $792**
Checked answer: $792 To check addition, add again from bottom to top.

14. **Week 7** After Week 6, the students have read 99 books. Therefore, the students have not read 100 books until Week 7.

PAGE 27

2. 653
$$\begin{array}{r} 957 \\ -304 \\ \hline 653 \end{array}$$
Check
$$\begin{array}{r} 653 \\ +304 \\ \hline 957 \end{array}$$

3. 474
$$\begin{array}{r} 899 \\ -425 \\ \hline 474 \end{array}$$
Check
$$\begin{array}{r} 474 \\ +425 \\ \hline 899 \end{array}$$

4. 743
$$\begin{array}{r} 786 \\ -\ 43 \\ \hline 743 \end{array}$$
Check
$$\begin{array}{r} 743 \\ +\ 43 \\ \hline 786 \end{array}$$

5. 25
$$\begin{array}{r} {}^{4\,13} \\ \cancel{53} \\ -28 \\ \hline 25 \end{array}$$
Check
$$\begin{array}{r} 1 \\ 25 \\ +28 \\ \hline 53 \end{array}$$

6. 155
$$\begin{array}{r} {}^{3\,12} \\ 4\cancel{2}6 \\ -271 \\ \hline 155 \end{array}$$
Check
$$\begin{array}{r} 1 \\ 155 \\ +271 \\ \hline 426 \end{array}$$

7. 179
$$\begin{array}{r} {}^{2\,12} \\ 3\cancel{2}9 \\ -150 \\ \hline 179 \end{array}$$
Check
$$\begin{array}{r} 1 \\ 179 \\ +150 \\ \hline 329 \end{array}$$

8. 467
$$\begin{array}{r} {}^{7\ ^{13}_{3}13} \\ 8\cancel{4}3 \\ -376 \\ \hline 467 \end{array}$$
Check
$$\begin{array}{r} 1\ 1 \\ 467 \\ +376 \\ \hline 843 \end{array}$$

9. 333

$$\begin{array}{r} \overset{9}{\cancel{5}}\overset{4\;10}{\cancel{0}0} \\ -167 \\ \hline 333 \end{array}$$ Check $$\begin{array}{r} {}^{1\,1} \\ 333 \\ +167 \\ \hline 500 \end{array}$$

10. 387

$$\begin{array}{r} \overset{6\;\overset{10}{0}15}{7\cancel{1}\cancel{5}} \\ -328 \\ \hline 387 \end{array}$$ Check $$\begin{array}{r} {}^{1\,1} \\ 387 \\ +328 \\ \hline 715 \end{array}$$

11. 877

$$\begin{array}{r} \overset{8\;\overset{9}{1}16}{9\cancel{0}\cancel{6}} \\ -\;\;29 \\ \hline 877 \end{array}$$ Check $$\begin{array}{r} {}^{1\,1} \\ 877 \\ +\;\;29 \\ \hline 906 \end{array}$$

12. 38

$$\begin{array}{r} \overset{4\;11\,10\,10}{\cancel{5}\cancel{2}\cancel{0}} \\ -482 \\ \hline 38 \end{array}$$ Check $$\begin{array}{r} {}^{1\,1} \\ 38 \\ +482 \\ \hline 520 \end{array}$$

13. 573

$$\begin{array}{r} \overset{7\;12}{8\cancel{2}7} \\ -254 \\ \hline 573 \end{array}$$ Check $$\begin{array}{r} {}^{1} \\ 573 \\ +254 \\ \hline 827 \end{array}$$

14. 318

$$\begin{array}{r} \overset{7\,10\,17}{8\cancel{1}\cancel{7}} \\ -499 \\ \hline 318 \end{array}$$ Check $$\begin{array}{r} {}^{1\,1} \\ 318 \\ +499 \\ \hline 817 \end{array}$$

15. 157

$$\begin{array}{r} \overset{6\;\overset{9}{1}10\,10}{7\cancel{0}\cancel{0}} \\ -543 \\ \hline 157 \end{array}$$ Check $$\begin{array}{r} {}^{1\,1} \\ 157 \\ +543 \\ \hline 700 \end{array}$$

16. 198

$$\begin{array}{r} \overset{5\;17\,14}{\cancel{6}\cancel{8}\cancel{4}} \\ -486 \\ \hline 198 \end{array}$$ Check $$\begin{array}{r} {}^{1\,1} \\ 198 \\ +486 \\ \hline 684 \end{array}$$

17. **Calculator answer: $358**
   Checked answer: $650 To check
   subtraction, add your answer and the
   numbers that were subtracted:
   $125 + $89 + $57 + $21 = $292
   Check: $358 + $292 = $650

18. 11 To find how many books Gayle needs
   to order, subtract the amount of books
   already ordered from the total amount she
   needs to order:
   $120 - 43 = 77 - 54 = 23 - 12 = 11.$

PAGE 29

1. (4) $450 − $225 To find how much
   more Paula pays for rent than for food,
   subtract what she pays for food from what
   she pays for rent. You do not need to
   know how much she earns.

2. (5) **Not enough information is given.**
   Rasheed delivers water **three** times a week.
   You know the number of bottles delivered
   on only two of the days. You do not know
   the number of bottles delivered on the
   third day.

3. (2) 45 × 5 To find how many miles
   Danielle could drive in 5 hours, multiply
   how many miles she drives in 1 hour by 5.
   You do not need to know how many miles
   to a gallon her car gets.

4. (3) 56 + 17 To find the total number of
   boxes, add the boxes he collected on
   Monday. You do not need to know how
   many boxes Kunio collected on Tuesday.

5. (5) **Not enough information is given.**
   To find how many pairs of sandals were
   sold, you must subtract the number of
   pairs of sandals left in stock (given as 27)
   from the number of pairs of sandals that
   were in stock (not given).

6. (2) 220 Use addition to find the total.
   Round the three numbers to the nearest
   ten and add to estimate the total. 67 is
   about 70, 73 is about 70, and 81 is
   about 80.
   70 + 70 + 80 = 220

7. Situations and details will vary. Sample
   answer: I would need to know the price of
   the item, how many items I was buying, if
   there was a discount or coupons, and any
   tax to be added.

PAGE 31

2. **Sports Vehicle** The tallest bar represents
   the greatest number of cars sold.

3. **225,000** The bar for mini-vans represents
   225,000 cars.

4. **200,000** The two smallest bars represent the two least popular types of cars sold. The bar for the 4-door is the smallest and represents about 75,000 cars. The next smallest bar is the convertible and it represents about 125,000 cars. Add to find the total.

$$\begin{array}{r} 75{,}000 \\ +125{,}000 \\ \hline 200{,}000 \end{array}$$

5. **300,000** The bar for sports vehicles represents 450,000 cars. The bar for the 2-door represents 150,000. Subtract to find the difference.

$$\begin{array}{r} 450{,}000 \\ -150{,}000 \\ \hline 300{,}000 \end{array}$$

6. **1,025,000** Find values for each bar. Sum the values. $150{,}000 + 75{,}000 + 450{,}000 + 225{,}000 + 125{,}000 = 1{,}025{,}000$

7.

TALL BUILDINGS IN THE UNITED STATES

8. **Height or Stories**
9. **Stories, by tens**
10. **Sample answer: Tall Buildings in the United States**
11. **One World Trade Center, 110 stories**
12. **First Interstate World Center**
13. **30** The bar for the One World Trade Center represents 110 stories. The bar for the Amoco Building represents 80 stories. Subtract to find the difference. $110 - 80 = 30$
14. **102 stories**
15. **Amoco Building, Empire State Building, and One World Trade Center**

PAGES 32–33

1. **733**

$$\begin{array}{r} \overset{1\ 1}{237} \\ +496 \\ \hline 733 \end{array}$$

2. **162**

$$\begin{array}{r} \overset{414}{\cancel{547}} \\ -385 \\ \hline 162 \end{array}$$

3. **$782**

$$\begin{array}{r} \overset{1\ 1}{\$295} \\ +\ 487 \\ \hline \$782 \end{array}$$

4. **279**

$$\begin{array}{r} \overset{5\ \overset{9}{\cancel{10}}10}{\cancel{600}} \\ -321 \\ \hline 279 \end{array}$$

5. **$312**

$$\begin{array}{r} \overset{5\ 10}{\$\cancel{6}07} \\ -\ 295 \\ \hline \$312 \end{array}$$

6. **928**

$$\begin{array}{r} \overset{1\ 1}{503} \\ 58 \\ +367 \\ \hline 928 \end{array}$$

7. **$689**

$$\begin{array}{r} \overset{1}{\$227} \\ 317 \\ +\ 145 \\ \hline \$689 \end{array}$$

8. **$699**

$$\begin{array}{r} \overset{8\ 18\ 13}{\$\cancel{993}} \\ -\ 294 \\ \hline \$699 \end{array}$$

9. **78**

$$\begin{array}{r} \overset{6\ 15\ 12}{\cancel{762}} \\ -684 \\ \hline 78 \end{array}$$

10. **1,250**

$$\begin{array}{r} \overset{1}{836} \\ +414 \\ \hline 1{,}250 \end{array}$$

11. **Calculator answer: $3,809**
12. **Calculator answer: 45,205**
13. **89 people** Subtract the number of deliveries made before Ed's break from the total number of deliveries he needs to make.

$$\begin{array}{r} \overset{1\ 12\ 18}{\cancel{238}} \\ -149 \\ \hline 89 \end{array}$$

14. **422 miles** Add the number of miles driven before lunch and the number of miles driven after lunch.

$$\begin{array}{r} \overset{1\ 1}{243} \\ +179 \\ \hline 422 \end{array}$$

15. **221 customers** Add the number of each type of customer.
$$\begin{array}{r} \overset{1\ 1}{153} \\ 26 \\ +\ 42 \\ \hline 221 \end{array}$$

16. **44** Subtract the number of tickets sold for the 5:00 show from the number of tickets sold for the 7:30 show.
$$\begin{array}{r} 127 \\ -83 \\ \hline 44 \end{array}$$

17. **(5) 20 + 20 + 14 + 14** You know that two sides measure 20 feet and two sides measure 14 feet. Add all four measurements to find the perimeter.

18. **(4) $92** Add to find the total.
$$\begin{array}{r} \overset{1}{\$28} \\ 53 \\ +\ 11 \\ \hline \$92 \end{array}$$

19. **(2) 208 − 185** The difference between Bill's present weight and the weight his doctor advised is the weight Bill should lose. Subtract the smaller amount from the larger one.

20. **(4) $452** Add the cost of tuition and the cost of books and supplies.
$$\begin{array}{r} \overset{1}{\$348} \\ +\ 104 \\ \hline \$452 \end{array}$$

21. **(3) comedy** The tallest bar represents the greatest number of rentals. The bar for comedy is the tallest.

22. **(4) 250** The bar for comedy is the tallest. It ends between the 4 and 5 marks and represents 450 rentals. The bar for drama is the smallest. It reaches the 2 mark and represents 200 rentals. Subtract to find the difference.
$$\begin{array}{r} 450 \\ -200 \\ \hline 250 \end{array}$$

23. **(2) 1,550** Find values for each bar. Sum the values.
$$350 + 200 + 250 + 450 + 300 = 1,550$$

SECTION 3

PAGE 35

1. **(1) 144 ÷ 12** To separate a large amount into dozens, divide the total number of cookies by the number in a dozen.

2. **(3) $849 ÷ 3** To separate an amount into equal payments, divide the full amount by the number of payments.

3. **(3) 4 × 10** To find the total number of feet, multiply the number of rolls by the number of feet in each roll.

4. **(4) 18 × 24** To find the total number of airplanes packed, multiply the number of airplanes that fit in each carton by the number of cartons packed.

5. **(5) 200 ÷ 10** To separate a large amount into equal parts, divide the number of miles driven by the number of gallons of gas used.

6. **(2) 18 + 30** To find the total, add the number originally on hand and the number in the shipment.

7. **Division is a shortcut for subtraction because it is repeated subtraction. It is like subtracting the same number more than once. Multiplication is a shortcut for addition because it is repeated addition. It is like adding the same number more than once. (Answer should include an example.)**

PAGE 37

2. 468
$$\begin{array}{r} 234 \\ \times\ 2 \\ \hline 468 \end{array}$$

3. 2,304
$$\begin{array}{r} \overset{5\ 2}{384} \\ \times\ 6 \\ \hline 2,304 \end{array}$$

4. 1,314
$$\begin{array}{r} \overset{2}{73} \\ \times 18 \\ \hline 584 \\ +73 \\ \hline 1,314 \end{array}$$

5. 1,992

$$\begin{array}{r} \overset{1}{}83 \\ \times 24 \\ \hline 332 \\ +1\ 66 \\ \hline 1{,}992 \end{array}$$

6. 12,173

$$\begin{array}{r} \overset{2}{4}\overset{3}{6} \\ 259 \\ \times\ 47 \\ \hline 1\ 813 \\ +10\ 36 \\ \hline 12{,}173 \end{array}$$

7. $4,275

$$\begin{array}{r} \overset{6\,4}{} \\ \$475 \\ \times\ \ \ 9 \\ \hline \$4{,}275 \end{array}$$

8. 19,200

$$\begin{array}{r} \overset{1}{}64 \\ \times 300 \\ \hline 19{,}200 \end{array}$$

9. 336,474

$$\begin{array}{r} \overset{1}{}\overset{2}{8}\overset{1}{37} \\ \times 402 \\ \hline 1\ 674 \\ +334\ 80 \\ \hline 336{,}474 \end{array}$$

10. 35,280

$$\begin{array}{r} \overset{2\,4}{1}\overset{1}{26} \\ \times 280 \\ \hline 10\ 080 \\ 25\ 2 \\ \hline 35{,}280 \end{array}$$

11. $2,723

$$\begin{array}{r} \overset{6\,6}{} \\ \$389 \\ \times\ \ \ 7 \\ \hline \$2{,}723 \end{array}$$

12. 1,920

$$\begin{array}{r} \overset{3}{}48 \\ \times 40 \\ \hline 1{,}920 \end{array}$$

13. 33,201

$$\begin{array}{r} \overset{1}{5}\overset{4}{2}7 \\ \times\ 63 \\ \hline 1\ 581 \\ +31\ 62 \\ \hline 33{,}201 \end{array}$$

14. 475,488

$$\begin{array}{r} \overset{1}{2}\overset{3}{4} \\ 936 \\ \times 508 \\ \hline 7\ 488 \\ +468\ 00 \\ \hline 475{,}488 \end{array}$$

15. **Calculator answer: 13,170**
$198 \times 55 = 10{,}890 \quad 57 \times 40 = 2{,}280$
$10{,}890 + 2{,}280 = 13{,}170$

16. **744** $45 + 48 = 93 \quad 93 \times 8 = 744$
17. **$1,500** $125 \times \$12 = \$1{,}500$
18. **14,280** $280 \times 51 = 14{,}280$
19. **2,016** $84 \times 24 = 2{,}016$
20. **4,900** $350 \times 14 = 4{,}900$

PAGE 38

2. **$215**

$$\begin{array}{r} \$215 \\ 7)\overline{\$1{,}505} \\ -14 \\ \hline 10 \\ -\ 7 \\ \hline 35 \\ -35 \\ \hline 0 \end{array} \qquad \begin{array}{r} \$215 \\ \times\ \ \ 7 \\ \hline \$1{,}505 \end{array}$$

3. **913**

$$\begin{array}{r} 913 \\ 6)\overline{5{,}478} \\ -5\ 4 \\ \hline 7 \\ -6 \\ \hline 18 \\ -18 \\ \hline 0 \end{array} \qquad \begin{array}{r} 913 \\ \times\ \ \ 6 \\ \hline 5{,}478 \end{array}$$

4. **165**

$$\begin{array}{r} 165 \\ 36)\overline{5{,}940} \\ -3\ 6 \\ \hline 2\ 34 \\ -2\ 16 \\ \hline 180 \\ -180 \\ \hline 0 \end{array} \qquad \begin{array}{r} 165 \\ \times\ 36 \\ \hline 990 \\ 4\ 95 \\ \hline 5{,}940 \end{array}$$

PAGE 39

2. **87 r11**

$$\begin{array}{r} 87\ \text{r}11 \\ 24)\overline{2{,}099} \\ -1\ 92 \\ \hline 179 \\ -168 \\ \hline 11 \end{array} \qquad \begin{array}{r} 87 \\ \times 24 \\ \hline 348 \\ 1\ 74 \\ \hline 2\ 088 \\ +\ \ \ 11 \\ \hline 2{,}099 \end{array}$$

3. 93 r6

$$\begin{array}{r} 93\ r6 \\ 9\overline{)843} \\ -81 \\ \hline 33 \\ -27 \\ \hline 6 \end{array}$$

$$\begin{array}{r} 93 \\ \times\ 9 \\ \hline 837 \\ +\ 6 \\ \hline 843 \end{array}$$

4. 246 r4

$$\begin{array}{r} 246\ r4 \\ 5\overline{)1,234} \\ -10 \\ \hline 23 \\ -20 \\ \hline 34 \\ -30 \\ \hline 4 \end{array}$$

$$\begin{array}{r} 246 \\ \times\ 5 \\ \hline 1\ 230 \\ +\ 4 \\ \hline 1,234 \end{array}$$

5. 96 r11

$$\begin{array}{r} 96\ r11 \\ 18\overline{)1,739} \\ -1\ 62 \\ \hline 119 \\ -108 \\ \hline 11 \end{array}$$

$$\begin{array}{r} 96 \\ \times 18 \\ \hline 768 \\ 96 \\ \hline 1\ 728 \\ +\ 11 \\ \hline 1,739 \end{array}$$

6. $35

$$\begin{array}{r} \$35 \\ 62\overline{)\$2,170} \\ -1\ 86 \\ \hline 310 \\ -310 \\ \hline 0 \end{array}$$

$$\begin{array}{r} \$35 \\ \times\ 62 \\ \hline 70 \\ 2\ 10 \\ \hline \$2,170 \end{array}$$

7. 46 r7

$$\begin{array}{r} 46\ r7 \\ 8\overline{)375} \\ -\ 32 \\ \hline 55 \\ -48 \\ \hline 7 \end{array}$$

$$\begin{array}{r} 46 \\ \times\ 8 \\ \hline 368 \\ +\ 7 \\ \hline 375 \end{array}$$

8. 47

$$\begin{array}{r} 47 \\ 27\overline{)\$1,269} \\ -1\ 08 \\ \hline 189 \\ -189 \\ \hline 0 \end{array}$$

$$\begin{array}{r} 47 \\ \times 27 \\ \hline 329 \\ 94 \\ \hline 1,269 \end{array}$$

PAGE 40

2. 101

$$\begin{array}{r} 101 \\ 26\overline{)2,626} \\ -2\ 6 \\ \hline 26 \\ -26 \\ \hline 0 \end{array}$$

$$\begin{array}{r} 101 \\ \times\ 26 \\ \hline 606 \\ 2\ 02 \\ \hline 2,626 \end{array}$$

3. $40

$$\begin{array}{r} \$40 \\ 8\overline{)\$320} \\ -32 \\ \hline 00 \\ -0 \\ \hline 0 \end{array}$$

$$\begin{array}{r} \$40 \\ \times\ 8 \\ \hline \$320 \end{array}$$

4. 304

$$\begin{array}{r} 304 \\ 6\overline{)1,824} \\ -18 \\ \hline 24 \\ -24 \\ \hline 0 \end{array}$$

$$\begin{array}{r} 304 \\ \times\ 6 \\ \hline 1,824 \end{array}$$

PAGE 41

6. 189

$$\begin{array}{r} 189 \\ 27\overline{)5,103} \\ -2\ 7 \\ \hline 2\ 40 \\ -2\ 16 \\ \hline 243 \\ -243 \\ \hline 0 \end{array}$$

$$\begin{array}{r} 189 \\ \times\ 27 \\ \hline 1\ 323 \\ 3\ 78 \\ \hline 5,103 \end{array}$$

7. 308

$$\begin{array}{r} 308 \\ 19\overline{)5,852} \\ -5\ 7 \\ \hline 152 \\ -152 \\ \hline 0 \end{array}$$

$$\begin{array}{r} 308 \\ \times\ 19 \\ \hline 2\ 772 \\ 3\ 08 \\ \hline 5,852 \end{array}$$

8. $276

$$\begin{array}{r} \$276 \\ 8\overline{)\$2,208} \\ -1\ 6 \\ \hline 60 \\ -56 \\ \hline 48 \\ -48 \\ \hline 0 \end{array}$$

$$\begin{array}{r} \$276 \\ \times\ 8 \\ \hline \$2,208 \end{array}$$

9. 800

$$\begin{array}{r} 800 \\ 6\overline{)4,800} \\ -4\ 8 \\ \hline 00 \\ -0 \\ \hline 00 \\ -0 \\ \hline 0 \end{array}$$

$$\begin{array}{r} 800 \\ \times\ 6 \\ \hline 4,800 \end{array}$$

10. 515

$$13\overline{)6{,}695}$$
$$\underline{-6\,5}$$
$$19$$
$$\underline{-13}$$
$$65$$
$$\underline{-65}$$
$$0$$

$$\begin{array}{r} 515 \\ \times\ 13 \\ \hline 1\ 545 \\ 5\ 15 \\ \hline 6{,}695 \end{array}$$

11. $580

$$4\overline{)\$2{,}320}$$
$$\underline{-2\,0}$$
$$32$$
$$\underline{-32}$$
$$00$$
$$\underline{-\ 0}$$
$$0$$

$$\begin{array}{r} \$580 \\ \times\ \ 4 \\ \hline \$2{,}320 \end{array}$$

12. 455

$$10\overline{)4{,}550}$$
$$\underline{-4\,0}$$
$$55$$
$$\underline{-50}$$
$$50$$
$$\underline{-50}$$
$$0$$

$$\begin{array}{r} 455 \\ \times\ 10 \\ \hline 4{,}550 \end{array}$$

14. 489 r6

$$7\overline{)3{,}429}$$
$$\underline{-2\,8}$$
$$62$$
$$\underline{-56}$$
$$69$$
$$\underline{-63}$$
$$6$$

$$\begin{array}{r} 489 \\ \times\ \ 7 \\ \hline 3{,}423 \\ +\ \ \ 6 \\ \hline 3{,}429 \end{array}$$

15. 632

$$9\overline{)5{,}688}$$
$$\underline{-5\,4}$$
$$28$$
$$\underline{-27}$$
$$18$$
$$\underline{-18}$$
$$0$$

$$\begin{array}{r} 632 \\ \times\ \ 9 \\ \hline 5{,}688 \end{array}$$

16. 93

$$21\overline{)1{,}953}$$
$$\underline{-1\,89}$$
$$63$$
$$\underline{-63}$$
$$0$$

$$\begin{array}{r} 93 \\ \times 21 \\ \hline 93 \\ 186 \\ \hline 1{,}953 \end{array}$$

17. 156 r4

$$8\overline{)1{,}252}$$
$$\underline{-\ 8}$$
$$45$$
$$\underline{-40}$$
$$52$$
$$\underline{-48}$$
$$4$$

$$\begin{array}{r} 156 \\ \times\ \ 8 \\ \hline 1{,}248 \\ +\ \ \ 4 \\ \hline 1{,}252 \end{array}$$

18. 107 r33

$$46\overline{)4{,}955}$$
$$\underline{-4\,6}$$
$$355$$
$$\underline{-322}$$
$$33$$

$$\begin{array}{r} 107 \\ \times\ 46 \\ \hline 642 \\ 4\ 28 \\ \hline 4{,}922 \\ +\ \ \ 33 \\ \hline 4{,}955 \end{array}$$

19. $917

$$4\overline{)\$3{,}668}$$
$$\underline{-3\,6}$$
$$6$$
$$\underline{-4}$$
$$28$$
$$\underline{-28}$$
$$0$$

$$\begin{array}{r} \$917 \\ \times\ \ 4 \\ \hline \$3{,}668 \end{array}$$

20. 346 r4

$$5\overline{)1{,}734}$$
$$\underline{-1\,5}$$
$$23$$
$$\underline{-20}$$
$$34$$
$$\underline{-30}$$
$$4$$

$$\begin{array}{r} 346 \\ \times\ \ 5 \\ \hline 1{,}730 \\ +\ \ \ 4 \\ \hline 1{,}734 \end{array}$$

21. Calculator Answer: 68 $1{,}632 \div 24$
    Checked Answer: 1,632 68×24

22. $1,455 $17{,}460 \div 12 = 1455$

23. 231 minutes $18{,}480 \div 80 = 231$

24. 14 hours $840 \div 60 = 14$

PAGE 43

2. $n + 16 = 20$

3. $5n = 50$

4. $27 - n = 12$

5. $\frac{n}{6} = 5$

6. $$\begin{array}{rcr} a\ +\ 12 & = & 32 \\ -\ 12 & = & -12 \\ \hline a & = & 20 \end{array}$$
    $a + 12 = 32$
    $20 + 12 = 32$

7. $$\begin{array}{rcr} d\ \div\ 7 & = & 8 \\ \times\ 7 & = & \times 7 \\ \hline d & = & 56 \end{array}$$
    $d \div 7 = 8$
    $56 \div 7 = 8$

8. $\begin{array}{rcr} n - 18 &=& 56 \\ + 18 &=& +18 \\ \hline n &=& 74 \end{array}$ $\qquad$ $\begin{array}{l} n - 18 = 56 \\ 74 - 18 = 56 \end{array}$

9. $\begin{array}{rcr} d - 47 &=& 24 \\ + 47 &=& +47 \\ \hline d &=& 71 \end{array}$ $\qquad$ $\begin{array}{l} d - 47 = 24 \\ 71 - 47 = 24 \end{array}$

10. $\begin{array}{rcr} n \div 4 &=& 15 \\ \times \ 4 &=& \times 4 \\ \hline n &=& 60 \end{array}$ $\qquad$ $\begin{array}{l} n \div 4 = 15 \\ 60 \div 4 = 15 \end{array}$

11. $9n$ is the same as $n \times 9$.
$$\begin{array}{rcl} n \times 9 &=& 108 \\ n \times 9 \div 9 &=& 108 \div 9 \\ n &=& 12 \end{array}$$

12. $3z$ is the same as $z \times 3$.
$$\begin{array}{rcl} z \times 3 &=& 54 \\ z \times 3 \div 3 &=& 54 \div 3 \\ z &=& 18 \end{array}$$

13. $\begin{array}{rcr} c + 13 &=& 13 \\ - 13 &=& -13 \\ \hline c &=& 0 \end{array}$ $\qquad$ $\begin{array}{l} c + 13 = 13 \\ 0 + 13 = 13 \end{array}$

14. $\begin{array}{rcr} y + 19 &=& 45 \\ - 19 &=& -19 \\ \hline y &=& 26 \end{array}$ $\qquad$ $\begin{array}{l} y + 19 = 45 \\ 26 + 19 = 45 \end{array}$

15. $10c$ is the same as $c \times 10$.
$$\begin{array}{rcl} c \times 10 &=& 100 \\ c \times 10 \div 10 &=& 100 \div 10 \\ c &=& 10 \end{array}$$

16. $45d = \$540$ and $45d$ is the same as $d \times 45$.
$$\begin{array}{rcl} d \times 45 &=& \$540 \\ d \times 45 \div 45 &=& \$540 \div 45 \\ d &=& \$12 \end{array}$$

17. $\begin{array}{rcr} p - 15 &=& 125 \\ + 15 &=& +15 \\ \hline p &=& 140 \end{array}$ $\qquad$ $\begin{array}{l} p - 15 = 125 \\ 140 - 15 = 125 \end{array}$

18. $\begin{array}{rcr} p + 6 &=& 18 \\ - 6 &=& -6 \\ \hline p &=& 12 \end{array}$ $\qquad$ $\begin{array}{l} p + 6 = 18 \\ 12 + 6 = 18 \end{array}$

19. $\begin{array}{rcr} n \div 26 &=& 3 \\ \times \ 26 &=& \times 26 \\ \hline n &=& 78 \end{array}$ $\qquad$ $\begin{array}{l} n \div 26 = 3 \\ 78 \div 26 = 3 \end{array}$

20. $\begin{array}{rcr} a + \$25 &=& \$560 \\ - 25 &=& -25 \\ \hline a &=& \$535 \end{array}$ $\qquad$ $\begin{array}{l} a + \$25 = \$560 \\ \$535 + \$25 = \$560 \end{array}$

PAGE 45

2. **Division** When given a total—here, total pay earned in a 40-hour work week—and asked <u>How much for each unit</u>—here, each hour—use division.

3. **Multiplication** When given an amount for each—here, admission per person—and the total (here, total number of people who paid), and asked for <u>the total amount</u>—here, money paid—use multiplication.

4. **Subtraction** When given two amounts—here, the budget and the cost of the computer—and asked <u>How much more</u>—here, money needed—use subtraction.

5. **addition; 18 pounds of fruit** $12 + 6 = 18$ Estimate to check: $10 + 5 = 15$. The answer 18 is reasonable.

6. **subtraction; 126 hamburgers** $198 - 72 = 126$ Estimate to check: $200 - 70 = 130$. The answer 126 is reasonable.

7. **2 videos** Two videos total $34. Three videos total $51. The customer can buy only two videos with $50. Estimate to check: $2 \times \$20 = \$40$; $3 \times \$20 = \$60$. The answer 2 videos is reasonable.

8. **$112** Add the amount of the bills: $\$53 + \$16 + \$27 + \$5 + \$11 = \$112$. Estimate to check: $\$50 + \$15 + \$30 + \$5 + \$10 = \$110$. The answer $112 is reasonable.

9. **$346** Subtract 25 from 371. Estimate to check: $\$375 - \$25 = \$350$. The answer $346 is reasonable.

10. **221** Add the three amounts: $67 + 73 + 81 = 221$ Estimate to check: $70 + 70 + 80 = 220$. The answer 221 is reasonable.

PAGE 47

2. **Mean: 376; Median: 368** To find the mean, first find the total of the numbers in the set. $410 + 350 + 368 = 1{,}128$ Then divide the total by the number of items. $1{,}128 \div 3 = 376$
To find the median, arrange the numbers in order. Then find the middle number. 350, <u>368</u>, 410

3. **Mean: 213; Median: 254** To find the mean, first find the total of the numbers in the set.
   261 + 254 + 105 + 280 + 165 = 1,065.
   Then divide the total by the number of items.
   1,065 ÷ 5 = 213
   To find the median, arrange the numbers in order. Then find the middle number.
   105, 165, <u>254</u>, 261, 280

4. **Mean: 2,350; Median: 2,430** To find the mean, first find the total of the numbers in the set. 2,450 + 2,100 + 1,970 + 2,430 + 2,840 + 1,800 + 2,860 = 16,450
   Then divide the total by the number of items. 16,450 ÷ 7 = 2,350
   To find the median, arrange the numbers in order. Then find the middle number.
   1,800; 1,970; 2,100; <u>2,430</u>; 2,450; 2,840; 2,860

5. **Median: 9** To find the median, arrange the numbers in order. Then find the average of the two middle numbers.
   7, 7, 8, <u>8</u>, <u>10</u>, 10, 12, 15
   8 + 10 = 18   18 ÷ 2 = 9

6. **Median: 202** To find the median, arrange the numbers in order. Then find the average of the two middle numbers.
   146, 162, 184, <u>198</u>, <u>206</u>, 212, 234, 523
   198 + 206 = 404   404 ÷ 2 = 202

7. **Median: 50** To find the median, arrange the numbers in order. Then find the average of the two middle numbers.
   26, 32, 41, <u>46</u>, <u>54</u>, 73, 81, 81
   46 + 54 = 100   100 ÷ 2 = 50

8. **Median: 3,881** To find the median, arrange the numbers in order. Then find the average of the two middle numbers.
   1,903; 2,950; <u>3,420</u>; <u>4,342</u>; 5,651; 7,418
   3,420 + 4,342 = 7,762
   7,762 ÷ 2 = 3,881

9. **Mean: 50** To find the mean, first find the total of the numbers in the set.
   25 + 42 + 45 + 37 + 102 + 86 + 46 + 38 + 27 = 448 Then divide the total by the number of items. 448 ÷ 9 = 49.77
   Round your answer to the nearest customer, 50 customers.

10. **Mean: 425** To find the mean, first find the total of the numbers in the set.
    502 + 147 + 425 + 454 + 518 + 504 = 2,550
    Then divide the total by the number of items.   2,550 ÷ 6 = 425

PAGE 49

1. **(1) 40 × 30** To find the area of the exhibit, multiply the length (40) by the width (30).

2. **(4) 13 × 27** To find the amount of carpeting, find the area of the garage by multiplying the length (27) by the width (13).

3. **(3) 6 × 4 ÷ 3** To find the total number of feet, multiply the number of feet needed for each panel (6) by the number of panels (4). Then divide by the number of feet in a yard (3).

4. **(3) 42 × 32** To find the amount of tile flooring, find the are of the basement by multiplying the length (42) by the width (32).

5. **(5) Not enough information is given.** To find the amount of lime needed, Nancy needs to find the area of the garden. Area = length × width, but Nancy only knows the length of the garden; she does not know the width of the garden.

6. **(4) 800 square inches** To find the area of a rectangle, multiply the length by the width. The tabletop is 40 inches long and 20 inches wide, so the area of the tabletop is 40 × 20 = 800 square inches.

7. **Area = length × width, or $A = l \times w$. Since, in a square, the length and width are equal, using s for this measurement, $l = s$ and $w = s$. Replacing l and w in the area formula with s, the formula for area of a square becomes $A = s \times s$, or s^2.**

PAGES 50–51

1. **1,368**

$$\begin{array}{r} {\overset{1\ 1}{456}} \\ \times\ \ \ 3 \\ \hline 1,368 \end{array}$$

2. 4,067

$$
\begin{array}{r}
83 \\
\times 49 \\
\hline
747 \\
+3\ 32 \\
\hline
4{,}067
\end{array}
$$

3. $1,782

$$
\begin{array}{r}
\$297 \\
\times\quad 6 \\
\hline
\$1{,}782
\end{array}
$$

4. 124,956

$$
\begin{array}{r}
178 \\
\times 702 \\
\hline
356 \\
+124\ 60 \\
\hline
124{,}956
\end{array}
$$

5. 199,165

$$
\begin{array}{r}
653 \\
\times 305 \\
\hline
3\ 265 \\
+195\ 90 \\
\hline
199{,}165
\end{array}
$$

6. 823 r2

$$
\begin{array}{r}
823\ \text{r2} \\
4\overline{)3{,}294} \\
-3\ 2 \\
\hline
09 \\
-\ 8 \\
\hline
14 \\
-12 \\
\hline
2
\end{array}
$$

7. $920

$$
\begin{array}{r}
\$920 \\
7\overline{)\$6{,}440} \\
-6\ 3 \\
\hline
14 \\
-14 \\
\hline
00
\end{array}
$$

8. 321 r16

$$
\begin{array}{r}
321\ \text{r16} \\
18\overline{)5{,}794} \\
-5\ 4 \\
\hline
39 \\
-36 \\
\hline
34 \\
-18 \\
\hline
16
\end{array}
$$

9. 201 r14

$$
\begin{array}{r}
201\ \text{r14} \\
43\overline{)8{,}657} \\
-8\ 6 \\
\hline
057 \\
-\ 43 \\
\hline
14
\end{array}
$$

10. 153

$$
\begin{array}{r}
153 \\
27\overline{)4{,}131} \\
-2\ 7 \\
\hline
1\ 43 \\
-1\ 35 \\
\hline
81 \\
-81 \\
\hline
0
\end{array}
$$

11. 31,200

$$
\begin{array}{r}
39 \\
\times 800 \\
\hline
31{,}200
\end{array}
$$

12. $4,210

$$
\begin{array}{r}
\$842 \\
\times\quad 5 \\
\hline
\$4{,}210
\end{array}
$$

13. $205

$$
\begin{array}{r}
\$205 \\
8\overline{)\$1{,}640} \\
-1\ 6 \\
\hline
040 \\
-\ 40 \\
\hline
0
\end{array}
$$

14. 143 r27

$$
\begin{array}{r}
143\ \text{r27} \\
52\overline{)7{,}463} \\
-5\ 2 \\
\hline
2\ 26 \\
-2\ 08 \\
\hline
183 \\
-156 \\
\hline
27
\end{array}
$$

15. **55 miles** To find the mean, first find the total of the numbers in the set.
$62 + 28 + 57 + 65 + 63 = 275$
Then divide the total by the number of items in the set. $275 \div 5 = 55$

16. **62 miles** To find the median, arrange the numbers in order. Find the middle value.
28, 57, <u>62</u>, 63, 65

17. $y = 42$ Use addition to solve for y.

$$
\begin{array}{rcr}
y - 13 &=& 29 \\
+ 13 &=& 13 \\
\hline
y &=& 42
\end{array}
$$

18. **15 customers** Divide the total collected for oil changes by the cost of an oil change.

$$
\begin{array}{r}
15 \\
\$22\overline{)\$330} \\
-22 \\
\hline
110 \\
-110 \\
\hline
0
\end{array}
$$

19. 18,075

20. 12,000
21. 460
22. 629
23. (2) 94 × 50 To find the area, multiply the length times the width.
24. (1) $56 Multiply Lena's hourly wage by the number of hours she worked. The hours Lena worked on Saturday and Sunday are extra information not needed to find the answer.
    $8 × 7 = $56
25. (3) $12 ÷ 6 To find the price of one quart, divide the total cost by the number of quarts bought.
26. (5) **Not enough information is given.** To find the area, multiply the length times the width. The problem does not give the width of the garden.
27. (1) $5p = $20 To find how many pairs of shoes Martin bought, multiply the price for each pair of shoes ($5) by the number of pairs of shoes (p) to get the total ($20).
28. (5) 42,000 To find how many miles, divide the total miles (126,000) by the number of years (3).
29. (4) 2,299 To find the total number of miles, add the numbers:
    244 + 996 + 1,059 = 2,299

SECTION 4

PAGE 53
1. (4) 15^2 The room is a square because all four sides have the same length. Use the formula $A = s^2$ and substitute 15 for s.
2. (5) 225 ÷ 9 Divide the number of square feet by the number of square feet in a square yard (9).
3. (1) 18^2 The room is a square because all four sides have the same length. Use the formula $A = s^2$ and substitute 18 for s.
4. (4) 12^2 The patio is a square because all four sides have the same length. Use the formula $A = s^2$ and substitute 12 for s.
5. (2) 36 ÷ 9 Divide the number of square feet by the number of square feet in a square yard (9).
6. (4) 4 × 18 Multiply the number of tiles in a square foot by the number of square feet.

7. You can find the area of a square room by measuring only one side because you can take that one measurement and multiply it by itself ($A = s^2 = s \times s$).

PAGE 55
2. 5 × 5 = 25
3. 1 × 1 = 1
4. 7 × 7 × 7 = 49 × 7 = 343
5. 64
6. 121
7. 324
8. 256
9. 125 $5^3 = 5 \times 5 \times 5 = 25 \times 5 = 125$
10. 512 $8^3 = 8 \times 8 \times 8 = 64 \times 8 = 512$
11. 64 $4^3 = 4 \times 4 \times 4 = 16 \times 4 = 64$
12. 1,000 $10^3 = 10 \times 10 \times 10 = 100 \times 10 = 1,000$
13. 11 $11^2 = 121$ Therefore $\sqrt{121} = 11$
14. 4 $4^2 = 16$ Therefore $\sqrt{16} = 4$
15. 17 $17^2 = 289$ Therefore $\sqrt{289} = 17$
16. 14 $14^2 = 196$ Therefore $\sqrt{196} = 14$
17. $21^2 = 441$ $31^2 = 961$
    $22^2 = 484$ $32^2 = 1,024$
    $23^2 = 529$ $33^2 = 1,089$
    $24^2 = 576$ $34^2 = 1,156$
    $25^2 = 625$ $35^2 = 1,225$
18. $\sqrt{441} = 21$ $\sqrt{3,481} = 59$
    $\sqrt{625} = 25$ $\sqrt{3,721} = 61$
    $\sqrt{841} = 29$ $\sqrt{4,225} = 65$
    $\sqrt{1,225} = 35$ $\sqrt{4,761} = 69$
    $\sqrt{1,521} = 39$ $\sqrt{5,041} = 71$

PAGE 57
1. (3) 3 × 2 × 5 To find the volume of a rectangular solid, multiply the length (3 feet) times the width (2 feet) times the height (5 feet).
2. (2) 6 × 4 × 1 The sandbox is a rectangular solid. Multiply the length (6 feet) times the width (4 feet) times the height (1 feet) to find the volume.
3. (3) 7^3 The box is a cube because each side has the same length. Use the formula $A = s^3$ and substitute 7 for s.
4. (4) 12 × 10 × 8 To find the volume of the shipping carton, multiply the length (12 inches) times the width (10 inches) times the height (8 inches).

5. **(5) Not enough information is given.** To find the volume of a carton, multiply the length by the width by the height. The problem does not give the height of the carton.

6. **(2) 736** You are asked to find the <u>space left in the truck</u>. You are given the volume of the cartons loaded in the truck. To find the volume of a truck (a rectangular container), multiply the length by the width by the height. $16 \times 8 \times 14 = 1,792$ cubic feet. To find the space left in the truck, subtract the volume of the cartons from the volume of the truck. $1,792 - 1,056 = 736$

7. **$V = s^3 = s \times s \times s$ You can find the volume of a cube by measuring only one side because each side has the same length. Therefore, you already know the measures of the other sides without actually measuring them.**

PAGE 59

1. **(3) $2 \times 15 + 2 \times 12$** Substitute the numbers into the formula:
$P = 2 \times \underline{15} + 2 \times \underline{12}$

2. **(1) 45×5** The distance formula is $d = r \times t$. Substitute 45 for r and 5 for t.

3. **(2) 4×20** The formula for perimeter of a square is $P = 4 \times s$. Substitute 20 for s.

4. **(3) $150 \div 12$** Use the distance formula $r = d \div t$. Replace 150 for d and 12 (6 hours $\times$ 2 days) for t.

5. **(5) Not enough information is given.** To find how many miles Kathy drove, you need to use the distance formula. You know that $t = 7$ hours. The problem does not give r, the rate at which she drove. The time she left and the number of gallons of gas she put in the car are extra information not needed to solve the problem asked.

6. **(4) 960** $v = l \times w \times h$
$\quad\quad = 12 \times 10 \times 8$
$\quad\quad = 960$

7. **To rewrite the distance formula in terms of t, divide both sides by r.**
$\quad\quad d = rt$
$\quad d \div r = rt \div r$
$\quad\quad d \div r = t$

PAGES 60–61

1. $8 \times 8 = 64$

2. $1 \times 1 \times 1 = 1$
3. $3 \times 3 \times 3 = 27$
4. $9 \times 9 = 81$
5. 289
6. 225
7. 5
8. 7
9. 46,225
10. 6,859
11. 18
12. 33
13. **400 sq. ft.** The room is a square because all four sides have the same length. Use the formula $A = s^2$ and substitute 20 for s. $20 \times 20 = 400$
14. **900 cu. ft.**
$\quad V = l \times w \times h$
$\quad\quad = 15 \times 12 \times 5$
$\quad\quad = 180 \times 5$
$\quad\quad = 900$
15. **720 cu. ft.**
$\quad V = l \times w \times h$
$\quad\quad = 30 \times 6 \times 4$
$\quad\quad = 180 \times 4$
$\quad\quad = 720$
16. **$126** Multiply the cost per yard by the number of yards bought.
$14 \times 9 = 126$
17. **64 sq. yd.** Divide the number of square feet by the number of square feet in a square yard (9). $576 \div 9 = 64$
18. **576 cu. ft.**
$\quad V = l \times w \times h$
$\quad\quad = 12 \times 8 \times 6$
$\quad\quad = 96 \times 6$
$\quad\quad = 576$
19. **(2) $6 \times 6 \times 6$** To find the volume of a cube, multiply the length (6 inches) times the width (6 inches) times the height (6 inches).
20. **(5) 15^2** The room is a square because all 4 sides have the same length. Use the formula $A = s^2$ and substitute 15 for s.
21. **(5) 160**
$\quad V = l \times w \times h$
$\quad\quad = 8 \times 5 \times 4$
$\quad\quad = 40 \times 4$
$\quad\quad = 160$ cu. ft.
22. **(3) $42^2 \div 9$** To find the area in square feet, use the formula $A = s^2$ and substitute 42 for s. Then divide the number of square feet by the number of square feet in a square yard (9).

23. **(3) 100 sq. ft.** The patio will be a square because all 4 sides are to be the same length. Use the formula $A = s^2$ and substitute 10 for s.

24. **(1) 18 × 12 × 3 cu. in.** To find the volume of a rectangular space, multiply the length (18 inches) times the width (12 inches) times the height (3 inches).

25. **(5) 6,480 cu. in.**
$$\begin{aligned} V &= l \times w \times h \\ &= 18 \times 18 \times 20 \\ &= 324 \times 20 \\ &= 6{,}480 \end{aligned}$$

26. **(2) 12 sq. yd.** Divide the number of square feet by the number of square feet in a square yard (9). $108 \div 9 = 12$

MATH AT WORK

PAGES 62–63

1. **(4) 198** You find this by looking at the mileage on Route 78 between Harrisburg, PA and New York, New York.

2. **a.** The northern route is
$126 + 50 + 416 = $ **592 miles.**
The southern route is
$129 + 50 + 169 + 198 = $ **546 miles.**
b. The **southern route** is the shorter.
c. Answers will vary. Here is one possible answer:
Even though the southern route is 46 miles shorter ($592 - 546$), I would suggest that he take the northern route. The one hour delay on the Route 76 will probably make travel on the southern route take longer.

UNIT 1 REVIEW

PAGES 64–65

1. ten thousands
2. millions
3. **417 < 1,740** 417 has fewer digits than 1,740.
4. **54,9_7_2 > 54,9_2_7** 7 is greater than 2.
5. **4,380,000** 4,3⑦5,429 The number to the right of 7 is 5; add 1 to the circled digit.
6. twenty-three thousand twelve
7. **$x = 11$**
$$\begin{aligned} x + 8 &= 19 \\ -8 &= -8 \\ \hline x &= 11 \end{aligned}$$

8. **$x = 5$**
$$\begin{aligned} 4x &= 20 \\ \div\ 4 &= \div 4 \\ \hline x &= 5 \end{aligned}$$

9. **763**
$$\begin{array}{r} {}^{1\,1} \\ 426 \\ 84 \\ +253 \\ \hline 763 \end{array}$$

10. **$992**
$$\begin{array}{r} {}^{1\,3} \\ \$248 \\ \times\ \ 4 \\ \hline \$992 \end{array}$$

11. **$576**
$$\begin{array}{r} {}^{8\,14\,14} \\ \$9\cancel{5}\cancel{5}\cancel{4} \\ -\ 378 \\ \hline \$576 \end{array}$$

12. **29 r2**
$$\begin{array}{r} 29\,\text{r}2 \\ 27\overline{)785} \\ -54 \\ \hline 245 \\ -243 \\ \hline 2 \end{array}$$

13. **223,608**
$$\begin{array}{r} {}^{2\,1}\cancel{4} \\ 726 \\ \times 308 \\ \hline 5\ 808 \\ +217\ 80 \\ \hline 223{,}608 \end{array}$$

14. **64** $4 \times 4 \times 4 = 16 \times 4 = 64$

15. **257**
$$\begin{array}{r} {}^{4}\,\cancel{10}^{9}\,10 \\ \cancel{5}\cancel{0}\cancel{0} \\ -243 \\ \hline 257 \end{array}$$

16. **8** $8 \times 8 = 64$

17. **$313**
$$\begin{array}{r} \$313 \\ 8\overline{)\$2{,}504} \\ -24 \\ \hline 10 \\ -\ 8 \\ \hline 24 \\ -24 \\ \hline 0 \end{array}$$

18. **305 r11**
$$\begin{array}{r} 305\,\text{r}11 \\ 13\overline{)3{,}976} \\ -3\ 9 \\ \hline 076 \\ -\ 65 \\ \hline 11 \end{array}$$

19. **78** $80 + 76 + 75 + 80 + 79 = 390$

$$
\begin{array}{r}
78 \\
5\overline{)390} \\
-35 \\
\hline
40 \\
-40 \\
\hline
0
\end{array}
$$

20. **31** Arrange the numbers in order from least to greatest. Find the middle number.
18, 21, 24, <u>31</u>, 35, 37, 42

21. **12,844,368**

22. **68**

23. **6,889**

24. **235**

25. **(5) $48 + 48 + 20 + 20$** Two sides measure 48 feet and two sides measure 20 feet. Add all four measurements to find the perimeter.

26. **(3) $90** Add to find a total. Round the three numbers to the nearest ten and estimate the total.

$$
\begin{array}{lll}
\$57 & \text{rounds to} & 60 \\
13 & \text{rounds to} & 10 \\
+\ 21 & \text{rounds to} & +\ 20 \\
\hline
& & \$90
\end{array}
$$

27. **(2) $14 × 43** To find the amount Ramon earned this week, multiply the amount he earns per hour ($14) times the number of hours he worked this week (43). The number of hours he worked last week is extra information that is not needed.

28. **(4) 350** The tallest bar is for December. It reaches the 5 mark and represents 500 sweat shirts. The bar for September is the smallest. It ends between the 1 and 2 marks and represents 150 sweat shirts. Subtract to find the difference.

$$
\begin{array}{r}
\overset{4\,10}{\cancel{5}\cancel{0}0} \\
-150 \\
\hline
350
\end{array}
$$

29. **(4) 196** The room is a square because all four sides have the same length. Use the formula $A = s^2$ and substitute 14 for s.

MATH CONNECTION: WHOLE NUMBERS AND MONEY

PAGES 66–67

1. **(5) Computers would make mistakes interpreting the "00" in 2000.** The other choices are not reflected in the reading.

2. **(2) they didn't want goods and services to be scarce.** The third paragraph describes possible scarcity of goods and services if there were a Y2K problem.

3.

Estimated Cost to States to Avoid Y2K Crisis
Top 12 States

State	Estimated Cost (in millions)	State	Estimated Cost (in millions)
New York	$250	Florida	$75–90
Texas	$207	Indiana	$87
California	$200	North Carolina	$82
Arizona	$100	Missouri	$75
Maryland	$100	Oregon	$75
Alabama	$92	Ohio	$60

4. Answers will vary. Your answer may include some of the points below:
   - If the power company computers shut down, homes would be without the power needed to supply heat and run appliances.
   - It is also used to power the wells in rural areas, which are used to supply water.
   - Power is also necessary for businesses to run their machines and computers. If power companies shut down, families, workers, and businesses would suffer greatly.

5. Answers will vary. The sample paragraph below illustrates one impact of computers on daily lives:
   My bank uses computers to check how much money is in people's accounts. If the Y2K problem shut down my bank, people's savings could be lost or their accounts could be in error. It would also make it very difficult to withdraw money from the bank, because the bank couldn't tell if there is enough money in someone's account. For many people, like myself, I need to be able to put in or take out money many times during the month. Computers shut down by the Y2K problem would not allow me to get the cash I need.

UNIT 2: FRACTIONS

SECTION 5

PAGE 71

1. **(4) D** Look at the ruler. The $3\frac{1}{16}$-inch mark is one small mark greater than (to the right of) the 3-inch mark.

2. **(2) $\frac{3}{4}$ is greater than $\frac{1}{2}$** Look at the measuring cup. The $\frac{3}{4}$-cup mark is above (greater than) the $\frac{1}{3}$-, $\frac{1}{2}$-, and $\frac{2}{3}$-cup marks and below (less than) the 1- and $1\frac{1}{4}$-cup marks.

3. **(2) $2\frac{1}{4}$ is equal to $2\frac{2}{8}$** The $\frac{1}{4}$ mark is the same as two $\frac{2}{8}$ marks. Thus, $\frac{1}{4}$ and $\frac{2}{8}$ are the same or equal.

4. **(1) below the $\frac{1}{4}$ mark** On the ruler, $\frac{1}{8}$ is smaller (less) than $\frac{1}{4}$, so, on the measuring cup, the $\frac{1}{8}$ mark would go below the $\frac{1}{4}$ mark.

5. **(4) $3\frac{1}{4}$ is greater than $3\frac{1}{8}$** $3\frac{1}{4} = 3\frac{2}{8}$ which is more than $3\frac{1}{8}$.

6. **(5) Not enough information is given.** To find the weight of all the fruit, Dean needs to know the weight of the *bananas* as well as the pears and grapes.

7. **Each fraction has 1 whole and a fraction amount. The fractions are equal because $\frac{2}{4}$ reduces to $\frac{1}{2}$. Draw two figures. Divide each figure into two equal parts. Shade $1\frac{1}{2}$ parts. Draw another set of two figures the same size as the first two figures. Divide each figure into four equal parts. Shade $1\frac{2}{4}$ parts. The shaded parts are the same, or equal.**

PAGE 73

2. $\frac{2}{3}$ There are 3 circles in the group, and 2 of the 3 circles (or $\frac{2}{3}$) are colored.

3. $\frac{1}{2}$ The figure is divided into 2 parts, and 1 of the 2 parts (or $\frac{1}{2}$) is colored.

4. $\frac{3}{5}$ The figure is divided into 5 parts, and 3 of the 5 parts (or $\frac{3}{5}$) are colored.

5. $\frac{3}{4}$ There are 4 figures in the group, and 3 of the 4 figures (or $\frac{3}{4}$) are colored.

6. $\frac{5}{8}$ The figure is divided into 8 parts, and 5 of the 8 parts (or $\frac{5}{8}$) are colored.

7. 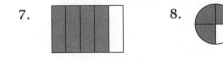   8. ⊘

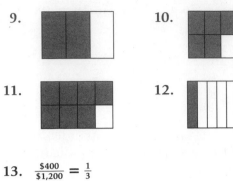

9.   10.

11.   12.

13. $\frac{\$400}{\$1{,}200} = \frac{1}{3}$

14. $\frac{8}{25}$

15. $\frac{\$18}{\$24} = \frac{3}{4}$

16. $\frac{8}{15}$

PAGE 75

2. $\frac{5}{2}$, $2\frac{1}{2}$ Look at the figures. Two whole figures and one-half of the third figure are colored.

3. $\frac{19}{8}$, $2\frac{3}{8}$ Look at the figures. Two whole figures and three-eighths of the third figure are colored.

4. $1\frac{3}{4}$
$$\begin{array}{r} 1 \\ 4\overline{)7} \\ -4 \\ \hline 3 \end{array}$$
$\rightarrow$ $1\frac{3}{4}$

5. 3
$$\begin{array}{r} 3 \\ 6\overline{)18} \\ -18 \\ \hline 0 \end{array}$$

6. $2\frac{4}{5}$
$$\begin{array}{r} 2 \\ 5\overline{)14} \\ -10 \\ \hline 4 \end{array}$$
$\rightarrow$ $2\frac{4}{5}$

7. $3\frac{1}{3}$
$$\begin{array}{r} 3 \\ 3\overline{)10} \\ -9 \\ \hline 1 \end{array}$$
$\rightarrow$ $3\frac{1}{3}$

8. 5
$$\begin{array}{r} 5 \\ 4\overline{)20} \\ -20 \\ \hline 0 \end{array}$$

9. $1\frac{7}{8}$
$$\begin{array}{r} 1 \\ 8\overline{)15} \\ -8 \\ \hline 7 \end{array}$$
$\rightarrow$ $1\frac{7}{8}$

10. $\frac{11}{6}$ $1\frac{5}{6} = \frac{(1 \times 6)}{6} + \frac{5}{6}$
$$= \frac{6}{6} + \frac{5}{6}$$
$$= \frac{11}{6}$$

11. $\frac{5}{2}$ $2\frac{1}{2} = \frac{(2 \times 2) + 1}{2}$
$= \frac{4 + 1}{2}$
$= \frac{5}{2}$

12. $\frac{7}{1}$ $7 = \frac{7}{1}$

13. $\frac{19}{5}$ $3\frac{4}{5} = \frac{(5 \times 3) + 4}{5}$
$= \frac{15 + 4}{5}$
$= \frac{19}{5}$

14. $\frac{12}{1}$ $12 = \frac{12}{1}$

15. $\frac{43}{4}$ $10\frac{3}{4} = \frac{(4 \times 10) + 3}{4}$
$= \frac{40 + 3}{4}$
$= \frac{43}{4}$

16. $\frac{7}{2}$ $3\frac{1}{2} = \frac{(3 \times 2) + 1}{2}$
$= \frac{6 + 1}{2}$
$= \frac{7}{2}$

17. $6\frac{1}{3}$

PAGE 76

2. $\frac{2}{3} \bowtie \frac{12}{18}$ $2 \times 18 = 36$
$3 \times 12 = 36$
$\frac{2}{3}$ and $\frac{12}{18}$ are equivalent

3. $\frac{1}{2} \bowtie \frac{5}{10}$ $1 \times 10 = 10$
$2 \times 5 = 10$
$\frac{1}{2}$ and $\frac{5}{10}$ are equivalent

4. $\frac{3}{6} \bowtie \frac{5}{12}$ $3 \times 12 = 36$
$6 \times 5 = 30$
$\frac{3}{6}$ and $\frac{5}{12}$ are not equivalent

5. $\frac{4}{6} \bowtie \frac{8}{12}$ $4 \times 12 = 48$
$6 \times 8 = 48$
$\frac{4}{6}$ and $\frac{8}{12}$ are equivalent

6. $\frac{4}{7} \bowtie \frac{8}{14}$ $4 \times 14 = 56$
$7 \times 8 = 56$
$\frac{4}{7}$ and $\frac{8}{14}$ are equivalent

7. $\frac{5}{6} \bowtie \frac{25}{30}$ $5 \times 30 = 150$
$6 \times 25 = 150$
$\frac{5}{6}$ and $\frac{25}{30}$ are equivalent

8. $\frac{3}{5} \bowtie \frac{6}{15}$ $3 \times 15 = 45$
$5 \times 6 = 30$
$\frac{3}{5}$ and $\frac{6}{15}$ are not equivalent

9. $\frac{6}{10} \bowtie \frac{24}{40}$ $6 \times 40 = 240$
$10 \times 24 = 240$
$\frac{6}{10}$ and $\frac{24}{40}$ are equivalent

10. $\frac{3}{5} \bowtie \frac{20}{25}$ $3 \times 25 = 75$
$5 \times 20 = 100$
$\frac{3}{5}$ and $\frac{20}{25}$ are not equivalent

11. $\frac{1}{2} \bowtie \frac{4}{8}$ $1 \times 8 = 8$
$2 \times 4 = 8$
$\frac{1}{2}$ and $\frac{4}{8}$ are equivalent

12. $\frac{2}{3} \bowtie \frac{6}{9}$ $2 \times 9 = 18$
$3 \times 6 = 18$
$\frac{2}{3}$ and $\frac{6}{9}$ are equivalent

13. $\frac{7}{8} \bowtie \frac{21}{24}$ $7 \times 24 = 168$
$8 \times 21 = 168$
$\frac{7}{8}$ and $\frac{21}{24}$ are equivalent

14. $\frac{3}{8} \bowtie \frac{6}{16}$ $3 \times 16 = 48$
$8 \times 6 = 48$
$\frac{3}{8}$ and $\frac{6}{16}$ are equivalent

15. $\frac{8}{9} \bowtie \frac{16}{18}$ $8 \times 18 = 144$
$9 \times 16 = 144$
$\frac{8}{9}$ and $\frac{16}{18}$ are equivalent

16. $\frac{1}{4} \bowtie \frac{8}{16}$ $1 \times 16 = 16$
$4 \times 8 = 32$
$\frac{1}{4}$ and $\frac{8}{16}$ are not equivalent

PAGE 77

2. $\frac{2}{3}$ $\frac{8}{12} = \frac{8 \div 4}{12 \div 4} = \frac{2}{3}$

3. $\frac{2}{5}$ $\frac{4}{10} = \frac{4 \div 2}{10 \div 2} = \frac{2}{5}$

4. $\frac{3}{4}$ $\frac{6}{8} = \frac{6 \div 2}{8 \div 2} = \frac{3}{4}$

5. $\frac{4}{5}$ $\frac{20}{25} = \frac{20 \div 5}{25 \div 5} = \frac{4}{5}$

6. $\frac{1}{3}$ $\frac{12}{36} = \frac{12 \div 12}{36 \div 12} = \frac{1}{3}$

7. $\frac{2}{3}$ $\frac{6}{9} = \frac{6 \div 3}{9 \div 3} = \frac{2}{3}$

8. $\frac{1}{2}$ $\frac{8}{16} = \frac{8 \div 8}{16 \div 8} = \frac{1}{2}$

9. $\frac{4}{5}$ $\frac{16}{20} = \frac{16 \div 4}{20 \div 4} = \frac{4}{5}$

10. $\frac{5}{7}$ $\frac{10}{14} = \frac{10 \div 2}{14 \div 2} = \frac{5}{7}$

11. $\frac{1}{3}$ $\frac{5}{15} = \frac{5 \div 5}{15 \div 5} = \frac{1}{3}$

12. $\frac{5}{6}$ $\frac{20}{24} = \frac{20 \div 4}{24 \div 4} = \frac{5}{6}$

13. $\frac{2}{3}$ $\frac{18}{27} = \frac{18 \div 9}{27 \div 9} = \frac{2}{3}$

14. $\frac{3}{4}$ $\frac{18}{24} = \frac{18 \div 6}{24 \div 6} = \frac{3}{4}$

15. $\frac{2}{3}$ $\frac{22}{33} = \frac{22 \div 11}{33 \div 11} = \frac{2}{3}$

16. $\frac{3}{4}$ $\frac{21}{28} = \frac{21 \div 7}{28 \div 7} = \frac{3}{4}$

PAGE 78

2. $\frac{8}{12}$ $\frac{2}{3} = \frac{2 \times 4}{3 \times 4} = \frac{8}{12}$

3. $\frac{3}{18}$ $\frac{1}{6} = \frac{1 \times 3}{6 \times 3} = \frac{3}{18}$

4. $\frac{14}{20}$ $\frac{7}{10} = \frac{7 \times 2}{10 \times 2} = \frac{14}{20}$

5. $\frac{15}{24}$ $\frac{5}{8} = \frac{5 \times 3}{8 \times 3} = \frac{15}{24}$

6. $\frac{12}{30}$ $\frac{2}{5} = \frac{2 \times 6}{5 \times 6} = \frac{12}{30}$

7. $\frac{8}{18}$ $\frac{4}{9} = \frac{4 \times 2}{9 \times 2} = \frac{8}{18}$

8. $\frac{3}{12}$ $\frac{1}{4} = \frac{1 \times 3}{4 \times 3} = \frac{3}{12}$

9. $\frac{28}{32}$ $\frac{7}{8} = \frac{7 \times 4}{8 \times 4} = \frac{28}{32}$

10. $\frac{15}{25}$ $\frac{3}{5} = \frac{3 \times 5}{5 \times 5} = \frac{15}{25}$

11. $\frac{21}{36}$ $\frac{7}{12} = \frac{7 \times 3}{12 \times 3} = \frac{21}{36}$

12. $\frac{12}{40}$ $\frac{3}{10} = \frac{3 \times 4}{10 \times 4} = \frac{12}{40}$

13. $\frac{20}{45}$ $\frac{4}{9} = \frac{4 \times 5}{9 \times 5} = \frac{20}{45}$

14. $\frac{18}{32}$ $\frac{9}{16} = \frac{9 \times 2}{16 \times 2} = \frac{18}{32}$

15. $\frac{24}{56}$ $\frac{3}{7} = \frac{3 \times 8}{7 \times 8} = \frac{24}{56}$

16. $\frac{72}{100}$ $\frac{18}{25} = \frac{18 \times 4}{25 \times 4} = \frac{72}{100}$

PAGE 79

2. $>$ $\frac{3}{4} \bowtie \frac{3}{8}$ $3 \times 8 = 24$ $24 > 12$
                      $4 \times 3 = 12$

3. $<$ Since both fractions have the same denominator (16), compare the numerators: $9 < 12$, therefore $\frac{9}{16} < \frac{12}{16}$.

4. $<$ $\frac{2}{4} \bowtie \frac{2}{3}$ $2 \times 3 = 6$ $6 < 8$
                      $4 \times 2 = 8$

5. $>$ $\frac{2}{3} \bowtie \frac{3}{5}$ $2 \times 5 = 10$ $10 > 9$
                      $3 \times 3 = 9$

6. $=$ $\frac{8}{10} \bowtie \frac{4}{5}$ $8 \times 5 = 40$ $40 = 40$
                      $10 \times 4 = 40$

7. $>$ $\frac{7}{8} \bowtie \frac{3}{4}$ $7 \times 4 = 28$ $28 > 24$
                      $8 \times 3 = 24$

8. $<$ Since both fractions have the same denominator (9), compare the numerators: $5 < 8$, therefore $\frac{5}{9} < \frac{8}{9}$.

9. $>$ Since both fractions have the same denominator (12), compare the numerators: $11 > 8$, therefore $\frac{11}{12} > \frac{8}{12}$.

10. $<$ $\frac{5}{6} \bowtie \frac{7}{8}$ $5 \times 8 = 40$ $40 < 42$
                      $6 \times 7 = 42$

11. $<$ $\frac{3}{8} \bowtie \frac{4}{5}$ $3 \times 5 = 15$ $15 < 32$
                      $8 \times 4 = 32$

12. $>$ $\frac{3}{4} \bowtie \frac{2}{3}$ $3 \times 3 = 9$ $9 > 8$
                      $4 \times 2 = 8$

13. $=$ $\frac{1}{3} \bowtie \frac{3}{9}$ $1 \times 9 = 9$ $9 = 9$
                      $3 \times 3 = 9$

14. $>$ Since both fractions have the same denominator (21), compare the numerators: $15 > 13$, therefore $\frac{15}{21} > \frac{13}{21}$.

15. $<$ Since both fractions have the same denominator (24), compare the numerators: $15 < 18$, therefore $\frac{15}{24} < \frac{18}{24}$.

16. $=$ $\frac{3}{4} \bowtie \frac{18}{24}$ $3 \times 24 = 72$ $72 = 72$
                      $4 \times 18 = 72$

PAGES 80–81

1. $\frac{5}{6}$ The figure is divided into 6 parts, and 5 parts are shaded.

2. $\frac{3}{5}$ There are 5 circles in the group, and 3 are shaded.

3. $\frac{7}{2}$ Each figure is divided into 2 parts, and 7 parts are shaded.

4. $\frac{14}{9}$ Each figure is divided into 9 parts, and 14 parts are shaded.

5. $1\frac{1}{3}$ Each figure is divided into 3 parts. One figure is completely shaded, and 1 of the 3 parts of the last figure is shaded.

6. $2\frac{3}{4}$ Each figure is divided into 4 parts. Two figures are completely shaded, and 3 of the 4 parts of the last figure are shaded.

7. $1\frac{5}{6}$ $6\overline{)11}$ $\dfrac{1}{}$ $\dfrac{-6}{5}$ $\rightarrow$ $1\frac{5}{6}$

8. 5 $5\overline{)25}$ $\dfrac{5}{}$ $\dfrac{-25}{0}$

9. $2\frac{3}{8}$ $8\overline{)19}$ $\dfrac{2}{}$ $\dfrac{-16}{3}$ $\rightarrow$ $2\frac{3}{8}$

10. $4\frac{4}{7}$ $7\overline{)32}$ $\dfrac{4}{}$ $\dfrac{-28}{4}$ $\rightarrow$ $4\frac{4}{7}$

11. $\frac{7}{1}$ $7 = \frac{7}{1}$

12. $\frac{15}{8}$ $\quad 1\frac{7}{8} = \frac{(8 \times 1) + 7}{8} = \frac{8 + 7}{8} = \frac{15}{8}$

13. $\frac{23}{5}$ $\quad 4\frac{3}{5} = \frac{(5 \times 4) + 3}{5} = \frac{20 + 3}{5} = \frac{23}{5}$

14. $\frac{20}{3}$ $\quad 6\frac{2}{3} = \frac{(6 \times 3) + 2}{3} = \frac{18 + 2}{3} = \frac{20}{3}$

15. $\frac{1}{2}$ $\quad \frac{5}{10} = \frac{5 \div 5}{10 \div 5} = \frac{1}{2}$

16. $\frac{3}{4}$ $\quad \frac{9}{12} = \frac{9 \div 3}{12 \div 3} = \frac{3}{4}$

17. $\frac{2}{3}$ $\quad \frac{16}{24} = \frac{16 \div 8}{24 \div 8} = \frac{2}{3}$

18. $\frac{3}{4}$ $\quad \frac{24}{32} = \frac{24 \div 8}{32 \div 8} = \frac{3}{4}$

19. $\frac{9}{18}$ $\quad \frac{1}{2} = \frac{1 \times 9}{2 \times 9} = \frac{9}{18}$

20. $\frac{20}{24}$ $\quad \frac{5}{6} = \frac{5 \times 4}{6 \times 4} = \frac{20}{24}$

21. $\frac{24}{32}$ $\quad \frac{3}{4} = \frac{3 \times 8}{4 \times 8} = \frac{24}{32}$

22. $\frac{16}{28}$ $\quad \frac{4}{7} = \frac{4 \times 4}{7 \times 4} = \frac{16}{28}$

23. $>$ $\quad \frac{4}{5} \times \frac{3}{4}$ $\quad 4 \times 4 = 16$, $5 \times 3 = 15$ $\quad 16 > 15$

24. $>$ $\quad \frac{7}{5} \times \frac{5}{6}$ $\quad 7 \times 6 = 42$, $5 \times 5 = 25$ $\quad 42 > 25$

25. $=$ $\quad \frac{4}{8} \times \frac{12}{24}$ $\quad 4 \times 24 = 96$, $8 \times 12 = 96$ $\quad 96 = 96$

26. $>$ $\quad \frac{8}{9} \times \frac{9}{12}$ $\quad 8 \times 12 = 96$, $9 \times 9 = 81$ $\quad 96 > 81$

27. $\frac{13}{40}$ $\quad$ 13 out of 40

28. $\frac{5}{36}$ $\quad$ 5 out of 36

29. (3) $\frac{3}{4}$ $\quad \frac{6}{8} = \frac{6 \div 2}{8 \div 2} = \frac{3}{4}$

30. (3) $3\frac{3}{4}$

$$\begin{array}{r} 3 \\ 4\overline{)15} \\ -12 \\ \hline 3 \end{array} \quad \rightarrow \quad 3\frac{3}{4}$$

31. (1) $\frac{7}{1}$ $\quad 7 = \frac{7}{1}$

32. (4) $\frac{3}{4}$ $\quad$ 6 ounces $= \frac{3}{4}$ cup

33. (2) **Stock 1 dropped more.**

$\frac{3}{4} \times \frac{5}{8}$ $\quad 3 \times 8 = 24$, $4 \times 5 = 20$ $\quad \frac{3}{4} > \frac{5}{8}$

34. (2) $\frac{1}{4}$ $\quad \frac{8}{32} = \frac{8 \div 8}{32 \div 8} = \frac{1}{4}$

SECTION 6

PAGE 83

1. **(3) $1\frac{3}{4} + 2$** Use the row for jumper. Add the number of yards needed for size 4 ($1\frac{3}{4}$) and for size 6 (2).
2. **(2) $1\frac{7}{8} + 1\frac{1}{8} + \frac{3}{4}$** Use the column for size 5. Add the number of yards needed for the jumper ($1\frac{7}{8}$); blouse B ($1\frac{1}{8}$); and facing, blouse B ($\frac{3}{4}$).
3. **(5) $5 - 1\frac{3}{4}$** Subtract the yards needed for the jumper ($1\frac{3}{4}$) from the number of yards of fabric Russ has (5).
4. **(1) $1\frac{3}{4} + 1\frac{3}{8}$** Add the number of yards needed for the jumper in size 4 ($1\frac{3}{4}$) and blouse A in size 6 ($1\frac{3}{8}$) to find the total number of yards of fabric.
5. **(5) jumper, size 5** $1\frac{3}{4} = 1\frac{6}{8}$ Jumper size 5 needs $1\frac{7}{8}$ yards of fabric, which is $\frac{1}{8}$ yard more than $1\frac{3}{4}$.
6. **You should have selected a table in the sports section of a newspaper and explained three row and/or column labels. For example, a table about a baseball game might have column headings of AB (meaning At Bats), AVG (meaning Average), and RBI (meaning Runs Batted In).**

PAGE 85

2. **36**
   multiples of 4: 4, 8, 12, 16, 20, 24, 28, 32, (36)
   multiples of 9: 9, 18, 27, (36)
3. **18**
   multiples of 2: 2, 4, 6, 8, 10, 12, 14, 16, (18)
   multiples of 9: 9, (18)
4. **21**
   multiples of 3: 3, 6, 9, 12, 15, 18, (21)
   multiples of 7: 7, 14, (21)
5. **12**
   multiples of 4: 4, 8, (12)
   multiples of 12: (12), 24
6. **15**
   multiples of 3: 3, 6, 9, 12, (15)
   multiples of 5: 5, 10, (15)
7. **10**
   multiples of 2: 2, 4, 6, 8, (10)
   multiples of 5: 5, (10)
   multiples of 10: (10), 20

24
multiples of 3: 3, 6, 9, 12, 15, 18, 21, (24)
multiples of 4: 4, 8, 12, 16, 20, (24)
multiples of 8: 8, 16, (24)

9. 20
multiples of 4: 4, 8, 12, 16, (20)
multiples of 5: 5, 10, 15, (20)
multiples of 10: 10, (20)

10. $\frac{18}{30}, \frac{25}{30}$

11. $\frac{7}{35}, \frac{30}{35}$

12. $\frac{5}{10}, \frac{2}{10}$

13. $\frac{32}{40}, \frac{35}{40}$

14. $\frac{3}{4}, \frac{1}{4}$

15. $\frac{16}{24}, \frac{18}{24}$

16. $\frac{2}{3}, \frac{2}{3}$

17. $\frac{16}{40}, \frac{35}{40}$

18. $\frac{10}{15}, \frac{9}{15}$

19. $\frac{21}{28}, \frac{22}{28}$

20. $\frac{40}{48}, \frac{9}{48}$

21. $\frac{20}{24}, \frac{21}{24}$

22. $\frac{15}{24}$ $\frac{5}{8} = \frac{5 \times 3}{8 \times 3} = \frac{15}{24}$

23. $\frac{21}{28}$ $\frac{3}{4} = \frac{3 \times 7}{4 \times 7} = \frac{21}{28}$

24. $\frac{16}{36}$ $\frac{4}{9} = \frac{4 \times 4}{9 \times 4} = \frac{16}{36}$

25. $\frac{35}{50}$ $\frac{7}{10} = \frac{7 \times 5}{10 \times 5} = \frac{35}{50}$

26. $\frac{12}{30}$ $\frac{2}{5} = \frac{2 \times 6}{5 \times 6} = \frac{12}{30}$

27. $\frac{12}{18}$ $\frac{2}{3} = \frac{2 \times 6}{3 \times 6} = \frac{12}{18}$

28. $\frac{40}{48}$ $\frac{5}{6} = \frac{5 \times 8}{6 \times 8} = \frac{40}{48}$

29. $\frac{20}{35}$ $\frac{4}{7} = \frac{4 \times 5}{7 \times 5} = \frac{20}{35}$

30. $\frac{20}{44}$ $\frac{5}{11} = \frac{5 \times 4}{11 \times 4} = \frac{20}{44}$

PAGE 87

2. $1\frac{3}{5}$ $\frac{7}{10} + \frac{9}{10} = \frac{16}{10} = 1\frac{6}{10} = 1\frac{3}{5}$

3. $\frac{17}{20}$ $\frac{1}{4} + \frac{3}{5} = \frac{5}{20} + \frac{12}{20} = \frac{17}{20}$

4. $1\frac{1}{2}$ $\frac{5}{6} + \frac{2}{3} = \frac{5}{6} + \frac{4}{6} = \frac{9}{6} = 1\frac{3}{6} = 1\frac{1}{2}$

5. $1\frac{1}{24}$ $\frac{3}{8} + \frac{2}{3} = \frac{9}{24} + \frac{16}{24} = \frac{25}{24} = 1\frac{1}{24}$

6. $1\frac{3}{35}$ $\frac{2}{7} + \frac{4}{5} = \frac{10}{35} + \frac{28}{35} = \frac{38}{35} = 1\frac{3}{35}$

7. $5\frac{3}{5}$

$$\begin{array}{r} 2\frac{1}{5} \\ +3\frac{2}{5} \\ \hline 5\frac{3}{5} \end{array}$$

8. 7

$$\begin{array}{r} 4\frac{5}{12} \\ +2\frac{7}{12} \\ \hline 6\frac{12}{12} = 6 + 1 = 7 \end{array}$$

9. $4\frac{11}{12}$

$$\begin{array}{l} 1\frac{1}{4} = 1\frac{3}{12} \\ +3\frac{2}{3} = 3\frac{8}{12} \\ \hline \quad\quad 4\frac{11}{12} \end{array}$$

10. $8\frac{1}{15}$

$$\begin{array}{l} 5\frac{2}{3} = 5\frac{10}{15} \\ +2\frac{2}{5} = 2\frac{6}{15} \\ \hline 7\frac{16}{15} = 7 + 1\frac{1}{15} = 8\frac{1}{15} \end{array}$$

11. $6\frac{19}{24}$

$$\begin{array}{l} 6\frac{5}{8} = 6\frac{15}{24} \\ + \frac{1}{6} = \frac{4}{24} \\ \hline \quad\quad 6\frac{19}{24} \end{array}$$

12. $6\frac{5}{12}$

$$\begin{array}{l} 3\frac{2}{3} = 3\frac{8}{12} \\ +2\frac{3}{4} = 2\frac{9}{12} \\ \hline 5\frac{17}{12} = 5 + 1\frac{5}{12} = 6\frac{5}{12} \end{array}$$

13. $6\frac{11}{16}$

$$\begin{array}{l} 4\frac{3}{8} = 4\frac{6}{16} \\ +2\frac{5}{16} = 2\frac{5}{16} \\ \hline \quad\quad 6\frac{11}{16} \end{array}$$

14. $21\frac{7}{16}$

$$\begin{array}{l} 12\frac{1}{4} = 12\frac{4}{16} \\ + 9\frac{3}{16} = 9\frac{3}{16} \\ \hline \quad\quad 21\frac{7}{16} \end{array}$$

15. $9\frac{5}{12}$

$$\begin{array}{l} 5\frac{3}{4} = 5\frac{9}{12} \\ +3\frac{2}{3} = 3\frac{8}{12} \\ \hline 8\frac{17}{12} = 9\frac{5}{12} \end{array}$$

16. $4\frac{1}{6}$

$$\begin{array}{l} 1\frac{3}{4} = 1\frac{9}{12} \\ 1\frac{1}{4} = 1\frac{3}{12} \\ \frac{1}{2} = \frac{6}{12} \\ + \frac{2}{3} = \frac{8}{12} \\ \hline 2\frac{26}{12} = 2 + 2 + \frac{2}{12} = 4\frac{1}{6} \end{array}$$

PAGE 89

2. $\frac{11}{24}$ $\frac{5}{6} - \frac{3}{8} =$

$$\frac{20}{24} - \frac{9}{24} = \frac{11}{24}$$

3. $\frac{5}{16}$ $\frac{9}{16} - \frac{1}{4} =$

$$\frac{9}{16} - \frac{4}{16} = \frac{5}{16}$$

4. $3\frac{1}{3}$ $6 - 2\frac{2}{3} =$

$$5\frac{3}{3} - 2\frac{2}{3} = 3\frac{1}{3}$$

5. $\frac{5}{24}$ $\frac{7}{8} - \frac{2}{3} =$

$$\frac{21}{24} - \frac{16}{24} = \frac{5}{24}$$

6. $\frac{1}{20}$ $\frac{4}{5} - \frac{3}{4} =$

$$\frac{16}{20} - \frac{15}{20} = \frac{1}{20}$$

7. $3\frac{1}{2}$
$$\begin{array}{r} 4\frac{7}{8} \\ -1\frac{3}{8} \\ \hline 3\frac{4}{8} = 3\frac{1}{2} \end{array}$$

8. $5\frac{1}{12}$
$$\begin{array}{r} 7\frac{3}{4} = 7\frac{9}{12} \\ -2\frac{2}{3} = 2\frac{8}{12} \\ \hline 5\frac{1}{12} \end{array}$$

9. $2\frac{7}{9}$
$$\begin{array}{r} 8 = \overset{7}{\cancel{8}}\frac{9}{9} \\ -5\frac{2}{9} = 5\frac{2}{9} \\ \hline 2\frac{7}{9} \end{array}$$

10. $2\frac{1}{3}$
$$\begin{array}{r} 6\frac{1}{6} = \overset{5}{\cancel{6}}\frac{7}{6} \\ -3\frac{5}{6} = 3\frac{5}{6} \\ \hline 2\frac{2}{6} = 2\frac{1}{3} \end{array}$$

11. $\frac{5}{8}$
$$\begin{array}{r} 9\frac{1}{4} = 9\frac{2}{8} = \overset{8}{\cancel{9}}\overset{10}{\cancel{\frac{2}{8}}} \\ -8\frac{5}{8} = 8\frac{5}{8} = 8\frac{5}{8} \\ \hline \frac{5}{8} \end{array}$$

12. $5\frac{11}{15}$
$$\begin{array}{r} 7\frac{2}{5} = 7\frac{6}{15} = \overset{6}{\cancel{7}}\overset{21}{\cancel{\frac{6}{15}}} \\ -1\frac{2}{3} = 1\frac{10}{15} = 1\frac{10}{15} \\ \hline 5\frac{11}{15} \end{array}$$

13. $6\frac{1}{2}$ $9\frac{3}{4} - 3\frac{1}{4} = 6\frac{2}{4} = 6\frac{1}{2}$

14. $26\frac{7}{16}$
$$\begin{array}{r} 35 = 34\frac{16}{16} \\ -8\frac{9}{16} = 8\frac{9}{16} \\ \hline 26\frac{7}{16} \end{array}$$

15. $2\frac{1}{4}$
$$\begin{array}{r} 4\frac{3}{4} = 4\frac{3}{4} \\ -2\frac{1}{2} = 2\frac{2}{4} \\ \hline 2\frac{1}{4} \end{array}$$

16. $3\frac{5}{6}$
$$\begin{array}{r} 8\frac{1}{2} = \overset{7}{\cancel{8}}\overset{9}{\frac{3}{6}} \\ -4\frac{2}{3} = 4\frac{4}{6} \\ \hline 3\frac{5}{6} \end{array}$$

PAGE 91

1. **(5)** $17\frac{1}{2}$ Add each distance to find the total.
$$\begin{array}{r} 6\frac{3}{10} = 6\frac{3}{10} \\ 6\frac{4}{5} = 6\frac{8}{10} \\ +4\frac{2}{5} = 4\frac{4}{10} \\ \hline 16\frac{15}{10} \end{array}$$
$$= 16 + 1\frac{5}{10} = 17\frac{5}{10} = 17\frac{1}{2} \text{ miles}$$

2. **(4)** $1\frac{1}{10}$ One-Stop Gas and Food is located between the city pool and the hospital. Jose drives $3\frac{3}{10}$ miles from the city pool to One-Stop Gas and Food and $4\frac{2}{5}$ miles from the city pool to the hospital. Subtract to find the distance from One-Stop Gas and Food to the hospital.
$$\begin{array}{r} 4\frac{2}{5} = 4\frac{4}{10} \\ -3\frac{3}{10} = 3\frac{3}{10} \\ \hline 1\frac{1}{10} \end{array}$$

3. **(1)** $8\frac{7}{10}$ Add each distance to find the total. See the explanation for Item 4.
$$\begin{array}{r} 1\frac{1}{10} \\ 1\frac{9}{10} \\ +5\frac{7}{10} \\ \hline 7\frac{17}{10} = 7 + 1\frac{7}{10} = 8\frac{7}{10} \end{array}$$

4. **(4)** $\frac{3}{5}$ Subtract each distance.
$$6\frac{3}{10} - 5\frac{7}{10} = 5\frac{13}{10} - 5\frac{7}{10} = \frac{6}{10} = \frac{3}{5}$$

5. **(5) Not enough information is given.** The map does not show where José lives, therefore you do not know the distance he lives from work.

6. $6\frac{3}{10}$ rounds to 6

   $6\frac{4}{5}$ rounds to 7

   $4\frac{2}{5}$ rounds to 4

   $1\frac{1}{10}$ rounds to 1

   $1\frac{9}{10}$ rounds to 2

   $5\frac{7}{10}$ rounds to 6

   Jose's route is about 26 miles
   $(6 + 7 + 4 + 1 + 2 + 6)$.

PAGE 93

1. **(3) 8** Kareem adds $4\frac{1}{4}$ and $3\frac{3}{4}$ hours to find the total.

$$
\begin{array}{r}
4\frac{1}{4} \\
+3\frac{3}{4} \\
\hline
7\frac{4}{4} = 7 + 1 = 8
\end{array}
$$

2. **(3) $3\frac{3}{4}$** Since 50 is close to 45, think of 8:50 as $8\frac{3}{4}$. Since 32 is close to 30, think of 12:32 as $12\frac{1}{2}$. Then subtract.

$$
\begin{array}{ccc}
12\frac{1}{2} = & 12\frac{2}{4} & \overset{11}{\cancel{12}}\overset{6}{\cancel{\frac{2}{4}}} \\
- 8\frac{3}{4} = & - 8\frac{3}{4} & - 8\frac{3}{4} \\
\hline
& & 3\frac{3}{4} \text{ hours}
\end{array}
$$

3. **(1) $3\frac{3}{4}$** Since 24 is close to 30, think of 1:24 as $1\frac{1}{2}$. Since 10 is close to 15, think of 5:10 as $5\frac{1}{4}$. Then subtract.

$$
\begin{array}{ccc}
5\frac{1}{4} = & 5\frac{1}{4} = & \overset{4}{\cancel{5}}\overset{5}{\cancel{\frac{1}{4}}} \\
-1\frac{1}{2} = & 1\frac{2}{4} = & -1\frac{2}{4} \\
\hline
& & 3\frac{3}{4} \text{ hours}
\end{array}
$$

4. **(1) $7\frac{1}{2}$** Add the hours Paul worked in the morning and the hours he worked in the afternoon.

$$
\begin{array}{r}
3\frac{3}{4} \\
+3\frac{3}{4} \\
\hline
6\frac{6}{4} = 6 + 1\frac{2}{4} = 7\frac{2}{4} = 7\frac{1}{2}
\end{array}
$$

5. **(5) Not enough information is given.** You need to know what time Paul went to lunch to answer the question.

6. **(3) $2\frac{1}{2}$** $40\frac{3}{4} - 38\frac{1}{4} = 2\frac{2}{4} = 2\frac{1}{2}$

7. Create a fraction with a numerator based on the number of minutes being discussed and a denominator based on the number of minutes in an hour.

   20 minutes = $\frac{1}{3}$ hour
   20 minutes = $\frac{20}{60} = \frac{20 \div 20}{60 \div 20} = \frac{1}{3}$

   40 minutes = $\frac{2}{3}$ hour
   40 minutes = $\frac{40}{60} = \frac{40 \div 20}{60 \div 20} = \frac{2}{3}$

   50 minutes = $\frac{5}{6}$ hour
   50 minutes = $\frac{50}{60} = \frac{50 \div 10}{60 \div 10} = \frac{5}{6}$

PAGES 94–95

1. $\frac{3}{8}$ $\frac{15}{16} - \frac{9}{16} = \frac{6}{16} = \frac{3}{8}$

2. $\frac{9}{10}$ $\frac{1}{5} + \frac{7}{10} =$
   $\frac{2}{10} + \frac{7}{10} = \frac{9}{10}$

3. $5\frac{5}{6}$

$$
\begin{array}{r}
8 = \overset{7}{\cancel{8}}\frac{6}{6} \\
-2\frac{1}{6} = 2\frac{1}{6} \\
\hline
5\frac{5}{6}
\end{array}
$$

4. $1\frac{5}{24}$ $\frac{5}{6} + \frac{3}{8} =$
   $\frac{20}{24} + \frac{9}{24} = \frac{29}{24} = 1\frac{5}{24}$

5. 8

$$
\begin{array}{r}
3\frac{7}{9} \\
+4\frac{2}{9} \\
\hline
7\frac{9}{9} = 7 + 1 = 8
\end{array}
$$

6. $2\frac{3}{5}$

$$
\begin{array}{r}
4\frac{9}{10} \\
-2\frac{3}{10} \\
\hline
2\frac{6}{10} = 2\frac{3}{5}
\end{array}
$$

7. $5\frac{3}{8}$

$$
\begin{array}{r}
2\frac{5}{8} = 2\frac{5}{8} \\
+2\frac{3}{4} = 2\frac{6}{8} \\
\hline
4\frac{11}{8} = 4 + 1\frac{3}{8} = 5\frac{3}{8}
\end{array}
$$

8. $\frac{3}{5}$

$$
\begin{array}{r}
9\frac{1}{5} = \overset{8}{\cancel{9}}\overset{6}{\cancel{\frac{1}{5}}} \\
-8\frac{3}{5} = 8\frac{3}{5} \\
\hline
\frac{3}{5}
\end{array}
$$

9. $2\frac{1}{8}$

$$
\begin{array}{r}
3 = \overset{2}{\cancel{3}}\frac{8}{8} \\
-\frac{7}{8} = \frac{7}{8} \\
\hline
2\frac{1}{8}
\end{array}
$$

10. $3\frac{7}{12}$ $\begin{aligned} 6\frac{1}{3} &= 6\frac{4}{12} = \cancel{6}\,\overset{5}{}\frac{\overset{16}{4}}{12} \\ -2\frac{3}{4} &= 2\frac{9}{12} = 2\frac{9}{12} \\ \hline & \phantom{2\frac{9}{12}=} 3\frac{7}{12} \end{aligned}$

11. **(3) $7\frac{1}{8}$ yards** Use the column for size 12. Add the number of yards of fabric in the main color ($5\frac{5}{8}$) and the number of yards of contrasting fabric ($1\frac{1}{2}$) needed for dress B.

$$\begin{aligned} 5\frac{5}{8} &= 5\frac{5}{8} \\ +1\frac{1}{2} &= 1\frac{4}{8} \\ \hline 6\frac{9}{8} &= 6 + 1\frac{1}{8} = 7\frac{1}{8} \text{ yards} \end{aligned}$$

12. **(3) $\frac{3}{8}$ yard** Use the column for size 8. Subtract the number of yards of lace needed for dress B ($2\frac{1}{8}$) from the number of yards of lace needed for dress A ($2\frac{1}{2}$).

$$\begin{aligned} 2\frac{1}{2} &= 2\frac{4}{8} \\ -2\frac{1}{8} &= 2\frac{1}{8} \\ \hline & \frac{3}{8} \text{ yard} \end{aligned}$$

13. **(3) $5\frac{1}{8}$ yards** Add the amounts of lace needed for dress A in size 8 ($2\frac{1}{2}$) and size 10 ($2\frac{5}{8}$).

$$\begin{aligned} 2\frac{1}{2} &= 2\frac{4}{8} \\ +2\frac{5}{8} &= 2\frac{5}{8} \\ \hline 4\frac{9}{8} &= 4 + 1\frac{1}{8} = 5\frac{1}{8} \end{aligned}$$

14. **6 miles** Sundra drives $3\frac{2}{5}$ miles from the courthouse to the art museum, $1\frac{1}{5}$ miles from the art museum to the science museum, and $1\frac{9}{10}$ miles from the science museum to the shopping mall. To estimate the total distance, round each distance and add.

$$\begin{aligned} 3\frac{2}{5} \text{ rounds to } & 3 \\ 1\frac{1}{5} \text{ rounds to } & 1 \\ +1\frac{9}{10} \text{ rounds to } & 2 \\ \hline & 6 \end{aligned}$$

15. **$6\frac{1}{2}$ miles** Add each distance to find the total distance. See the explanation for Item 14.

$$\begin{aligned} 3\frac{2}{5} &= 3\frac{4}{10} \\ 1\frac{1}{5} &= 1\frac{2}{10} \\ +1\frac{9}{10} &= 1\frac{9}{10} \\ \hline 5\frac{15}{10} &= 5 + 1\frac{5}{10} = 6\frac{5}{10} = 6\frac{1}{2} \end{aligned}$$

16. **$8\frac{3}{5}$ miles** Use your answer from Item 15 ($6\frac{1}{2}$). Add the distance from the mall to the bus depot.

$$\begin{aligned} 6\frac{1}{2} &= 6\frac{5}{10} \\ +2\frac{1}{10} &= 2\frac{1}{10} \\ \hline 8\frac{6}{10} &= 8\frac{3}{5} \end{aligned}$$

17. **12 miles** Sundra drives $5\frac{3}{10}$ miles from the bus depot to the library. The entire route is $17\frac{1}{5}$ miles. Round each distance and subtract.

$$\begin{aligned} 17\frac{1}{5} \text{ rounds to } & 17 \\ -5\frac{3}{10} \text{ rounds to } & 5 \\ \hline & 12 \end{aligned}$$

18. **$3\frac{3}{4}$ hours** Since 24 is close to 30, think of 8:24 as $8\frac{1}{2}$. Since 13 is close to 15, think of 12:13 as $12\frac{1}{4}$. Then subtract.

$$\begin{aligned} 12\frac{1}{4} &= 12\frac{1}{4} = \cancel{12}\,\overset{11}{}\overset{5}{}\frac{1}{4} \\ -8\frac{1}{2} &= 8\frac{2}{4} = 8\frac{2}{4} \\ \hline & \phantom{8\frac{2}{4}=} 3\frac{3}{4} \end{aligned}$$

19. **$4\frac{3}{4}$ hours** Since 58 is close to 60, think of 12:58 as 1. Since 43 is close to 45, think of 5:43 as $5\frac{3}{4}$. Then subtract.

$$\begin{aligned} 5\frac{3}{4} \\ -1 \\ \hline 4\frac{3}{4} \end{aligned}$$

20. **$8\frac{1}{2}$ hours** Add the hours Joy worked in the morning ($3\frac{3}{4}$) and the hours she worked in the afternoon ($4\frac{3}{4}$).

$$\begin{aligned} 3\frac{3}{4} \\ +4\frac{3}{4} \\ \hline 7\frac{6}{4} = 7 + 1\frac{2}{4} = 8\frac{2}{4} = 8\frac{1}{2} \end{aligned}$$

SECTION 7

PAGE 97

1. **(5) $5\frac{1}{2} \div 3$** Vince wants to make one third of the recipe, so divide the amount of potatoes by 3.

2. **(3) $7\frac{1}{4} \times \frac{1}{2}$** Kasia wants to make half of the recipe, so multiply the amount of flour by $\frac{1}{2}$. (This is the same as dividing by 2.)

3. **(4) $4\frac{2}{3} \div \frac{1}{3}$** Divide the total amount of cocoa by the amount needed for each pound.

4. **(5)** $1\frac{1}{4} \times 2$ Luis is planning to double the recipe, so multiply the amount of milk by 2.

5. **(5) Not enough information is given.** You need to know the number of cups of chiles needed in the recipe.

6. **(3)** $1\frac{1}{2} \times 5$ Since Armand needs $1\frac{1}{2}$ cups of cleaning solution for every gallon of water, he would need five times the amount of cleaning solution for 5 gallons of water.

7. **When you invert a number, you turn it upside down. Thus when 2 (or $\frac{2}{1}$) is inverted, you get $\frac{1}{2}$. When you invert a number used as a divisor, you change division to multiplication. Instead of dividing by 2, you would multiply by $\frac{1}{2}$. So, multiplying by $\frac{1}{2}$ is the same as dividing by 2.**

PAGE 99

2. $\frac{25}{48}$ $\frac{5}{6} \times \frac{5}{8} = \frac{25}{48}$ (canceling cannot be used here.)

3. $\frac{3}{10}$ $\frac{3}{8} \times \frac{4}{5} = \frac{3}{\cancel{8}_2} \times \frac{\cancel{4}^1}{5} = \frac{3}{10}$

4. $\frac{2}{21}$ $\frac{\cancel{3}^1}{7} \times \frac{2}{\cancel{9}_3} = \frac{2}{21}$

5. $2\frac{1}{4}$ $\frac{3}{8} \times 6 = \frac{3}{\cancel{8}_4} \times \frac{\cancel{6}^3}{1} = \frac{9}{4} = 2\frac{1}{4}$

6. $11\frac{1}{2}$ $5 \times 2\frac{3}{10} = \frac{\cancel{5}^1}{1} \times \frac{23}{\cancel{10}_2} = \frac{23}{2} = 11\frac{1}{2}$

7. $1\frac{1}{2}$ $\frac{5}{8} \times 2\frac{2}{5} = \frac{\cancel{5}^1}{\cancel{8}_2} \times \frac{\cancel{12}^3}{\cancel{5}_1} = \frac{3}{2} = 1\frac{1}{2}$

8. $3\frac{1}{9}$ $3\frac{1}{2} \times \frac{8}{9} = \frac{7}{\cancel{2}_1} \times \frac{\cancel{8}^4}{9} = \frac{28}{9} = 3\frac{1}{9}$

9. $3\frac{1}{2}$ $2\frac{1}{3} \times 1\frac{1}{2} = \frac{7}{\cancel{3}_1} \times \frac{\cancel{3}^1}{2} = \frac{7}{2} = 3\frac{1}{2}$

10. 12 $1\frac{4}{5} \times 6\frac{2}{3} = \frac{\cancel{9}^3}{\cancel{5}_1} \times \frac{\cancel{20}^4}{\cancel{3}_1} = \frac{12}{1} = 12$

11. **18 people** $\frac{3}{4} \times 24 = \frac{3}{\cancel{4}_1} \times \frac{\cancel{24}^6}{1} = \frac{18}{1} = 18$

12. $5\frac{1}{4}$ **miles** $3\frac{1}{2} \times 1\frac{1}{2} = \frac{7}{2} \times \frac{3}{2} = \frac{21}{4} = 5\frac{1}{4}$

13. $10\frac{1}{2}$ **pounds** $15\frac{3}{4} \times \frac{2}{3} = \frac{\cancel{63}^{21}}{\cancel{4}_2} \times \frac{\cancel{2}^1}{\cancel{3}_1} = \frac{21}{2} = 10\frac{1}{2}$

14. $25\frac{1}{2}$ **pints** $4\frac{1}{4} \times 6 = \frac{17}{\cancel{4}_2} \times \frac{\cancel{6}^3}{1} = \frac{51}{2} = 25\frac{1}{2}$

PAGE 101

2. $1\frac{1}{3}$ $\frac{5}{6} \div \frac{5}{8} = \frac{\cancel{5}^1}{\cancel{6}_3} \times \frac{\cancel{8}^4}{\cancel{5}_1} = \frac{4}{3} = 1\frac{1}{3}$

3. $\frac{1}{3}$ $\frac{4}{15} \div \frac{4}{5} = \frac{\cancel{4}^1}{\cancel{15}_3} \times \frac{\cancel{5}^1}{\cancel{4}_1} = \frac{1}{3}$

4. $\frac{4}{5}$ $\frac{2}{3} \div \frac{5}{6} = \frac{2}{\cancel{3}_1} \times \frac{\cancel{6}^2}{5} = \frac{4}{5}$

5. $\frac{7}{36}$ $\frac{7}{12} \div 3 = \frac{7}{12} \div \frac{3}{1} = \frac{7}{12} \times \frac{1}{3} = \frac{7}{36}$

6. $1\frac{3}{4}$ $2 \div 1\frac{1}{7} = \frac{2}{1} \div \frac{8}{7} = \frac{\cancel{2}^1}{1} \times \frac{7}{\cancel{8}_4} = \frac{7}{4} = 1\frac{3}{4}$

7. $\frac{15}{44}$ $\frac{5}{8} \div 1\frac{5}{6} = \frac{5}{8} \div \frac{11}{6} = \frac{5}{\cancel{8}_4} \times \frac{\cancel{6}^3}{11} = \frac{15}{44}$

8. $12\frac{1}{2}$ $3\frac{3}{4} \div \frac{3}{10} = \frac{15}{4} \div \frac{3}{10} = \frac{\cancel{15}^5}{\cancel{4}_2} \times \frac{\cancel{10}^5}{\cancel{3}_1} = \frac{25}{2} = 12\frac{1}{2}$

9. $\frac{28}{45}$ $1\frac{2}{5} \div 2\frac{1}{4} = \frac{7}{5} \div \frac{9}{4} = \frac{7}{5} \times \frac{4}{9} = \frac{28}{45}$

10. $2\frac{5}{6}$ $4\frac{1}{4} \div 1\frac{1}{2} = \frac{17}{4} \div \frac{3}{2} = \frac{17}{\cancel{4}_2} \times \frac{\cancel{2}^1}{3} = \frac{17}{6} = 2\frac{5}{6}$

11. **24 building lots** $15 \div \frac{5}{8} = \frac{\cancel{15}^3}{1} \times \frac{8}{\cancel{5}_1} = 24$

12. **22 patties**
$5\frac{1}{2} \div \frac{1}{4} = \frac{11}{2} \div \frac{1}{4} = \frac{11}{\cancel{2}_1} \times \frac{\cancel{4}^2}{1} = \frac{22}{1} = 22$

13. $2\frac{1}{2}$ **hours** $6\frac{1}{4} \div 2\frac{1}{2} = \frac{25}{4} \div \frac{5}{2} = \frac{\cancel{25}^5}{\cancel{4}_2} \times \frac{\cancel{2}^1}{\cancel{5}_1} = \frac{5}{2} = 2\frac{1}{2}$

14. **10 tapes**
$12\frac{1}{2} \div 1\frac{1}{4} = \frac{25}{2} \div \frac{5}{4} = \frac{\cancel{25}^5}{\cancel{2}_1} \times \frac{\cancel{4}^2}{\cancel{5}_1} = \frac{10}{1} = 10$

PAGE 103

1. **(2) 100 inches** Multiply the length of each shelf ($1\frac{2}{3}$ feet) by the number of shelves (5). Then multiply by 12 to change feet to inches.

$1\frac{2}{3} \times 5 = \frac{5}{3} \times \frac{5}{1} = \frac{25}{3} = 8\frac{1}{3}$ feet

$8\frac{1}{3} \times 12 = \frac{25}{3} \times \frac{\cancel{12}^4}{1} = \frac{100}{1} = 100$ inches

2. **(4) 4 cuts**

$\frac{3}{16}"$	$\frac{3}{16}"$	$\frac{3}{16}"$	$\frac{3}{16}"$	waste

$1\frac{2}{3}$	$1\frac{2}{3}$	$1\frac{2}{3}$	$1\frac{2}{3}$	$1\frac{2}{3}$	5 shelves
1	2	3	4		cuts

3. **(5) $\frac{3}{4}$ inch** Multiply the waste per cut ($\frac{3}{16}$ inch) by the number of cuts (4; see the diagram for Item 2).

$$\frac{3}{16} \times 4 = \frac{3}{\cancel{16}_{4}} \times \frac{\cancel{4}^{1}}{1} = \frac{3}{4} \text{ inch}$$

4. **(4) $27\frac{1}{2}$ feet** Multiply the length of each board ($6\frac{7}{8}$ feet) by the number of shelves (4).

$$6\frac{7}{8} \times 4 = 24\frac{28}{8} \text{ or } 24\frac{7}{2} = 24 + 3\frac{1}{2} = 27\frac{1}{2}.$$

5. **(2) $\frac{3}{4}$ inch** Divide the total waste ($2\frac{1}{4}$ inches) by the number of cuts (3).

$$2\frac{1}{4} \div 3 = \frac{9}{4} \div 3 = \frac{\cancel{9}^{3}}{4} \times \frac{1}{\cancel{3}_{1}} = \frac{3}{4}.$$

6. **Your diagram should show a board with five cuts separating the board into six equal pieces.**

PAGES 104–105

1. $\frac{4}{15}$ $\frac{4}{5} \times \frac{1}{3} = \frac{4}{15}$

2. $2\frac{2}{3}$ $\frac{4}{9} \div \frac{1}{6} = \frac{4}{\cancel{9}_{3}} \times \frac{\cancel{6}^{2}}{1} = \frac{8}{3} = 2\frac{2}{3}$

3. $1\frac{1}{24}$ $\frac{5}{8} \div \frac{3}{5} = \frac{5}{8} \times \frac{5}{3} = \frac{25}{24} = 1\frac{1}{24}$

4. $\frac{5}{42}$ $\frac{\cancel{7}^{1}}{7} \times \frac{5}{\cancel{12}_{6}} = \frac{5}{42}$

5. 8 $7 \div \frac{7}{8} = \frac{7}{1} \div \frac{7}{8} = \frac{\cancel{7}^{1}}{1} \times \frac{8}{\cancel{7}_{1}} = \frac{8}{1} = 8$

6. $4\frac{1}{2}$ $6 \times \frac{3}{4} = \frac{\cancel{6}^{3}}{1} \times \frac{3}{\cancel{4}_{2}} = \frac{9}{2} = 4\frac{1}{2}$

7. $\frac{6}{25}$ $\frac{\cancel{4}^{2}}{\cancel{15}_{5}} \times \frac{\cancel{9}^{3}}{\cancel{10}_{5}} = \frac{6}{25}$

8. $\frac{9}{10}$ $\frac{2}{5} \div \frac{4}{9} = \frac{\cancel{2}^{1}}{5} \times \frac{9}{\cancel{4}_{2}} = \frac{9}{10}$

9. $7\frac{1}{3}$ $\frac{11}{12} \times 8 = \frac{11}{\cancel{12}_{3}} \times \frac{\cancel{8}^{2}}{1} = \frac{22}{3} = 7\frac{1}{3}$

10. $11\frac{1}{3}$ $4 \times 2\frac{5}{6} = \frac{\cancel{4}^{2}}{1} \times \frac{17}{\cancel{6}_{3}} = \frac{34}{3} = 11\frac{1}{3}$

11. $\frac{3}{32}$ $\frac{3}{8} \div 4 = \frac{3}{8} \div \frac{4}{1} = \frac{3}{8} \times \frac{1}{4} = \frac{3}{32}$

12. $4\frac{1}{2}$ $5 \div 1\frac{1}{9} = \frac{5}{1} \div \frac{10}{9} = \frac{\cancel{5}^{1}}{1} \times \frac{9}{\cancel{10}_{2}} = \frac{9}{2} = 4\frac{1}{2}$

13. $2\frac{1}{10}$ $\frac{7}{12} \times 3\frac{3}{5} = \frac{7}{\cancel{12}_{2}} \times \frac{\cancel{18}^{3}}{5} = \frac{21}{10} = 2\frac{1}{10}$

14. $\frac{1}{3}$ $\frac{5}{6} \div 2\frac{1}{2} = \frac{5}{6} \div \frac{5}{2} = \frac{\cancel{5}^{1}}{\cancel{6}_{3}} \times \frac{\cancel{2}^{1}}{\cancel{5}_{1}} = \frac{1}{3}$

15. $1\frac{5}{6}$ $2\frac{3}{4} \times \frac{2}{3} = \frac{11}{\cancel{4}_{2}} \times \frac{\cancel{2}^{1}}{3} = \frac{11}{6} = 1\frac{5}{6}$

16. $3\frac{3}{4}$ $4\frac{1}{2} \div 1\frac{1}{5} = \frac{9}{2} \div \frac{6}{5} = \frac{\cancel{9}^{3}}{2} \times \frac{5}{\cancel{6}_{2}} = \frac{15}{4} = 3\frac{3}{4}$

17. $1\frac{1}{3}$ $3\frac{1}{5} \div 2\frac{2}{5} = \frac{16}{5} \div \frac{12}{5} = \frac{\cancel{16}^{4}}{\cancel{5}} \times \frac{\cancel{5}^{1}}{\cancel{12}_{3}} = \frac{4}{3} = 1\frac{1}{3}$

18. $4\frac{1}{7}$ $2\frac{5}{12} \times 1\frac{5}{7} = \frac{29}{\cancel{12}_{1}} \times \frac{\cancel{12}^{1}}{7} = \frac{29}{7} = 4\frac{1}{7}$

19. $3\frac{3}{4}$ $2\frac{6}{7} \times 1\frac{5}{16} = \frac{\cancel{20}^{5}}{7} \times \frac{\cancel{21}^{3}}{\cancel{16}_{4}} = \frac{15}{4} = 3\frac{3}{4}$

20. $\frac{4}{5}$ $3\frac{1}{3} \div 4\frac{1}{6} = \frac{10}{3} \div \frac{25}{6} = \frac{\cancel{10}^{2}}{\cancel{3}_{1}} \times \frac{\cancel{6}^{2}}{\cancel{25}_{5}} = \frac{4}{5}$

21. **$2\frac{1}{2}$ hours** Multiply the hours worked by the part of the day spent answering the phone.

$$7\frac{1}{2} \times \frac{1}{3} = \frac{\cancel{15}^{5}}{2} \times \frac{1}{\cancel{3}_{1}} = \frac{5}{2} = 2\frac{1}{2}$$

22. **4 dozen plants** Since $\frac{1}{7}$ of the plants were lost, $\frac{6}{7}$ survived. Multiply the total number of plants by the fraction that survived.

$$4\frac{2}{3} \times \frac{6}{7} = \frac{\cancel{14}^{2}}{\cancel{3}_{1}} \times \frac{\cancel{6}^{2}}{\cancel{7}_{1}} = \frac{4}{1} = 4$$

23. **8 batches** Divide the total amount of pectin by the amount needed for each batch.

$$14 \div 1\frac{3}{4} = \frac{14}{1} \div \frac{7}{4} = \frac{\cancel{14}^{2}}{1} \times \frac{4}{\cancel{7}_{1}} = \frac{8}{1} = 8$$

24. **40 strips** Divide the total width by the width of each strip.

$$45 \div 1\frac{1}{8} = \frac{45}{1} \div \frac{9}{8} = \frac{\cancel{45}^{5}}{1} \times \frac{8}{\cancel{9}_{1}} = \frac{40}{1} = 40$$

25. **(3) $\frac{3}{4} \times 2$** To double the recipe, multiply the amount of bread crumbs by 2.

26. **(2) $\frac{5}{8}$** Since 6 is one third of 18, Carla needs to make one third of the recipe. Multiply the amount of mayonnaise by $\frac{1}{3}$.

$$1\frac{7}{8} \times \frac{1}{3} = \frac{\cancel{15}^{5}}{8} \times \frac{1}{\cancel{3}_{1}} = \frac{5}{8}$$

27. **(4) 6**

| $2\frac{1}{2}$ | $2\frac{1}{2}$ | $2\frac{1}{2}$ | $2\frac{1}{2}$ | $2\frac{1}{2}$ | $2\frac{1}{2}$ | $2\frac{1}{2}$ | 7 shelves |

  1 2 3 4 5 6 cuts

28. **(5) 42** Multiply 3 feet by 12 inches to get 36 inches. Add 6 inches ($\frac{1}{2}$ ft). $36 + 6 = 42$ or $3\frac{1}{2} \times 12 = \frac{7}{2} \times \frac{\cancel{12}^{6}}{1} = 42$

MATH AT WORK
PAGE 107

1. **(3) $6\frac{1}{8}$ feet** The length of the whole wall with the door is $15\frac{1}{4}$ feet, the same as the length of the opposite wall. Subtract the sum of the door (3 ft.) and the part of the wall to the left of the door ($6\frac{1}{8}$ ft.) from the length of the whole wall to get the part of the wall to the right of the door.
$15\frac{1}{4} - (3 + 6\frac{1}{8}) = 15\frac{1}{4} - 9\frac{1}{8} = 15\frac{2}{8} - 9\frac{1}{8} = 6\frac{1}{8}$

2. **(2) $46\frac{1}{2}$ feet** Jake needs baseboard for the perimeter minus the door. (There is no baseboard on the door.)
$P = (2l + 2w) - \text{door}$
$= (2 \times 15\frac{1}{4} + 2 \times 9\frac{1}{2}) - 3$
$= (\frac{\cancel{2}^{1}}{1} \times \frac{61}{\cancel{4}_{2}} + \frac{\cancel{2}^{1}}{1} \times \frac{19}{\cancel{2}_{1}}) - 3$
$= (\frac{61}{2} + 19) - 3$
$= 30\frac{1}{2} + 19 - 3 = 46\frac{1}{2}$

3. **Jake needs a total of 96 feet of stripping material for this room.**
$15\frac{1}{4} \times 2 = 30\frac{1}{2}$ feet (two lengths of room)
$9\frac{1}{2} \times 2 = 19$ feet (two widths of room)
$30\frac{1}{2} + 19 = 49\frac{1}{2}$ feet (2 lengths + two widths of ceiling molding)
$49\frac{1}{2} - 3 = 46\frac{1}{2}$ feet (perimeter of ceiling minus the door area = baseboard)
$49\frac{1}{2} + 46\frac{1}{2} = 96$ feet (baseboard and ceiling molding lengths)

UNIT 2 REVIEW
PAGES 108–109

1. **$1\frac{7}{12}$** $\frac{5}{6} + \frac{3}{4} =$
$\frac{10}{12} + \frac{9}{12} = \frac{19}{12} = 1\frac{7}{12}$

2. **$\frac{7}{30}$** $\frac{9}{10} - \frac{2}{3} =$
$\frac{27}{30} - \frac{20}{30} = \frac{7}{30}$

3. **$7\frac{5}{24}$** $\begin{aligned}4\frac{7}{8} &= 4\frac{21}{24}\\ +2\frac{1}{3} &= 2\frac{8}{24}\\ \hline &6\frac{29}{24} = 6 + 1\frac{5}{24} = 7\frac{5}{24}\end{aligned}$

4. **$1\frac{13}{20}$** $\begin{aligned}3\frac{2}{5} &= 3\frac{8}{20} = \cancel{3}^{2}\frac{\cancel{8}^{28}}{20}\\ -1\frac{3}{4} &= 1\frac{15}{20} = 1\frac{15}{20}\\ \hline &1\frac{13}{20}\end{aligned}$

5. **$\frac{14}{15}$** $\frac{7}{12} \div \frac{5}{8} = \frac{7}{\cancel{12}} \times \frac{\cancel{8}^{2}}{5} = \frac{14}{15}$

6. **$\frac{5}{6}$** $\frac{\cancel{15}^{5}}{\cancel{16}_{2}} \times \frac{\cancel{8}^{1}}{\cancel{9}_{3}} = \frac{5}{6}$

7. **$8\frac{1}{3}$** $3\frac{1}{3} \times 2\frac{1}{2} = \frac{\cancel{10}^{5}}{3} \times \frac{5}{\cancel{2}_{1}} = \frac{25}{3} = 8\frac{1}{3}$

8. **$\frac{15}{16}$** $3\frac{3}{8} \div 3\frac{3}{5} = \frac{27}{8} \div \frac{18}{5} = \frac{\cancel{27}^{3}}{8} \times \frac{5}{\cancel{18}_{2}} = \frac{15}{16}$

9. **$\frac{3}{20}$**   3 out of 20

10. **40 packets** Divide the weight of a box by the weight of a packet.
$30 \div \frac{3}{4} = \frac{30}{1} \div \frac{3}{4} = \frac{\cancel{30}^{10}}{1} \times \frac{4}{\cancel{3}_{1}} = \frac{40}{1} = 40$

11. **$4\frac{5}{8}$ feet** Since there are 12 inches in one foot, and inches are smaller than feet, divide the number of inches by 12.
$55\frac{1}{2} \div 12 = \frac{111}{2} \div \frac{12}{1} = \frac{111}{2} \times \frac{1}{12}$
$= \frac{111}{24} = 4\frac{15}{24} = 4\frac{5}{8}$

12. **25 feet** Since there are 3 feet in one yard, and yards are larger than feet, multiply the number of yards by 3.
$8\frac{1}{3} \times 3 = \frac{25}{\cancel{3}} \times \frac{\cancel{3}^{1}}{1} = \frac{25}{1} = 25$

13. **$5\frac{3}{10}$ miles** Henry drives $1\frac{3}{10}$ miles from point C to point D, $1\frac{1}{5}$ miles from point D to point E, and $2\frac{4}{5}$ miles from point E to the garage. To find the total distance, add each distance.
$\begin{aligned}1\frac{3}{10} &= 1\frac{3}{10}\\ 1\frac{1}{5} &= 1\frac{2}{10}\\ +2\frac{4}{5} &= 2\frac{8}{10}\\ \hline &4\frac{13}{10} = 4 + 1\frac{3}{10} = 5\frac{3}{10}\end{aligned}$

14. **$7\frac{1}{2}$ hours** Find the number of hours Geri worked in the morning. Since 51 is close to 45, think of 8:51 as $8\frac{3}{4}$. Since 24 is close to 30, think of 12:24 as $12\frac{1}{2}$. Then subtract.

$$12\frac{1}{2} = 12\frac{2}{4} = \overset{11}{\cancel{12}}\overset{6}{\frac{2}{4}}$$
$$-\ 8\frac{3}{4} =\ \ 8\frac{3}{4} =\ \ 8\frac{3}{4}$$
$$\rule{2cm}{0.4pt}$$
$$3\frac{3}{4}$$

Find the number of hours Geri worked in the afternoon. Since 8 is close to 15, think of 1:08 as $1\frac{1}{4}$. Since 3 is close to zero, think of 5:03 as 5. Then subtract.

$$5\ \ =\ \overset{4}{\cancel{5}}\frac{4}{4}$$
$$-1\frac{1}{4} = 1\frac{1}{4}$$
$$\rule{2cm}{0.4pt}$$
$$3\frac{3}{4}$$

Add the number of hours Geri worked in the morning and in the afternoon.

$$3\frac{3}{4}$$
$$+3\frac{3}{4}$$
$$\rule{2cm}{0.4pt}$$
$$6\frac{6}{4} = 6 + 1\frac{2}{4} = 7\frac{2}{4} = 7\frac{1}{2}$$

15. **(3) $4\frac{1}{8}$ yards** Use the column for size 12. Add the number of yards needed for blouse A ($2\frac{7}{8}$) and the number of yards needed for the skirt ($1\frac{1}{4}$).

$$2\frac{7}{8} = 2\frac{7}{8}$$
$$+1\frac{1}{4} = 1\frac{2}{8}$$
$$\rule{2cm}{0.4pt}$$
$$3\frac{9}{8} = 3 + 1\frac{1}{8} = 4\frac{1}{8}$$

16. **(4) $3\frac{1}{4} - 1\frac{5}{8}$** Use the column for size 14 to find the number of yards needed for blouse B ($3\frac{1}{4}$) and the skirt ($1\frac{5}{8}$). Subtract to find how many more yards of fabric Pearl needs for the blouse than for the skirt.

17. **(2) $2\frac{1}{3} \times 2$** Since 16 is 2 times 8, Clara needs to double the recipe. Multiply the amount of milk ($2\frac{1}{3}$ cups) by 2.

MATH CONNECTION: FRACTIONS AND DRAMA

PAGE 111

1. **(2) $\frac{4}{10}$** $\frac{4 \text{ movies released before 1990}}{10 \text{ movies}} = \frac{4}{10} = \frac{2}{5}$

2. **(3) $\frac{1}{4}$** Round the earnings of the top 5 movies and add.
$600 + $500 + $400 + $400 + $300 = $2,200
Divide the earnings of *Titanic* by the earnings of the top 5.
$\frac{Titanic}{\text{earnings of the top 5}} = \frac{\$600}{\$2,200} = \frac{6}{22} = \frac{3}{11} \sim \frac{3}{12} = \frac{1}{4}$

3. **(2) $\frac{1}{3}$** Add the earnings of the *Star Wars* Trilogy. $500 + $300 + $300 = $1,100
Divide by the earnings of the top 10 movies.
$\frac{Star\ Wars\ \text{Trilogy}}{\text{earnings of the top 10}} = \frac{\$1,100}{\$3,700} = \frac{1}{3}$

4. **(2) *Home Alone/Titanic*** *Home Alone* time is 102, or about 100 minutes; *Titanic* time is 201, or about 200 minutes. $\frac{100}{200} = \frac{1}{2}$
The fractions for the other answers are:
(1) $\frac{200}{125}$ or about $\frac{200}{100}$ or about $\frac{2}{1}$.
(3) $\frac{88}{120}$ or about $\frac{100}{100}$ or about $\frac{1}{1}$.
(4) $\frac{145}{201}$ or about $\frac{150}{200}$ or about $\frac{1.5}{2}$.
(5) $\frac{135}{127}$ or about $\frac{100}{100}$ or about $\frac{1}{1}$.

5. *Answers may vary.* A sample response is: I saw *E.T.—The Extra-Terrestrial*. I loved this movie and so did my children. We all liked how the children in the movie worked together to save E.T. and return him to his spaceship.

CUMULATIVE REVIEW

PAGES 112–113

1. $32,774 > 3,274$
$32,774$ has more digits than $3,274$.

2. $948,\underline{5}26 > 948,\underline{2}56$ 5 is greater than 2.

3. $x = 34$
$$x + 19 = 53$$
$$-\ 19 = -19$$
$$x = 34$$

4. $x = 36$ $\frac{9}{1} \times \frac{x}{9} = 4 \times \frac{9}{1}$
$$x = 36$$

5. 454
$$\overset{14}{\overset{7\ \cancel{4}12}{\$\cancel{\$}\cancel{5}\cancel{2}}}$$
$$-\ 398$$
$$\rule{1.5cm}{0.4pt}$$
$$\$454$$

6. $360,558$
$$594$$
$$\times 607$$
$$\rule{1.5cm}{0.4pt}$$
$$4\ 158$$
$$356\ 40$$
$$\rule{1.5cm}{0.4pt}$$
$$360,558$$

7. 483
$$\begin{array}{r} \overset{8\ 9\ 10}{\cancel{9}\cancel{0}\cancel{0}} \\ -417 \\ \hline 483 \end{array}$$

8. 206 r15
$$\begin{array}{r} 206 \text{ r}15 \\ 38\overline{)7{,}843} \\ -7\,6 \\ \hline 243 \\ -228 \\ \hline 15 \end{array}$$

9. 976
$$\begin{array}{r} \overset{1\ 1}{684} \\ 35 \\ +257 \\ \hline 976 \end{array}$$

10. $423
$$\begin{array}{r} \$423 \\ 9\overline{)\$3807} \\ -\ 36 \\ \hline 20 \\ -18 \\ \hline 27 \\ -27 \\ \hline 0 \end{array}$$

11. $1\frac{7}{24}$ $\frac{2}{3} + \frac{5}{8} =$
$$\frac{16}{24} + \frac{15}{24} = \frac{31}{24} = 1\frac{7}{24}$$

12. $\frac{2}{3}$ $\frac{5}{9} \div \frac{5}{6} = \frac{\cancel{5}^1}{\cancel{9}_3} \times \frac{\cancel{6}^2}{\cancel{5}_1} = \frac{2}{3}$

13. $9\frac{2}{15}$ $3\frac{4}{5} = 3\frac{12}{15}$
$$+5\frac{1}{3} = 5\frac{5}{15}$$
$$8\frac{17}{15} = 8 + 1\frac{2}{15} = 9\frac{2}{15}$$

14. $3\frac{11}{12}$ $6\frac{2}{3} = 6\frac{8}{12} = \cancel{6}^{5}\cancel{\frac{8}{12}}^{20}$
$$-2\frac{3}{4} = 2\frac{9}{12} = 2\frac{9}{12}$$
$$\overline{3\frac{11}{12}}$$

15. $3\frac{3}{4}$ $6\frac{2}{3} \div 1\frac{7}{9} = \frac{20}{3} \div \frac{16}{9} = \frac{\cancel{20}^5}{\cancel{3}_1} \times \frac{\cancel{9}^3}{\cancel{16}_4} = \frac{15}{4} = 3\frac{3}{4}$

16. $4\frac{1}{2}$ $2\frac{5}{8} \times 1\frac{5}{7} = \frac{\cancel{21}^3}{\cancel{8}_2} \times \frac{\cancel{12}^3}{\cancel{7}_1} = \frac{9}{2} = 4\frac{1}{2}$

17. **18** Because $18^2 = 324$, the square root of 324 is 18.

18. **5,600,000** Look at the 6, the digit in the hundred thousands place. 5,⑥48,312 The number to the right of 6 is 4. Do not change the circled digit.

19. **1,500 parts** The tallest bar, Thursday, represents the highest production. This bar ends between the 3 and 4 marks and represents 3,500 parts. The smallest bar, Tuesday, represents the lowest production. This bar reaches the 2 mark and represents 2,000 parts. Subtract to find the difference.
$$\begin{array}{r} 3{,}500 \\ -2{,}000 \\ \hline 1{,}500 \end{array}$$

20. **60 feet** Two sides measure 12 feet and two sides measure 18 feet. Add all four sides together to find the perimeter.
$$\begin{array}{r} \overset{2}{12} \\ 12 \\ 18 \\ +18 \\ \hline 60 \end{array}$$

21. **3,240 cubic inches**
$$\begin{aligned} V &= l \times w \times h \\ &= 18 \times 12 \times 15 \\ &= 216 \times 15 \\ &= 3{,}240 \end{aligned}$$

22. **(4) \$36 + \$15** Add the amount Lupe paid for the overalls and the amount he paid to hold the jacket. You do not need the total cost of the jacket.

23. **(3) 200** Add to find a total. Round the three numbers to the nearest ten and add to estimate the total.
$$\begin{array}{l} 63 \text{ rounds to } 60 \\ 89 \text{ rounds to } 90 \\ +45 \text{ rounds to } \underline{50} \\ \phantom{+45 \text{ rounds to } 0}200 \end{array}$$

24. **(2) $54\frac{3}{4} \div 12$** Since there are 12 inches in one foot and inches are smaller than feet, divide the number of inches by 12.

25. **(3) 35** To find the mean, first find the total of the numbers of the set of data.
$$38 + 42 + 35 + 33 + 27 = 175$$
Then divide the total by the number of items in the set.
$$175 \div 5 = 35$$

UNIT 3: DECIMALS

SECTION 8

PAGE 117

1. **(3) 6 dollars** The 6 is in the dollars column.
2. **(3) 40 cents** The 4 equals 4 dimes, or 40 cents.
3. **(3) 20 dollars** The 2 is in the tens column, so you have 2 tens or 20 dollars.
4. **(3) 49 dollars and 55 cents** $49.55 means 49 dollars and 55 cents.
5. **(4) six ten thousandths** The 6 is in the ten thousandths column.
6. **(3) two thousandths** The 2 is in the thousandths column.
7. **(1) one tenth** The 1 is in the tenths column.
8. **10 has one zero, so the tenths place is one place to the right of the decimal point; 100 has two zeros, so the hundredths place is two places to the right of the decimal point; 1,000 has three zeros, so the thousandths place is three places to the right of the decimal point; and so on.**

PAGE 118

2. two hundred fifty-six thousandths
3. two and nine hundredths
4. six and eight hundred five thousandths
5. 24.356
6. 3.78
7. 0.491
8. 267.3
9. 14.1736
10. 5.7084

PAGE 119

2. $\frac{9}{20}$ $0.45 = \frac{45}{100} = \frac{45 \div 5}{100 \div 5} = \frac{9}{20}$
3. $\frac{3}{20}$ $0.15 = \frac{15}{100} = \frac{15 \div 5}{100 \div 5} = \frac{3}{20}$
4. $\frac{16}{25}$ $0.64 = \frac{64}{100} = \frac{64 \div 4}{100 \div 4} = \frac{16}{25}$
5. $\frac{1}{8}$ $0.125 = \frac{125}{1000} = \frac{125 \div 25}{1000 \div 25} = \frac{5}{40} = \frac{1}{8}$
6. $1\frac{4}{5}$ $1.8 = 1\frac{8}{10} = 1\frac{8 \div 2}{10 \div 2} = 1\frac{4}{5}$
7. $\frac{1}{10}$ $0.10 = \frac{10}{100} = \frac{10 \div 10}{100 \div 10} = \frac{1}{10}$
8. $\frac{1}{200}$ $0.005 = \frac{5}{1000} = \frac{5 \div 5}{1000 \div 5} = \frac{1}{200}$
9. $\frac{8}{25}$ $0.32 = \frac{32}{100} = \frac{32 \div 4}{100 \div 4} = \frac{8}{25}$
10. $\frac{3}{8}$ $0.375 = \frac{375}{1000} = \frac{375 \div 25}{1000 \div 25} = \frac{15 \div 5}{40 \div 5} = \frac{3}{8}$
11. $2\frac{319}{1,000}$ $2.319 = 2\frac{319}{1,000}$
12. $\frac{1}{250}$ $0.004 = \frac{4}{1000} = \frac{4 \div 4}{1000 \div 4} = \frac{1}{250}$
13. $3\frac{7}{250}$ $3.028 = 3\frac{28}{1000} = 3\frac{28 \div 4}{1000 \div 4} = 3\frac{7}{250}$

14. $\frac{17}{20}$ $0.85 = \frac{85}{100} = \frac{85 \div 5}{100 \div 5} = \frac{17}{20}$
15. $\frac{5}{8}$ $0.625 = \frac{625}{1000} = \frac{625 \div 25}{1000 \div 25} = \frac{25 \div 5}{40 \div 5} = \frac{5}{8}$
16. $4\frac{1}{250}$ $4.0024 = 4\frac{24}{10,000} = 4\frac{24 \div 8}{10,000 \div 8} = 4\frac{3}{1,250}$

PAGE 121

2. **0.43 > 0.09**
   0.4̲3   0.0̲9   4 is greater than 0.
3. **0.73 > 0.542**
   Add a zero: 0.7̲30   0.5̲42
   Then compare: 7 is greater than 5.
4. **8.058 < 8.58**
   Add a zero: 8.0̲58   8.5̲80
   Then compare: 0 is less than 5.
5. **2.58 = 2.580**
   Add a zero: 2.580   2.580
   The numbers are the same.
6. **53.005 > 52.008**
   53.̲005   52.̲008   3 is greater than 2.
7. **0.863 < 0.9**
   Add two zeros: 0.8̲63   0.9̲00
   Then compare: 8 is less than 9.
8. **2.001 < 2.01**
   Add a zero: 2.0̲01   2.0̲10
   Then compare: 0 is less than 1.
9. **1.32 > 1.319**
   Add a zero: 1.32̲0   1.31̲9
   Then compare: 2 is greater than 1.
10. **5 = 5.00**
    Add two zeros: 5.00   5.00
    The numbers are the same.
11. **0.95 < 0.954**
    Add a zero: 0.95̲0   0.95̲4
    Then compare: 0 is less than 4.
12. **0.005 < 0.05**
    Add a zero: 0.0̲05   0.0̲50
    Then compare: 0 is less than 5.
13. **0.43 > 0.34**
    0.4̲3   0.3̲4   4 is greater than 3.
14. **0.54 > 0.054**
    Add a zero: 0.5̲40   0.0̲54
    Then compare: 5 is greater than 0.
15. **5.2 = 5.200**
    Add two zeros: 5.200   5.200
    The numbers are the same.
16. **6.38** 6.3̲782 Since 8 is greater than 5, add 1 to the underlined digit.
17. **8** 8̲.276 Since 2 is less than 5, do not change the underlined digit.
18. **46.4** 46.3̲518 The number to the right of 3 is 5; add 1 to the underlined digit.

19. **71.048**   71.04$\underline{8}$3 Since 3 is less than 5, do not change the underlined digit.

20. **56.28**   56.2$\underline{8}$19 Since 1 is less than 5, do not change the underlined digit.

21. **0.5**   0.$\underline{5}$462 Since 4 is less than 5, do not change the underlined digit.

22. **102**   10$\underline{1}$.667 Since 6 is greater than 5, add 1 to the underlined digit.

23. **25.024**   25.02$\underline{3}$5 The number to the right of 3 is 5, add 1 to the underlined digit.

24. **Adrian**
    0.21$\underline{5}$ > 0.21$\underline{3}$ 5 is greater than 3.

25. **96**   9$\underline{5}$.7 Since 7 is greater than 5, add 1 to the underlined digit.

26. **Brand B**
    $2.$\underline{7}$9 < $2.$\underline{8}$3   7 is less than 8.

27. **0.5, 0.57, 0.69, 0.75**
    0.5 < 0.57 < 0.69 < 0.75

PAGE 123

1. **(4) 5.8 + 5.8 + 2.7 + 2.7** To find the perimeter, add the lengths of all four sides.

2. **(2) 2.7 − 0.7** Find the difference between the length of the side and the opening.

3. **(5) Not enough information is given.** To find the perimeter, you need the length and width of the swimming pool.

4. **(3) 60.2 + 3.1 + 3.1** Add the lengths of the three sides that do not touch the side of the building.

5. **(3) 60.2 − 15.8** Find the difference between the length of the side and the length of the yellow strip.

6. Answers will vary.

PAGES 124–125

1. **seven hundredths**
2. **two ten thousandths**
3. **six thousandths**
4. **five tenths**
5. **eighty-two hundredths**
6. **three and five hundred thirteen thousandths**
7. **eighteen and three thousand four hundred six ten thousandths**
8. **seven hundred ninety and four thousandths**
9. **2.43**
10. **0.2343**
11. **37.621**
12. **429.09**

13. $\frac{1}{20}$ $0.05 = \frac{5}{100} = \frac{1}{20}$

14. $\frac{7}{25}$ $0.28 = \frac{28}{100} = \frac{7}{25}$

15. $\frac{3}{4}$ $0.75 = \frac{75}{100} = \frac{3}{4}$

16. $\frac{1}{40}$ $0.025 = \frac{25}{1000} = \frac{1}{40}$

17. $\frac{24}{25}$ $0.96 = \frac{96}{100} = \frac{24}{25}$

18. $3\frac{41}{200}$ $3.205 = 3\frac{205}{1000} = 3\frac{41}{200}$

19. $5\frac{1}{1,250}$ $5.0008 = 5\frac{8}{10,000} = 5\frac{1}{1,250}$

20. $9\frac{7}{20}$ $9.35 = 9\frac{35}{100} = 9\frac{7}{20}$

21. **0.759 < 0.795**
    0.7$\underline{5}$9   0.7$\underline{9}$5
    5 is less than 9.

22. **0.326 < 0.54**
    Add a zero: 0.$\underline{3}$26   0.$\underline{5}$40
    Then compare: 3 is less than 5.

23. **0.6 = 0.60**
    Add a zero: 0.60 0.60
    The numbers are the same.

24. **32.574 > 32.547**
    32.5$\underline{7}$4   32.5$\underline{4}$7
    7 is greater than 4.

25. **3.20**   3.1$\underline{9}$6 Since 6 is greater than 5, add 1 to the underlined digit.

26. **6.5**   6.$\underline{4}$53 The number to the right of 4 is 5; add 1 to the underlined digit.

27. **1** $\underline{0}$.724 Since 7 is greater than 5, add 1 to the underlined digit.

28. **42.936**   42.93$\underline{5}$8 Since 8 is greater than 5, add 1 to the underlined digit.

29. **(2) 0.016**

30. **(4) 5.2 + 5.2 + 3.4 + 3.4** To find the perimeter, add the lengths of all four sides.

31. **(1) 4.2 − 2.7** Subtract the number of meters of fabric Calvin used from the number of meters of fabric he had.

32. **(4) Ron finished before Al.**
    Ron = 10.$\underline{0}$6 < 10.$\underline{2}$5 = Al
    0 is less than 2.

33. **(2) three and fifty-nine thousandths**

34. **(4) 24.6350 > 24.6305**
    24.63$\underline{5}$0 > 24.63$\underline{0}$5
    5 is greater than 0.

35. **(4) $1.33**   1.3$\underline{2}$5 The number to the right of 2 is 5; add 1 to the underlined digit.

36. **(2)** $\frac{2}{5}$ $0.4 = \frac{4}{10} = \frac{2}{5}$

SECTION 9

PAGE 127

1. **(1) $39.99 + 58.00** To find the subtotal, add the cost of the two jackets.

2. **(3) subtotal + $6.86** To find the total including tax, add the tax to the subtotal.

3. **(1) total − $45.00** To find how much Mrs. Wilson still owes on the coats, subtract the layaway cost from the total.

4. **(2) $30.00 + $24.00 + $12.00** To find the subtotal, add the cost of the three items.

5. **(5) (3 × $18) + (2 × $5.99)** To find the total of the sale before tax, find the cost of the three shirts, the cost of the two containers, and add their totals.

6. **(5) Not enough information is given.** To find the total including tax, you need to add the sales tax to the subtotal. The sales tax was not given.

7. **Step 1: Find the subtotal by adding the costs of all three items. Step 2: Find the total by adding the sales tax to the subtotal. Step 3: Find the change by subtracting the total from the amount the customer gives the sales clerk.**

PAGE 129

2. 8.953
$$\begin{array}{r} {}^{0\,16\,13}\\ 1\cancel{7}.\cancel{3}68\\ -\ 8.415\\ \hline 8.953 \end{array}$$

3. 10.58
$$\begin{array}{r} {}^{3\ 12}\\ 2\cancel{4}.\cancel{2}8\\ -13.70\\ \hline 10.58 \end{array}$$

4. 65.757
$$\begin{array}{r} {}^{1\quad 1}\\ 28.467\\ +37.290\\ \hline 65.757 \end{array}$$

5. 5.25
$$\begin{array}{r} {}^{0\,11\,10}\\ 1\cancel{2}.\cancel{0}5\\ -\ 6.80\\ \hline 5.25 \end{array}$$

6. 20.24
$$\begin{array}{r} {}^{1\ 1}\\ 3.70\\ 14.24\\ +\ 2.30\\ \hline 20.24 \end{array}$$

7. 88.784
$$\begin{array}{r} {}^{1\ 1}\\ 24.810\\ 35.700\\ +28.274\\ \hline 88.784 \end{array}$$

8. 8.09
$$\begin{array}{r} {}^{3\,10}\\ 8.\cancel{4}\cancel{0}\\ -0.31\\ \hline 8.09 \end{array}$$

9. 8.047
$$\begin{array}{r} {}^{0\,14\ 7\ 9\,10}\\ 1\cancel{4}.\cancel{8}\cancel{0}\cancel{0}\\ -\ 6.753\\ \hline 8.047 \end{array}$$

10. 17.88
$$\begin{array}{r} 12.00\\ +\ 5.88\\ \hline 17.88 \end{array}$$

11. 6.18
$$\begin{array}{r} {}^{1\quad 1}\\ 2.43\\ 0.57\\ +3.18\\ \hline 6.18 \end{array}$$

12. 17.484
$$\begin{array}{r} {}^{5\,16\ 12\,10\,10}\\ \cancel{6}\cancel{7}.\cancel{3}\cancel{1}\cancel{0}\\ -49.826\\ \hline 17.484 \end{array}$$

13. 27.93
$$\begin{array}{r} {}^{4\,16\ 14}\\ \cancel{5}\cancel{7}.\cancel{4}3\\ -29.50\\ \hline 27.93 \end{array}$$

14. 25.208
$$\begin{array}{r} {}^{1}\\ 0.300\\ 21.508\\ +\ 3.400\\ \hline 25.208 \end{array}$$

15. 17.75
$$\begin{array}{r} {}^{1}\\ 4.26\\ 5.10\\ +8.39\\ \hline 17.75 \end{array}$$

16. 3.703
$$\begin{array}{r} {}^{0\,11\ 15\,9\,10}\\ 1\cancel{2}.\cancel{6}\cancel{0}\cancel{0}\\ -\ 8.897\\ \hline 3.703 \end{array}$$

17. $364.20

$$
\begin{array}{r}
\$256.81 \\
-\quad 2.80 \\
\hline 254.01 \\
-\quad 34.33 \\
\hline 219.68 \\
-\quad 62.18 \\
\hline 157.50 \\
+122.77 \\
\hline 280.27 \\
+\quad 83.93 \\
\hline \$364.20
\end{array}
$$

18. $116.00

$$
\begin{array}{r}
{\scriptstyle 1\;1} \\
\$43.50 \\
+\;72.50 \\
\hline \$116.00
\end{array}
$$

Add the two amounts Roberto was paid for overtime that weekend.

OR

3+5=8

$$
\begin{array}{r}
\$14.50 \\
\times \qquad 8 \\
\hline \$116.00
\end{array}
$$

Multiply the number of overtime hours (8) times the hourly wage ($14.50) for total overtime pay.

PAGE 131

1. **(4) $14.50 × (38 + 43)** Work the parentheses first to find the total number of hours worked. Then multiply the total hours by the hourly wage.
2. **(2) $60 − (2 × $7.50) − (2 × $3.25)** First, find the total for the tickets and the drinks. Then subtract the two totals from $60.
3. **(5) (205 + 165 + 104 + 190) ÷ 4 + 40** Find the mean of the bowlers: (205 + 165 + 104 + 190) ÷ 4. Then add 40 to the mean.
4. **(5) ($6.50 + $9.25) × ($9.25 − $6.50)** First, add the two amounts. Then find the difference of the two amounts. Last, multiply the two totals.
5. **(1) (3 × $11.00) + (5.5 × $8.40)** To find the cost of the sale before tax, work the parentheses first to find the cost of 3 yards of fabric and the cost of 5.5 yards of fabric. Then sum the two totals.
6. **(5) (24 × $9.95) + (18 × $11.75)** To find how much Glen makes a week, work the parentheses first to find how much he earns for 24 hours and how much he earns for 18 hours. Then sum the two totals.

7. 13 $(6 \times 2) + 4 - 3 = 12 + 4 - 3 = 13$
   33 $6 \times (2 + 4) - 3 = 6 \times 6 - 3 = 36 - 3 = 33$
   18 $6 \times (2 + 4 - 3) = 6 \times 3 = 18$

PAGE 133

1. **(2) 2.9** The point for week 4 is at the ninth small line above 2 on the scale.
2. **(3) from week 3 to week 4** The steepest part of the graph is between the points for week 3 and week 4.
3. **(2) 2.5** Subtract the number of miles Ahmad walked each day during week 1 (1.3) from the number of miles he walked each day during week 6 (3.8).

$$
\begin{array}{r}
3.8 \\
-1.3 \\
\hline 2.5
\end{array}
$$

4. **(3) $2.6 million** The point for 1995 is at the sixth small line above 2 on the scale.
5. **(3) 17.3 million** To find the total sales from 1994 to 1999, sum the sales for each year. (1.1 + 2.6 + 3.4 + 3.7 + 3.1 + 3.4).
6. **The lines show the increases and decreases in the amount of sales from 1994 to 1999. The steeper the line, the greater the increase or decrease in sales. The steepest part of the graph is between points for 1997 and 1998. This means that the greatest increase in sales was from 1997 to 1998. When the line is flat this means that there was the least increase or decrease in sales.**

PAGES 134–135

1. 77.04

$$
\begin{array}{r}
{\scriptstyle 1} \\
35.20 \\
+41.84 \\
\hline 77.04
\end{array}
$$

2. 22.88

$$
\begin{array}{r}
{\scriptstyle 3\;\;1210} \\
7\cancel{4}.\cancel{3}\cancel{0} \\
-51.42 \\
\hline 22.88
\end{array}
$$

3. 78.924

$$
\begin{array}{r}
{\scriptstyle 8\,12\;12\,5\,10} \\
\cancel{9}\cancel{3}.\cancel{2}\cancel{6}\cancel{0} \\
-14.336 \\
\hline 78.924
\end{array}
$$

4. 20.825

$$
\begin{array}{r}
{\scriptstyle 1\,1\;\;1} \\
15.930 \\
+\;4.895 \\
\hline 20.825
\end{array}
$$

5. 23.766

$$\overset{1}{}8.450$$
$$+15.316$$
$$\overline{23.766}$$

6. 8.207

$$\overset{1\ \ 1}{}4.260$$
$$+3.947$$
$$\overline{8.207}$$

7. 6.899

$$\overset{6\ \ 1218}{7.389}$$
$$-0.490$$
$$\overline{6.899}$$

8. 1.12

$$\overset{8\ 10}{5.90}$$
$$-4.78$$
$$\overline{1.12}$$

9. 21.673

$$\overset{1\ 1}{}3.810$$
$$12.463$$
$$+\ 5.400$$
$$\overline{21.673}$$

10. 178.307

$$\overset{1\ \ 1}{42.180}$$
$$53.700$$
$$+82.427$$
$$\overline{178.307}$$

11. 2.899

$$\overset{0\ 11\ 121612}{12.372}$$
$$-\ 9.473$$
$$\overline{2.899}$$

12. 1.477

$$\overset{6\ \ 121610}{7.370}$$
$$-5.893$$
$$\overline{1.477}$$

13. 20.78

$$\overset{1}{}8.25$$
$$9.47$$
$$+3.06$$
$$\overline{20.78}$$

14. 22.046

$$\overset{1\ 2}{15.430}$$
$$2.800$$
$$+\ 3.816$$
$$\overline{22.046}$$

15. 9.015

$$\overset{2\ \ 1}{}4.245$$
$$1.970$$
$$+2.800$$
$$\overline{9.015}$$

16. 1.876

$$\overset{2\ 17910}{3.800}$$
$$-1.924$$
$$\overline{1.876}$$

17. 148.7734

18. 1,336.2735

19. **(4) $100 − ($30 + $24 + $15.00 + $5.18)**
First, sum the expenses. Then subtract this amount from $100.

20. **(3) 10** $4 + (2 \times 6) − (12 \div 6)$
$4 + 12 \quad − \quad 6$
$16 \quad − \quad 6 \quad = 10$

21. **$30** Round the cost of the tape player and the cost of the tape and add.

$$\$24.99 \text{ rounds to } \overset{1}{}\$25$$
$$+\ \ 4.99 \text{ rounds to } +\ \ 5$$
$$\overline{\$30}$$

22. **$29.98** Add the cost of the tape player and the cost of the tape.

$$\overset{1\ \ 1}{\$24.99}$$
$$+\ \ \ 4.99$$
$$\overline{\$29.98}$$

23. **$32.08** Add the tax to the subtotal (see the explanation for Item 22.)

$$\overset{1\ 1}{\$29.98}$$
$$+\ \ 2.10$$
$$\overline{\$32.08}$$

24. **$7.92** Subtract the total (see the explanation for Item 23) from the amount the customer gave George.

$$\overset{3\ 9\ \ 910}{\$40.00}$$
$$-\ \ 32.08$$
$$\overline{\$7.92}$$

25. **(2) 2,300** The point for 1995 is at the third small line above 2 on the scale.

26. **(5) an increase from 1997 to 1998** The steepest part of the graph is between the points for 1997 and 1998. This part of the graph slants upward, indicating an increase.

27. **(1) 1.3 thousand** Subtract the number of students enrolled in 1994 (1.7 thousand) from the number enrolled in 1999 (3.0 thousand).

$$\overset{2\ 10}{3,000}$$
$$-1,700$$
$$\overline{1,300}$$

SECTION 10

PAGE 137

1. **(4)** $0.99 × 6 Find the total cost by multiplying the unit cost ($0.99) by the number of calendar refills (6).

2. **(3)** $19.95 ÷ 5 Find the unit cost by dividing the total cost ($19.95) by the number of tapes (5).

3. **(1)** $47.94 ÷ 6 Find the unit cost by dividing the total cost ($47.94) by the number of reams of paper (6).

4. **(5)** $0.19 × 25 Find the total cost by multiplying the unit cost ($0.19) by the number of pencils (25).

5. **(5) Not enough information is given.** Find the total cost by multiplying the unit cost of the paper towels (unknown) by the number of rolls of paper towels (25).

6. **(4)** $1.69 ÷ 4 Find the unit cost by dividing the total cost ($1.69) by the number of bulbs (4).

7. **Recall division is the inverse operation of multiplication. You know total cost = number of units × unit cost. To find unit cost, divide both sides by the number of units: total cost ÷ number of units = unit cost.**

PAGE 138

2. 72.8

$$\begin{array}{r} 9.1 \\ \times\ 8 \\ \hline 72.8 \end{array}$$

3. 6.15

$$\begin{array}{r} 12.3 \\ \times\ 0.5 \\ \hline 6.15 \end{array}$$

4. 0.0728

$$\begin{array}{r} 1.04 \\ \times 0.07 \\ \hline 0.0728 \end{array}$$

5. 0.375

$$\begin{array}{r} 0.75 \\ \times\ 0.5 \\ \hline 0.375 \end{array}$$

6. 0.816

$$\begin{array}{r} 136 \\ \times 0.006 \\ \hline 0.816 \end{array}$$

7. 25.6

$$\begin{array}{r} 128 \\ \times\ 0.2 \\ \hline 25.6 \end{array}$$

8. 27.68

$$\begin{array}{r} 17.3 \\ \times\ 1.6 \\ \hline 10\ 38 \\ +\ 1\ 73 \\ \hline 27.68 \end{array}$$

9. 0.0126

$$\begin{array}{r} 0.42 \\ \times 0.03 \\ \hline 0.0126 \end{array}$$

10. 1.845

$$\begin{array}{r} 2.05 \\ \times\ 0.9 \\ \hline 1.845 \end{array}$$

11. 18.972

$$\begin{array}{r} 5.27 \\ \times\ 3.6 \\ \hline 3\ 162 \\ +15\ 81 \\ \hline 18.972 \end{array}$$

12. 0.496

$$\begin{array}{r} 6.2 \\ \times 0.08 \\ \hline 0.496 \end{array}$$

13. 152.928

$$\begin{array}{r} 28.32 \\ \times\ 5.4 \\ \hline 11\ 328 \\ +141\ 60 \\ \hline 152.928 \end{array}$$

14. 1.0395

$$\begin{array}{r} 16.5 \\ \times 0.063 \\ \hline 495 \\ +990 \\ \hline 1.0395 \end{array}$$

15. 31.80

$$\begin{array}{r} 21.2 \\ \times\ 1.5 \\ \hline 10\ 6\ 0 \\ +21\ 2 \\ \hline 31.8\ 0 \end{array}$$

16. 0.2075

$$\begin{array}{r} 8.3 \\ \times 0.025 \\ \hline 415 \\ +166 \\ \hline .2075 \end{array}$$

PAGE 139

2. 1.5 × $0.79

3.

Weight lb	Unit Price $	Total Price $
02.25 lb	$ 02.69	$ 6.05

4.

Weight lb	Unit Price $	Total Price $
00.75 lb	$ 00.69	$ 0.52

PAGE 141

2. 0.63

```
    0.63
9)5.67
  −5 4
    27
   −27
     0
```

3. 5.9

```
    5.9
7)41.3
 −35
   6 3
  −6 3
     0
```

4. 2.8

```
     2.8
12)33.6
  −24
    9 6
   −9 6
      0
```

5. 8.5

```
      8.5
13)110.5
  −104
     6 5
    −6 5
       0
```

6. 0.306

```
    0.306
4)1.224
 −1 2
     24
    −24
      0
```

7. 0.042

```
     0.042
21)0.882
  −  84
      42
     −42
       0
```

8. 0.002

```
        0.002
0.5.)0.0.010
     − 10
        0
```

9. 0.31

```
        0.31
0.08.)0.02.48
     −  24
          8
         −8
          0
```

10. 8.53

```
        8.53
0.12.)1.02.36
     −  96
         6 3
        −6 0
          36
         −36
           0
```

11. .0175

```
     0.0175
6)0.1050
 − 6
    45
   −42
    30
   −30
     0
```

12. 42.5

```
          42.5
0.012.)0.510.0
      −48
        30
       −24
         6 0
        −6 0
           0
```

13. 3.5

```
         3.5
3.6.)12.6.0
    −10 8
      1 80
     −1 80
         0
```

14. 0.06

```
          0.06
0.25.)0.01.50
     −  1 50
           0
```

15. 16,000

$$0.\underset{\smile}{004}\,)\overline{64.\underset{\smile}{000.}}\;\;\;\overset{16{,}000.}{}$$

$$
\begin{array}{r}
-4 \\
\hline
24 \\
-24 \\
\hline
0
\end{array}
$$

16. 5.151

$$5.\underset{\smile}{1}\,)\overline{26.\underset{\smile}{2}.701}\;\;\;\overset{5.151}{}$$

$$
\begin{array}{r}
-25\;5 \\
\hline
77 \\
-51 \\
\hline
260 \\
-255 \\
\hline
51 \\
-51 \\
\hline
0
\end{array}
$$

17. $17.30
18. $16.50

PAGE 143

1. **(5) 1,750 cm** Multiply by 100 to convert meters to centimeters.
    $17.5 \times 100 = 17.\underset{\smile}{50}. = 1{,}750$

2. **(4) 5,500 mg** Multiply by 1,000 to convert grams to milligrams.
    $5.5 \times 1{,}000 = 5.\underset{\smile}{500}. = 5{,}500$

3. **(2) 160 cm** First multiply 0.4 by 4 to find the perimeter of the square.
    $0.4 \times 4 = 1.6$ m
    Then multiply by 100 to convert meters to centimeters.
    $1.6 \times 100 = 1.\underset{\smile}{60}. = 160$ cm

4. **(3) 0.1 L** Divide by 1,000 to convert milliliters to liters.
    $100 \div 1{,}000 = 0.\underset{\smile}{100}. = 0.1$ L

5. **(5) Not enough information is given.** To find how many meters long the race is, you need to know the number of kilometers, and then convert kilometers to meters.

6. **(1) 298,000 mg** Multiply by 1,000 to convert grams to milligrams.
    $298 \times 1{,}000 = 298.\underset{\smile}{000}. = 298{,}000$

7. **Answers will vary. To convert amounts from grams to kilograms, divide by 1,000. To convert amounts from liters to milliliters, multiply by 1,000.**

PAGE 145

1. **(4) $149.97** Find the total amount by multiplying the unit price ($49.99) by the quantity (3).

2. **(2) 4 × $135** Find the total amount by multiplying the unit price ($135) by the quantity (4).

3. **(1) 48** Sum the numbers in the quantity column ($12 + 8 + 15 + 3 + 6 + 4$) to find the total (48).

4. **(4) $117.80** Find the total amount of the washcloths ($4.49 \times 12 = \$53.88$) and the total amount of the hand towels ($7.99 \times 8 = \$63.92$). Sum the two totals ($53.88 + \$63.92 = \$117.80$).

5. **(5) $1,236.77** Find the TOTAL by finding the total amount of each unit item and finding their sum
    ($53.88 + \$63.92 + \$375 + \$149.97 + \$54 + \$540 = \$1{,}236.77$)

6. **(5) Not enough information is given.** To find the total amount of the blenders, you would need to multiply the unit price (unknown) by the quantity (4).

7. **(3) $189** To find the total amount for the mugs, find the amount for the first order ($9 \times 6 = \$54$) and the amount for the second order ($9 \times 15 = \$135$). Then sum the amounts ($54 + \$135 = \$189$).

8. **$696.77** Subtract the total amount of the tote bags ($4 \times \$135 = \$540$) from the TOTAL found in Item 5 ($1,236.77).

PAGES 146–147

1. 19.2

$$
\begin{array}{r}
3.2 \\
\times\;6 \\
\hline
19.2
\end{array}
$$

2. 0.0824

$$
\begin{array}{r}
2.06 \\
\times 0.04 \\
\hline
0.0824
\end{array}
$$

3. 1.251

$$
\begin{array}{r}
4.17 \\
\times\;0.3 \\
\hline
1.251
\end{array}
$$

4. 33.48

$$
\begin{array}{r}
12.4 \\
\times\;2.7 \\
\hline
8\;68 \\
+24\;8 \\
\hline
33.48
\end{array}
$$

5. **0.13536**

$$\begin{array}{r} 0.752 \\ \times\ 0.18 \\ \hline 6016 \\ +752\ \ \\ \hline 0.13536 \end{array}$$

6. **0.76**

$$\begin{array}{r} 0.76 \\ 6\overline{)4.56} \\ -4\,2\ \ \\ \hline 36 \\ -36 \\ \hline 0 \end{array}$$

7. **0.825**

$$\begin{array}{r} 0.825 \\ 4\overline{)3.300} \\ -3\,2\ \ \ \\ \hline 10 \\ -\ 8 \\ \hline 20 \\ -20 \\ \hline 0 \end{array}$$

8. **3.87**

$$\begin{array}{r} 3.87 \\ 0.5.\overline{)1\,9.35} \\ -1.5\ \ \ \\ \hline 4\,3 \\ -4\,0 \\ \hline 35 \\ -35 \\ \hline 0 \end{array}$$

9. **67.75**

$$\begin{array}{r} 67.75 \\ 0.08.\overline{)5.42.00} \\ -4.8\ \ \ \ \\ \hline 62 \\ -56 \\ \hline 6\,0 \\ -5\,6 \\ \hline 40 \\ -40 \\ \hline 0 \end{array}$$

10. **9.3**

$$\begin{array}{r} 9.3 \\ 1.5.\overline{)13.9.5} \\ -13.5\ \ \\ \hline 4\,5 \\ -4\,5 \\ \hline 0 \end{array}$$

11. **0.252**
12. **130.19**
13. **12,500 milligrams** Multiply by 1,000 to convert grams to milligrams.
$12.5 \times 1,000 = 12.500. = 12,500$

14. **14.9 miles** Divide the total distance by the number of days.
$74.5 \div 5 = 14.9$

15.

16. **$0.40** Divide the total cost ($4.82) by the number of servings (12).
$\$4.82 \div 12 = \$0.402$, which rounds to $0.40

17. **(2) $197.00 ÷ 4** Find the unit cost by dividing the total cost ($197.00) by the number of tires (4).

18. **(3) $2.40** Multiply the cost of each pound by the number of pounds.
$\$0.60 \times 4 = \$2.40$

19. **(3) $644.93** Find the total amount for each of the two items. Then find their sum.
$(4 \times \$149) + (7 \times \$6.99) =$
$\$596 + \$48.93 = \$644.93$

20. **(4) 24.3 miles per gallon** Divide the total miles by the number of gallons of gas.
$291.6 \div 12 = 24.3$

21. **(4) $1.29 × 31** Find the total cost by multiplying the unit cost ($1.29) by the number of notebooks (31).

22. **(1) 5 ÷ 1.2** Divide the total length (5 meters) by the length of each piece (1.2 meters).

23. **(5) Not enough information is given.** You need to know the number of doses in each bottle to find the total amount.

24. **(2) 3.98 meters** Divide by 100 to convert centimeters to meters.
$398 \div 100 = 3.98$

MATH AT WORK
PAGE 149
1. **(5) $1,512.24** $(3 \times \$345.23) + \$476.55$
2. **(3) Subtotal**
3. **(4) $1,386.74** Subtotal − cash received
$\$1,512.24 - \$125.50$

UNIT 3 REVIEW

PAGES 150–151

1. **six ten thousandths**
2. **eight hundredths**
3. **0.25 > 0.025** Add a zero: 0.2<u>5</u>0 0.0<u>2</u>5
   Then compare: 2 is greater than 0.
4. **0.97 = 0.970** Add a zero: 0.970 0.970
   Then compare: 0.970 and 0.970 are the same.
5. **5.36** 5.3⑥2 Since 2 is less than 5, do not change the circled digit.
6. **7.4** 7.③<u>5</u>1 The number to the right of 3 is 5; add 1 to the circled digit.
7. $\frac{21}{50}$ $\quad \frac{42 \div 2}{100 \div 2} = \frac{21}{50}$

8. **0.76**
$$\begin{array}{r} 0.76 \\ 25\overline{)19.0} \\ -17\ 5 \\ \hline 1\ 50 \\ -1\ 50 \\ \hline 0 \end{array}$$

9. **11.785**
$$\begin{array}{r} {}^{1} \\ 3.925 \\ 4.600 \\ +3.260 \\ \hline 11.785 \end{array}$$

10. **7.86**
$$\begin{array}{r} {}^{11\,4\,13} \\ 2\cancel{5}.\cancel{3}\cancel{6} \\ -17.50 \\ \hline 7.86 \end{array}$$

11. **5.143**
$$\begin{array}{r} {}^{3\,9\,10} \\ 8.4\cancel{0}\cancel{0} \\ -3.257 \\ \hline 5.143 \end{array}$$

12. **9.4**
$$\begin{array}{r} 9.4 \\ 4\overline{)37.6} \\ -36 \\ \hline 1\ 6 \\ -1\ 6 \\ \hline 0 \end{array}$$

13. **50.4**
$$\begin{array}{r} 6.3 \\ \times\ \ 8 \\ \hline 50.4 \end{array}$$

14. **2.286**
$$\begin{array}{r} 2.54 \\ \times\ 0.9 \\ \hline 2.286 \end{array}$$

15. **45**
$$\begin{array}{r} 45. \\ 0.06.\overline{)2.70.} \\ -2\ 4 \\ \hline 30 \\ -30 \\ \hline 0 \end{array}$$

16. **468**
$$\begin{array}{r} 468 \\ 0.016.\overline{)7.488.} \\ -6\ 4 \\ \hline 1\ 08 \\ -\ 96 \\ \hline 128 \\ -128 \\ \hline 0 \end{array}$$

17. **15.56**
$$\begin{array}{r} {}^{1\ \ 1} \\ 3.54 \\ 6.83 \\ +5.19 \\ \hline 15.56 \end{array}$$

18. **0.51**
$$\begin{array}{r} 0.51 \\ 0.07.\overline{)0.03.57} \\ -3\ 5 \\ \hline 07 \\ -\ 7 \\ \hline 0 \end{array}$$

19. **42.48**
$$\begin{array}{r} {}^{4\ 1310} \\ 4\cancel{5}.\cancel{4}0 \\ -\ 2.92 \\ \hline 42.48 \end{array}$$

20. **0.0285**
$$\begin{array}{r} 0.57 \\ \times 0.05 \\ \hline 0.0285 \end{array}$$

21. **$5.51** Add the cost of the shoes and the socks to find the subtotal. Then add the tax to find the total. Finally, subtract the total from the amount Bryant gave the clerk.

$$\begin{array}{r} {}^{1\,1\ \ 1} \\ \$64.95 \\ +\ \ 5.99 \\ \hline \$70.94 \end{array} \qquad \begin{array}{r} {}^{1} \\ \$70.94 \\ +\ \ 3.55 \\ \hline \$74.49 \end{array} \qquad \begin{array}{r} {}^{7\,9\ 9\,10} \\ \$\cancel{8}\cancel{0}.\cancel{0}\cancel{0} \\ -\ 74.49 \\ \hline \$5.51 \end{array}$$

22. **11** $9 \div (4 - 1) + 8$
   $9 \div 3 + 8 = 3 + 8 = 11$

23. **$0.40** Find the unit cost, then multiply by 5.
   $\$0.96 \div 12 = \$0.08$
   $\$0.08 \times 5 = \$0.40$

24. **$1.08** $0.40 \times \$2.69 = \$1.076$, which rounds to $1.08

25. **$151.96** Sum the total amounts of the two items. $(4 \times \$12.99) + (5 \times \$20) = \$51.96 + \$100 = \$151.96$

26. **(4) $42.50 × 4** Find the total cost by multiplying the unit cost ($42.50) by the number of tires (4).

27. **(5) 6.1 thousand** Add the number of parts produced in November (3.2 thousand) and the number produced in December (2.9 thousand).

$$\begin{array}{r} \overset{1}{}3.2 \text{ thousand} \\ +2.9 \text{ thousand} \\ \hline 6.1 \text{ thousand} \end{array}$$

MATH CONNECTION: DECIMALS AND WORLD HISTORY

PAGES 152–153

1. **(4) 1916, 1940, and 1944**
2. **(4) 1 and 8 hundredths seconds**

$$\begin{array}{r} \overset{3\,10\ \ 2\,13}{4\cancel{0}.\cancel{3}\cancel{3}} \\ -39.25 \\ \hline 1.08 \end{array}$$

3. and 4.

Skaters	Marks								
	Judge 1	Judge 2	Judge 3	Judge 4	Judge 5	Judge 6	Judge 7	Judge 8	Judge 9
Chen Lu	5.5	5.5	5.6	5.5	5.5	5.5	5.7	5.5	5.5
	5.8	5.8	5.8	5.8	5.8	5.8	5.8	5.7	5.8
Total	11.3	11.3	11.4	11.3	11.3	11.3	11.5	11.2	11.3
Michelle Kwan	5.7	5.7	5.8	5.7	5.8	5.8	5.7	5.7	5.8
	5.9	5.9	5.9	5.9	5.9	5.9	5.9	5.9	5.9
Total	11.6	11.6	11.7	(11.6)	(11.7)	11.7	11.6	(11.6)	11.7
Tara Lipinski	5.9	5.9	5.9	5.8	5.8	5.9	5.9	5.8	5.9
	5.8	5.8	5.9	5.8	5.8	5.9	5.9	5.8	5.9
Total	(11.7)	(11.7)	(11.8)	(11.6)	11.6	(11.8)	(11.8)	(11.6)	(11.8)

5. a. **Tara Lipinski**
   b. **Michelle Kwan**
   c. **Chen Lu**

CUMULATIVE REVIEW

PAGES 154–155

1. **two ten thousands**
2. **seven millions**
3. **three tenths**
4. **four thousandths**

5. **348**
$$\begin{array}{r} \overset{6\ \ 9\,10}{7\cancel{0}\cancel{0}} \\ -352 \\ \hline 348 \end{array}$$

6. **200,438**
$$\begin{array}{r} 973 \\ \times 206 \\ \hline 5\,838 \\ +194\,60 \\ \hline 200{,}438 \end{array}$$

7. **208 r24**
$$\begin{array}{r} 208\ \text{r}24 \\ 42\overline{)8{,}760} \\ \underline{-8\,4} \\ 360 \\ \underline{-336} \\ 24 \end{array}$$

8. **15**
$$\begin{array}{r} 15 \\ 2.4.\overline{)36.0.} \\ \underline{-24} \\ 12\,0 \\ \underline{-12\,0} \\ 0 \end{array}$$

9. $6\frac{3}{8}$
$$\begin{array}{r} 2\frac{3}{4} = 2\frac{6}{8} \\ +3\frac{5}{8} = 3\frac{5}{8} \\ \hline 5\frac{11}{8} = 5 + 1\frac{3}{8} = 6\frac{3}{8} \end{array}$$

10. $4\frac{11}{15}$
$$\begin{array}{r} 8\frac{2}{5} = 8\frac{6}{15} = \overset{7\ \ 21}{8\frac{\cancel{6}}{15}} \\ -3\frac{2}{3} = 3\frac{10}{15} = 3\frac{10}{15} \\ \hline 4\frac{11}{15} \end{array}$$

11. $8\frac{1}{4}$
$$\begin{array}{r} 5\frac{3}{4} = 5\frac{3}{4} \\ +2\frac{1}{2} = 2\frac{2}{4} \\ \hline 7\frac{5}{4} = 7 + 1\frac{1}{4} \\ = 8\frac{1}{4} \end{array}$$

12. $5\frac{5}{12}$
$$\begin{array}{r} 6\frac{2}{3} = 6\frac{8}{12} \\ -1\frac{1}{4} = 1\frac{3}{12} \\ \hline 5\frac{5}{12} \end{array}$$

13. 453

$$\begin{array}{r} {}^{1\,2} \\ 28 \\ 376 \\ +\ 49 \\ \hline 453 \end{array}$$

14. $5.46

$$\begin{array}{r} {}^{8\ 9\,15} \\ \$9.0\cancel{5} \\ -\ 3.59 \\ \hline \$5.46 \end{array}$$

15. $2\frac{5}{12}$ $3\frac{5}{8} \div 1\frac{1}{2} = \frac{29}{8} \div \frac{3}{2} = \frac{29}{\cancel{8}}^{} \times \frac{\cancel{2}^{1}}{3} = \frac{29}{12} = 2\frac{5}{12}$

16. 21.758

$$\begin{array}{r} {}^{4\ 15\,9\,10} \\ 2\cancel{5}.\cancel{6}\cancel{0}\cancel{0} \\ -\ 3.842 \\ \hline 21.758 \end{array}$$

17. 729 $9^3 = 9 \times 9 \times 9 = 729$

18. 42.5

$$\begin{array}{r} 42.5 \\ 0.04.\overline{)1.70.0} \\ -1\ 6 \\ \hline 10 \\ -\ 8 \\ \hline 2\ 0 \\ -2\ 0 \\ \hline 0 \end{array}$$

19. $14\frac{2}{5}$ $5\frac{1}{3} \times 2\frac{7}{10} = \frac{\cancel{16}^{8}}{\cancel{3}_{1}} \times \frac{\cancel{27}^{9}}{\cancel{10}_{5}} = \frac{72}{5} = 14\frac{2}{5}$

20. 67.141

$$\begin{array}{r} {}^{1\ 1\ 1} \\ 3.270 \\ 54.800 \\ +\ 9.071 \\ \hline 67.141 \end{array}$$

21. 0.0441

$$\begin{array}{r} 0.063 \\ \times\ \ 0.7 \\ \hline 0.0441 \end{array}$$

22. **6** Because $6^2 = 36$, the square root of 36 is 6.

23. **8 miles** Monica drives $2\frac{1}{5}$ miles from the dry cleaner to point A, $2\frac{1}{10}$ miles from point A to point B, and $3\frac{7}{10}$ miles from point B to point C.

$$\begin{array}{r} 2\frac{1}{5} = \quad 2\frac{2}{10} \\ 2\frac{1}{10} = \quad 2\frac{1}{10} \\ +3\frac{7}{10} = +3\frac{7}{10} \\ \hline 7\frac{10}{10} = 8 \end{array}$$

24. **$18\frac{2}{5}$ miles** Subtract the distance from point E to the dry cleaner ($6\frac{3}{5}$ miles) from the total length of the route.

$$\begin{array}{r} 25\ \ = \cancel{25}^{24}\frac{5}{5} \\ -\ 6\frac{3}{5} = \ \ 6\frac{3}{5} \\ \hline 18\frac{2}{5} \end{array}$$

25. **(4) 480 sq. ft.** To find the area, multiply the length times the width.

$$\begin{array}{r} 24 \\ \times 20 \\ \hline 480 \end{array}$$

26. **(3) $\frac{2}{3} \times 3$** Rosa is planning to triple the recipe, so multiply the amount of cheese by 3.

27. **(5) $8\frac{1}{4}$ hours** First find the number of hours Mike worked in the morning. Since 28 is close to 30, think of 8:28 as $8\frac{1}{2}$. Since 21 is close to 15, think of 12:21 as $12\frac{1}{4}$. Then subtract.

$$\begin{array}{r} 12\frac{1}{4} = 12\frac{1}{4} = \cancel{12}^{11}\frac{5}{4} \\ -\ 8\frac{1}{2} = \ \ 8\frac{2}{4} = \ \ 8\frac{2}{4} \\ \hline 3\frac{3}{4} \end{array}$$

Then find the number of hours Mike worked in the afternoon. Since 6 is close to zero, think of 1:06 as 1. Since 36 is close to 30, think of 5:36 as $5\frac{1}{2}$. Then subtract.

$$\begin{array}{r} 5\frac{1}{2} \\ -1 \\ \hline 4\frac{1}{2} \end{array}$$

Finally, add the number of hours Mike worked in the morning and in the afternoon.

$$\begin{array}{r} 3\frac{3}{4} = 3\frac{3}{4} \\ +4\frac{1}{2} = 4\frac{2}{4} \\ \hline 7\frac{5}{4} = 7 + 1\frac{1}{4} = 8\frac{1}{4} \end{array}$$

28. **(3) $\frac{4}{7}$** Write a fraction for the number of hours Mario worked overtime, and reduce to lowest terms.

$$\frac{12}{21} = \frac{12 \div 3}{21 \div 3} = \frac{4}{7}$$

UNIT 4: RATIOS, PROPORTIONS, AND PERCENTS AND SPECIAL TOPICS

SECTION 11

PAGE 159

1. (2) $\dfrac{\text{1 cup sugar}}{\text{6 cups flour}}$

2. (4) $\dfrac{\text{1,267 nonfiction}}{\text{1,055 fiction}}$

3. (1) $\dfrac{\$0.24}{\text{1 oz.}}$

4. (4) $\dfrac{\$18}{\text{1 hr.}}$

5. (2) $\dfrac{\text{19 hours}}{\text{26 hours}}$ Subtract the 19 hours Daksha spends baking cakes from the total hours she spends baking cakes and cookies: $45 - 19 = 26$ hours baking cookies.

6. **Car 1 gets more miles per gallon.** You know Car 1 gets 15 miles to 1 gallon of gasoline. Car 2 gets 100 miles to 10 gallons. Divide 100 by 10 to find how many miles Car 2 gets to 1 gallon. $\frac{100}{10} = \frac{100 \div 10}{10 \div 10} = \frac{10}{1}$ or 10 miles to 1 gallon.

PAGE 161

2. $\frac{16}{21}$

$\dfrac{\text{16 employees}}{\text{21 employees}} = \dfrac{16}{21}$

3. $\dfrac{\text{1 gallon}}{\text{200 square feet}}$

$\dfrac{\text{2 gallons of paint}}{\text{400 square feet}} = \dfrac{2 \div 2}{400 \div 2} = \dfrac{1}{200}$

4. $\dfrac{\text{1 pound}}{\$0.75}$

$\dfrac{\text{2 pounds of apples}}{\$1.50} = \dfrac{2 \div 2}{\$1.50 \div 2} = \dfrac{1}{\$0.75}$

5. $\frac{2}{5}$

$\dfrac{\text{4 hours}}{\text{10 hours}} = \dfrac{4 \div 2}{10 \div 2} = \dfrac{2}{5}$

6. $\frac{2}{5}$

$\dfrac{\text{8 grams}}{\text{20 grams}} = \dfrac{8 \div 4}{20 \div 4} = \dfrac{2}{5}$

7. $\dfrac{\text{1 pair}}{\$4}$

$\dfrac{\text{3 pairs of socks}}{\$12} = \dfrac{3 \div 3}{\$12 \div 3} = \dfrac{\text{1 pair of socks}}{\$4}$

8. $\dfrac{\text{3 women}}{\text{2 men}}$

$\dfrac{\text{12 women}}{\text{8 men}} = \dfrac{12 \div 4}{8 \div 4} = \dfrac{3}{2}$

9. $\dfrac{\$14}{\text{1 hr.}}$

$\dfrac{\$280}{\text{20 hr.}} = \dfrac{\$280 \div 20}{20 \div 20} = \dfrac{\$14}{\text{1 hr.}}$

10. $\frac{8}{3}$

$\dfrac{\text{40 books}}{\text{15 books}} = \dfrac{40 \div 5}{15 \div 5} = \dfrac{8}{3}$

11. $\frac{9}{31}$

$\dfrac{\text{9 rainy days}}{\text{31 days in the month}} = \dfrac{9}{31}$

12. $\frac{2}{3}$

$\dfrac{\text{12 won}}{\text{18 played}} = \dfrac{12 \div 6}{18 \div 6} = \dfrac{2}{3}$

13. $\frac{5}{2}$

$\dfrac{\text{35 cars}}{\text{14 vans}} = \dfrac{35 \div 7}{14 \div 7} = \dfrac{5}{2}$

14. $\frac{23}{1}$

$\dfrac{\text{230 miles}}{\text{10 gallons}} = \dfrac{230 \div 10}{10 \div 10} = \dfrac{23}{1}$

15. $\frac{3}{5}$

$\dfrac{\text{15 cash}}{\text{25 charge}} = \dfrac{15 \div 5}{25 \div 5} = \dfrac{3}{5}$

16. $\dfrac{\$6}{1}$

$\dfrac{\$42}{\text{7 hours}} = \dfrac{\$42 \div 7}{7 \div 7} = \dfrac{\$6}{1}$

PAGE 163

2. 6 $4 \times 18 = 72; 72 \div 12 = 6$
3. 6 $18 \times 10 = 180; 180 \div 30 = 6$
4. 2 $7 \times 4 = 28; 28 \div 14 = 2$
5. 55 $20 \times 11 = 220; 220 \div 4 = 55$
6. 25 $30 \times 10 = 300; 300 \div 12 = 25$
7. 35 $42 \times 10 = 420; 420 \div 12 = 35$

8. **19.2** $24 \times 4 = 96; 96 \div 5 = 19.2$

$$
\begin{array}{r}
19.2 \\
5)\overline{96.0} \\
\underline{5} \\
46 \\
\underline{45} \\
1\,0 \\
\underline{1\,0} \\
\end{array}
$$

9. **25** $5 \times 40 = 200; 200 \div 8 = 25$

10. **$210** $\dfrac{40 \text{ hr.}}{\$240} = \dfrac{35 \text{ hr.}}{?}$

$240 \times 35 = 8{,}400; 8{,}400 \div 40 = 210$

11. **21** $\dfrac{35 \text{ laps}}{25 \text{ min.}} = \dfrac{?}{15 \text{ min.}}$

$35 \times 15 = 525; 525 \div 25 = 21$

12. **17** $\dfrac{5 \text{ rolls}}{15 \text{ rolls}} = \dfrac{?}{51 \text{ rolls}}$

$5 \times 51 = 255; 255 \div 15 = 17$

13. **6** $\dfrac{45 \text{ min.}}{1 \text{ day}} = \dfrac{?}{8 \text{ days}}$

$45 \times 8 = 360; 360 \div 1 =$
$360 \text{ min} \div 60 \text{ min./hr.} = 6 \text{ hr.}$

14. **98.425** $\dfrac{1 \text{ meter}}{39.37 \text{ inches}} = \dfrac{2.5}{?}$

$39.37 \times 2.5 = 98.425; 98.425 \div 1 = 98.425$

15. **1,250** $\dfrac{2{,}500 \text{ cars}}{4{,}000 \text{ spaces}} = \dfrac{?}{2{,}000 \text{ spaces}}$

$2{,}500 \times 2{,}000 = 5{,}000{,}000 \div 4{,}000 = 1{,}250$

PAGE 165

1. **(3)** $\dfrac{10}{\$22.50} = \dfrac{6}{?}$ Both ratios are in the same order. The first ratio means that 10 reams cost $22.50. The second ratio means that 6 reams cost an unknown amount.

2. **(4)** $\dfrac{12}{\$3.96} = \dfrac{18}{?}$ Both ratios are in the same order. The first ratio means that 12 pads cost $3.96. The second ratio means that 18 pads cost an unknown amount.

3. **(3) $3.32** $\dfrac{3}{\$2.49} = \dfrac{4}{?}$ $\$2.49 \times 4 = \$9.96$; $\$9.96 \div 3 = \$3.32$

4. **(2)** $\dfrac{2}{\$1.28} = \dfrac{5}{?}$ Both ratios are in the same order. The first ratio means that 2 heads cost $1.28. The second ratio means that 5 heads cost an unknown amount.

5. **(2) $0.81** $\dfrac{3}{\$1.22} = \dfrac{2}{?}$ $\$1.22 \times 2 = \$2.44$; $\$2.44 \div 3 = \$0.813$, about $0.81.

6. **Set up a proportion and solve for the unknown.**
$\dfrac{5}{\$3.15} = \dfrac{?}{\$12.60}$ $5 \times \$12.60 = \$63.00$;
$\$63.00 \div \$3.15 = \textbf{20 pounds}$

PAGE 167

1. **(3) 20** The distance on the map from Riverton to Plainview is 1 inch.

$\dfrac{2 \text{ in.}}{40 \text{ mi.}} = \dfrac{1 \text{ in.}}{? \text{ mi.}}$ $\quad 40 \times 1 = 40$
$\quad 40 \div 2 = 20$

2. **(2) 55** The distance on the map from Plainview to Rock Falls is $2\frac{3}{4}$ inches.

$\dfrac{2 \text{ in.}}{40 \text{ mi.}} = \dfrac{2\frac{3}{4} \text{ in.}}{? \text{ mi.}}$ $\quad 40 \times 2\frac{3}{4} = 110$
$\quad 110 \div 2 = 55$

3. **(3) 3**

$\dfrac{2 \text{ in.}}{40 \text{ mi.}} = \dfrac{? \text{ in.}}{60 \text{ mi.}}$ $\quad 2 \times 60 = 120$
$\quad 120 \div 40 = 3$

4. **(4) $7\frac{1}{2}$** The distance on the map from Mesa to Canyon City is $1\frac{1}{2}$ inches.

$\dfrac{3 \text{ in.}}{15 \text{ mi.}} = \dfrac{1\frac{1}{2} \text{ in.}}{? \text{ mi.}}$ $\quad 15 \times 1\frac{1}{2} = 22\frac{1}{2}$
$\quad 22\frac{1}{2} \div 3 = 7\frac{1}{2}$

5. **(4) $11\frac{1}{4}$** The distance on the map from Bluffton to Canyon City is $2\frac{1}{4}$ inches.

$\dfrac{3 \text{ in.}}{15 \text{ mi.}} = \dfrac{2\frac{1}{4} \text{ in.}}{? \text{ mi.}}$ $\quad 15 \times 2\frac{1}{4} = 33\frac{3}{4}$
$\quad 33\frac{3}{4} \div 3 = 11\frac{1}{4}$

6. The distance from Mesa to Canyon City is $7\frac{1}{2}$ miles. The distance from Canyon City to Bluffton is $11\frac{1}{4}$ miles. The total miles traveled is $7\frac{1}{2} + 11\frac{1}{4}$, or $18\frac{3}{4}$ miles. Round $18\frac{3}{4}$ miles to 20 miles. If Jerome drives 40 miles in 1 hour, then he will drive 20 miles in $\frac{1}{2}$ hour. Therefore, it takes Jerome about 30 minutes to drive from Mesa to Bluffton.

PAGES 168–169

1. $\frac{5}{7}$

$\dfrac{20 \text{ in office}}{28 \text{ in warehouse}} = \dfrac{20 \div 4}{28 \div 4} = \dfrac{5}{7}$

2. $\frac{5}{12}$

$\dfrac{15 \text{ with field goals}}{36 \text{ total points}} = \dfrac{15 \div 3}{36 \div 3} = \dfrac{5}{12}$

3. **36** $6 \times 42 = 252; 252 \div 7 = 36$

4. **20** $8 \times 30 = 240; 240 \div 12 = 20$

5. **265 miles**

$\dfrac{106 \text{ miles}}{2 \text{ hours}} = \dfrac{? \text{ miles}}{5 \text{ hours}}$ $\quad 106 \times 5 = 530$
$\quad 530 \div 2 = 265$

6. **12 parts**

$\dfrac{400 \text{ total parts}}{3 \text{ defective parts}} = \dfrac{1{,}600 \text{ total parts}}{? \text{ defective parts}}$

$3 \times 1{,}600 = 4{,}800$
$4{,}800 \div 400 = 12$

7. **$1.65**

$$\frac{3}{\$0.99} = \frac{5}{?} \qquad \$0.99 \times 5 = \$4.95$$
$$\$4.95 \div 3 = \$1.65$$

8. **$6\frac{2}{5}$ hours**

$$\frac{2 \text{ hours}}{5 \text{ pictures}} = \frac{? \text{ hours}}{16 \text{ pictures}} \qquad 2 \times 16 = 32$$
$$32 \div 5 = 6\frac{2}{5}$$

9. **$7\frac{1}{2}$ gallons**

$$\frac{3 \text{ parts blue}}{4 \text{ parts gray}} = \frac{? \text{ gallons blue}}{10 \text{ gallons gray}}$$
$$3 \times 10 = 30 \qquad 30 \div 4 = 7\frac{1}{2}$$

10. **400 miles**

$$\frac{300 \text{ miles}}{12 \text{ gallons}} = \frac{? \text{ miles}}{16 \text{ gallons}}$$
$$300 \times 16 = 4,800 \qquad 4,800 \div 12 = 400$$

11. **240 women**

$$\frac{5 \text{ women}}{9 \text{ employees}} = \frac{? \text{ women}}{432 \text{ employees}}$$
$$5 \times 432 = 2,160 \qquad 2,160 \div 9 = 240$$

12. **$10\frac{1}{2}$ miles** The distance on the map from Mountainview to Somerset is $1\frac{3}{4}$ inches.

$$\frac{2 \text{ in.}}{12 \text{ mi.}} = \frac{1\frac{3}{4} \text{ in.}}{? \text{ mi.}} \qquad 12 \times 1\frac{3}{4} = 21$$
$$21 \div 2 = 10\frac{1}{2}$$

13. **21 miles** The distance on the map from Somerset to Princeton is $3\frac{1}{2}$ inches.

$$\frac{2 \text{ in.}}{12 \text{ mi.}} = \frac{3\frac{1}{2} \text{ in.}}{? \text{ mi.}} \qquad 12 \times 3\frac{1}{2} = 42$$
$$42 \div 2 = 21$$

14. **5 inches**

$$\frac{2 \text{ in.}}{12 \text{ mi.}} = \frac{? \text{ in.}}{30 \text{ mi.}} \qquad 2 \times 30 = 60$$
$$60 \div 12 = 5$$

SECTION 12

PAGE 171

1. **(4) 30** 30% means "30 out of 100," so there are 30 pennies.

2. **(3) $\frac{3}{10}$** $\quad 30\% = \frac{30}{100} = \frac{3}{10}$

3. **(1) $0.30** $\quad 30\% = \frac{30}{100} = \$0.30$

4. **(2) $\frac{90}{100}$** 90% means "90 out of 100," so the score is $\frac{90}{100}$.

5. **(3) 0.85** $\quad \frac{85}{100} = 0.85$

6. **Count the number of shaded squares. Nine squares are shaded. Nine squares out of 100 are shaded, or $\frac{9}{100}$. The fraction $\frac{9}{100} = 0.09$. Percent means "out of 100," so 9% of the squares are shaded.**

2. **$\frac{3}{4}$, 0.75, 75%** 75 of 100 parts are colored. $\frac{75}{100}$ reduces to $\frac{3}{4}$.
$$\frac{75 \div 25}{100 \div 25} = \frac{3}{4}; \qquad \frac{75}{100} = 0.75;$$
$\frac{75}{100}$ is the same as 75%.

3. **$\frac{1}{5}$, 0.2, 20%** 20 of 100 parts are colored. $\frac{20}{100}$ reduces to $\frac{1}{5}$.
$$\frac{20 \div 20}{100 \div 20} = \frac{1}{5}; \qquad \frac{20}{100} = 0.2;$$
$\frac{20}{100}$ is the same as 20%.

4. **$\frac{1}{100}$, 0.01, 1%** 1 of 100 parts is colored. $\frac{1}{100}$ is in lowest terms; $\frac{1}{100} = 0.01;$
$\frac{1}{100}$ is the same as 1%.

PAGE 173

2. **0.04** $\quad .04. = 0.04$

3. **2.5** $\quad 2.50. = 2.5$

4. **0.065** $\quad 6\frac{1}{2}\% = 6.5\% = .06.5 = 0.065$

5. **0.0525** $\quad .05.25 = 0.0525$

6. **0.0014** $\quad .00.14 = 0.0014$

7. **0.9725** $\quad .97.25 = 0.9725$

8. **1.00** $\quad 1.00. = 1.00$

9. **260%** $\quad 2.60. = 260\%$

10. **3%** $\quad 0.03. = 3\%$

11. **62.5%** $\quad 0.62.5 = 62.5\%$

12. **0.08%** $\quad 0.00.08 = 0.08\%$

13. **463.5%** $\quad 4.63.5 = 463.5\%$

14. **20%** $\quad 0.20. = 20\%$

15. **586%** $\quad 5.86. = 586\%$

16. **34.5%** $\quad 0.34.\frac{1}{2} = 34\frac{1}{2}\%$ or 34.5%

PAGE 174

2. **$\frac{2}{25}$** $\quad 8\% = \frac{8}{100} = \frac{8 \div 4}{100 \div 4} = \frac{2}{25}$

3. **$\frac{13}{25}$** $\quad 52\% = \frac{52}{100} = \frac{52 \div 4}{100 \div 4} = \frac{13}{25}$

4. **$\frac{1}{100}$**

5. **$\frac{5}{2}$** $\quad \frac{250}{100} = \frac{250 \div 50}{100 \div 50} = \frac{5}{2}$

6. **$\frac{4}{25}$** $\quad \frac{16}{100} = \frac{16 \div 4}{100 \div 4} = \frac{4}{25}$

7. **1** $\quad \frac{100}{100} = \frac{100 \div 100}{100 \div 100} = \frac{1}{1} = 1$

8. $\dfrac{21}{50}$ $\dfrac{42}{100} = \dfrac{42 \div 2}{100 \div 2} = \dfrac{21}{50}$

9. 50% $\dfrac{1}{2} = 2\overline{)1.0}$ $\begin{array}{r} 0.5 = 50\% \\ \underline{-1\,0} \\ 0 \end{array}$

10. 60% $\dfrac{3}{5} = 5\overline{)3.0}$ $\begin{array}{r} 0.6 = 60\% \\ \underline{-3\,0} \\ 0 \end{array}$

11. 70% $10\overline{)7.0}$ $\begin{array}{r} 0.7 = 70\% \\ \underline{-7\,0} \\ 0 \end{array}$

12. 37.5% $8\overline{)3.0}$ $\begin{array}{r} 0.375 = 37.5\% \\ \underline{-2\,4} \\ 60 \\ \underline{-56} \\ 40 \\ \underline{-40} \\ 0 \end{array}$

13. 25% $4\overline{)1.0}$ $\begin{array}{r} 0.25 = 25\% \\ \underline{-8} \\ 20 \\ \underline{-20} \\ 0 \end{array}$

14. 62.5% $8\overline{)5.0}$ $\begin{array}{r} 0.625 = 62.5\% \\ \underline{-4\,8} \\ 20 \\ \underline{-16} \\ 40 \\ \underline{-40} \\ 0 \end{array}$

15. 650% $6\dfrac{5}{10} = \dfrac{65}{10} = 10\overline{)65}$ $\begin{array}{r} 6.5 = 650\% \\ \underline{-60} \\ 5\,0 \\ \underline{-5\,0} \\ 0 \end{array}$

16. 300% $\dfrac{15}{5} = 3 = 300\%$

PAGE 175

1. 0.37   To convert a percent to a decimal, divide by 100 (move decimal point 2 places to the left).
   37% = .37. = 0.37

2. 0.04   4% = .04. = 0.04

3. 2.25   225% = 2.25. = 2.25

4. 0.065 $6\frac{1}{2}\% = 6.5\% = .06.5 = 0.065$

5. 46%   To convert a decimal to a percent, multiply by 100 (move decimal point 2 places to the right).
   0.46 = 0.46. = 46%

6. 8%   0.08 = 0.08. = 8%

7. 250%   2.5 = 2.50. = 250%

8. 37.5%   0.375 = 0.37.5 = 37.5%

9. $\dfrac{1}{2}$ $50\% = \dfrac{50}{100} = \dfrac{50 \div 50}{100 \div 50} = \dfrac{1}{2}$

10. $\dfrac{2}{5}$ $40\% = \dfrac{40}{100} = \dfrac{40 \div 20}{100 \div 20} = \dfrac{2}{5}$

11. $\dfrac{1}{4}$ $25\% = \dfrac{25}{100} = \dfrac{25 \div 25}{100 \div 25} = \dfrac{1}{4}$

12. $\dfrac{9}{10}$ $90\% = \dfrac{90}{100} = \dfrac{90 \div 10}{100 \div 10} = \dfrac{9}{10}$

13. 70% $10\overline{)70}$ 0.7 $0.7 = 0.70. = 70\%$

14. 75% $4\overline{)3.00}$ 0.75 $0.75 = 0.75. = 75\%$

15. 80% $5\overline{)4.0}$ 0.8 $0.8 = 0.80. = 80\%$

16. 87.5% $8\overline{)7.000}$ 0.875 $0.875 = 0.87.5 = 87.5\%$

SECTION 13

PAGE 177

1. **(3) the part** base = 8 hours, rate = 20%, part = the hours it takes Hector to do the tasks

2. **(2) 8%** The rate is followed by the percent sign.

3. **(2) the rate** base = 64 people, part = 16 are women, rate = the percent of the employees that are women

4. **(1) 6** rate = 50%, part = 3 hours, base = the whole = 6 hours

5. **(3) the part** base = $24, rate = 20%, part = amount saved

6. **(1) the base** rate = 25%, part = $10, base = regular price of the sweater

7. **(4) 12** base = 16 games, rate = 75%, part = 12 games

8. Problems will vary. Possible answer: Fashions Plus has a 20% discount on goose down jackets. The original price is $500. The customer saves $100. Which number is the part?
Part = base × rate
$100 = $500 × 20%

PAGE 179

2. $18   40% = 0.40

$$\begin{array}{r} \$45 \\ \times\ .40 \\ \hline 00 \\ 180 \\ \hline \$18.00 \end{array}$$

3. 1.28   8% = 0.08

$$\begin{array}{r} 16 \\ \times .08 \\ \hline 1.28 \end{array}$$

4. 120   250% = 2.5

$$\begin{array}{r} 48 \\ \times 2.5 \\ \hline 240 \\ 96 \\ \hline 120.0 \end{array}$$

5. $2.25 $4\frac{1}{2}\%$ = 0.045

$$\begin{array}{r} \$50 \\ \times .045 \\ \hline 250 \\ 200 \\ \hline \$2.250 \end{array}$$

6. 10.44 $7\frac{1}{4}\%$ = 0.0725

$$\begin{array}{r} 144 \\ \times .0725 \\ \hline 720 \\ 288 \\ 1008 \\ \hline 10.4400 \end{array}$$

7. $8.00   25% = 0.25

$$\begin{array}{r} \$32 \\ \times .25 \\ \hline 160 \\ 64 \\ \hline \$8.00 \end{array}$$

8. 12   80% = 0.80

$$\begin{array}{r} 15 \\ \times .80 \\ \hline 00 \\ 120 \\ \hline 12.00 \end{array}$$

9. 400   5% = 0.05

$$\begin{array}{r} 8.000 \\ \times\ .05 \\ \hline 400.00 \end{array}$$

10. 108   45% = 0.45

$$\begin{array}{r} 240 \\ \times .45 \\ \hline 1200 \\ 960 \\ \hline 108.00 \end{array}$$

11. **$5,200**   16% = 0.16
$32,500 × 0.16 = $5,200

12. **$0.36**   27% = 0.27
$1.32 × 0.27 = $0.3564; $0.3564 rounded to the nearest cent is $0.36.

PAGE 181

1. **(2) $650 × 0.12 × 1**
interest = principal × rate × time
= $650 × 0.12 × 1

2. **(1) $75.60** $i = p \times r \times t$
   = $420 × 0.18 × 1
   = $75.60

3. **(4) $210.80** $i = p \times r \times t$
   = $1,240 × 0.085 × 2
   = $210.80

4. **(2) $252** $i = p \times r \times t$
   = $1,600 × 0.0525 × 3
   = $252

5. **(2) $410 × 0.16 × 2** Subtract the down payment from the cost to find the principal.
$560 − $150 = $410
Use the formula to find the interest.
$i = p \times r \times t = \$410 \times 0.016 \times 2$

6. **(3) $541.20** Add the principal and interest to find the amount to be paid back.
$410 + $131.20 = $541.20

7. **(4) $2,352** Use the formula to find the interest. $i = p \times r \times t$
   = $2,100 × 0.12 × 1
   = $252
Add the principal and interest to find the amount to be paid back.
$2,100 + $252 = $2,352

8. Estimate: Use the formula to find the interest. $i = p \times r \times t$ (Round .15 to .20)
$$= \$10{,}000 \times 0.20 \times 2$$
$$= \$4{,}000$$
Divide the amount to be paid back ($10,000 + $4,000) by the number of monthly payments. Round 24 months (in 2 years) to 25.
$14,000 ÷ 25 is 560 so the monthly payment is about $560.
Actual Calculations:
$i = p \times r \times t$
$$= \$10{,}000 \times 0.15 \times 2$$
$$= \$3{,}000$$
$$\$13{,}000 \div 24 = \$541.67$$

PAGE 183
1. **(3) 360° × 0.15** To find the degrees in the section, multiply the base (360°) times the rate (0.15).
2. **(3) $18.00** Multiply the base ($120) times the rate (15%).
$120 × 0.15 = $18.00
3. **(1) federal income tax** The largest section of the graph represents federal income tax.
4. **(4) 360° × 0.40** To find the degrees in the section, multiply the base (360°) times the rate (0.40).
5. **(5) Not enough information is given.** To find the cost of heat each month, you need to know the total cost (base) of Mr. Quan's electric bill.
6. **To find the degrees in each section, multiply the base (360°) times the rate for each item.**

Heating	360° × 0.4	=	144°
Refrigerator	360° × 0.2	=	72°
Water heater	360° × 0.1	=	36°
Cooking	360° × 0.05	=	18°
Other	360° × 0.25	=	90°

PAGES 184–186
1. 192   64% = 0.64
$$\begin{array}{r} 300 \\ \times 0.64 \\ \hline 192.00 \end{array}$$

2. 4.32   6% = 0.06
$$\begin{array}{r} 72 \\ \times 0.06 \\ \hline 4.32 \end{array}$$

3. $86.40   90% = 0.9
$$\begin{array}{r} \$96 \\ \times\ 0.9 \\ \hline \$86.40 \end{array}$$

4. $224   400% = 4
$$\begin{array}{r} \$56 \\ \times\ \ 4 \\ \hline \$224 \end{array}$$

5. 45   125% = 1.25
$$\begin{array}{r} 1.25 \\ \times\ 36 \\ \hline 45.00 \end{array}$$

6. 5.25 $3\frac{1}{2}\% = 0.035$
$$\begin{array}{r} 150 \\ \times 0.035 \\ \hline 5.250 \end{array}$$

7. **$116** Multiply the base ($145) times the rate (80%).
80% = 0.8
$$\begin{array}{r} \$145 \\ \times\ \ 0.8 \\ \hline \$116.0 \end{array}$$

8. **$41.25** Multiply the base ($125) times the rate (33%).
33% = 0.33
$$\begin{array}{r} \$125 \\ \times 0.33 \\ \hline 3\ 75 \\ +37\ 5 \\ \hline \$41.25 \end{array}$$

9. **1,650** Multiply the base (3,000) times the rate (55%).
55% = 0.55
$$\begin{array}{r} 3{,}000 \\ \times\ 0.55 \\ \hline 1{,}650.00 \end{array}$$

10. **$37.20** Multiply the base ($310) times the rate (12%).
12% = 0.12
$$\begin{array}{r} \$310 \\ \times 0.12 \\ \hline \$37.20 \end{array}$$

11. **$127.50** $i = p \times r \times t$
$$= \$1{,}500 \times 0.085 \times 1$$
$$= \$127.50$$

12. **$560** $i = p \times r \times t$
$$= \$2{,}000 \times 0.14 \times 2$$
$$= \$560.00$$

13. **$150** $i = p \times r \times t$
$$= \$1{,}200 \times 0.125 \times 1$$
$$= \$150$$

ANSWERS AND EXPLANATIONS

14. **$2,040** Subtract the down payment from the cost to find the principal.
$1,750 − $250 = $1,500
Use the formula to find the interest.
$i = p \times r \times t$
   $= \$1,500 \times 0.18 \times 2$
   $= \$540$
Add the principal and the interest to find the amount to be paid back.
$1,500 + $540 = $2,040

15. **$1,950** Use the formula to find the interest.
$i = p \times r \times t$
   $= \$1,500 \times 0.15 \times 2$
   $= \$450$
Add the principal and the interest to find the amount to be paid back.
$1,500 + $450 = $1,950

16. **$56.10** Use the formula to find the interest.
$i = p \times r \times t$
   $= \$990 \times 0.18 \times 2$
   $= \$356.40$
Add the principal and interest to find the amount to be paid back.
$990 + $356.40 = $1,346.40
Divide the amount to be paid back by the number of monthly payments.
$1,346.40 ÷ 24 = $56.10

17. **$14,960** The principal is $9,350.
Use the formula to find the interest.
$i = p \times r \times t$
   $= \$9,350 \times 0.12 \times 5$
   $= \$5,610$
Add the principal and interest to find the amount to be paid back.
$9,350 + $5,610 = $14,960

18. **$97.15** Add the cost of both appliances.
$475 + $680 = $1,155
Subtract the down payment from the cost to find the principal.
$1,155 − $150 = $1,005
Use the formula to find the interest.
$i = p \times r \times t$
   $= \$1,005 \times 0.16 \times 1$
   $= \$160.80$
Add the principal and interest to find the amount to be paid back.
$1,005 + $160.80 = $1,165.80
Divide the amount to be paid back by the number of monthly payments.
$1,165.80 ÷ 12 = $97.15

19. **(4) 40 × 15%** Multiply the base (40) times the rate (15%) to find the part.

20. **(2) 40% + 25%** Add the rates for supervising (40%) and planning (25%).

21. **(1) 8** Multiply the base (40) times the rate (20%).
20% = 0.2   40 × 0.2 = 8

22. To find the degrees in each section, multiply the base (360°) times the rate for each item.
Rent and utilities 360° × 0.35 = **126°**
Food   360° × 0.25 = **90°**
Transportation   360° × 0.2 = **72°**
Clothes   360° × 0.15 = **54°**
Other   360° × 0.05 = **18°**

23. **(5) Other** The smallest section of the graph represents *other*.

24. **(5) $770** Multiply the base ($2,200) times the rate (35%).
$2,200 × 0.35 = $770

SECTION 14
PAGE 188

1. **(3) $299 ÷ $1,150** To solve for the rate, divide the part ($299) by the base ($1,150).

2. **(1) $500 ÷ $4,500** Divide the part ($500) by the base ($4,500) to find the rate.

3. **(3) $30 ÷ $200** Divide the part ($30) by the base ($200) to find the rate.

4. **(3) 65 ÷ 110** To find the rate, divide the part (65) by the base (110).

5. **(4) $45 ÷ $380** Divide the part ($45) by the base ($380) to find the rate.

6. **(5) Not enough information is given.**
You are given the base ($420) but you do not have the part (how much Flavia spends a week).

7. **(4) 117 ÷ 300** Divide the part (117) by the base (300) to find the rate.

8. **The two equations are the inverse of each other. To find the rate, you divide both sides by the base.**
part = base × rate
part ÷ base = base ÷ base × rate
part ÷ base = rate

PAGE 190

1. **(2) $30.00 ÷ $600.00** Divide the part deducted for health insurance ($30.00) by the base ($600.00) to find the rate.

2. **(3) ($84.00 + $12.60) ÷ $600.00** Add the deductions for federal ($84.00) and state ($12.60) income taxes to find the part. Then divide the part by the base ($600.00) to find the rate.

3. **(3) $600.00 × 3%** Multiply the base ($600.00) times the rate (3%) to find the part.

4. **(1) $143.75 ÷ $1,250.00** Divide the part deducted for federal income tax ($143.75) by the base ($1,250.00) to find the rate.

5. **(4) $1,250.00 × 5%** Multiply the base ($1,250.00) times the rate (5%) to find the part.

6. **Add the deductions for the retirement fund ($62.50) and insurance ($25.00) to find the part. Then divide the part by the base ($1,250.00) to find the rate.** ($62.50 + $25.00) ÷ $1,250.00 = $87.50 ÷ $1,250.00 = 0.07 = 7%

PAGE 192

2. **10%**   0.1 = 10%

$$\frac{0.1}{360)36.0}$$

3. **400%**   4 = 400%

$$\frac{4}{45)180}$$

4. **60%**   0.6 = 60%

$$\frac{0.6}{95)57.0}$$

5. **2%**   0.02 = 2%

$$\frac{0.02}{200)4.00}$$

6. **250%**   2.5 = 250%

$$\frac{2.5}{6)15.0}$$

7. **80%** Divide the part (32) by the base (40) to find the rate.

$$\frac{0.8}{40)32.0} \quad 0.8 = 80\%$$

8. **57%** Divide the part (42) by the base (74) to find the rate.

$$\frac{0.567}{74)42.000} \quad \begin{array}{l} 0.567 \text{ rounds to } 0.57 \\ 0.57 = 57\% \end{array}$$

9. **25%** Divide the part ($9) by the base ($36) to find the rate.

$$\frac{0.25}{\$36)\$9.00} \quad 0.25 = 25\%$$

10. **16%** Divide the part ($73.60) by the base ($460) to find the rate.

$$\frac{0.16}{\$460)\$73.60} \quad 0.16 = 16\%$$

11. **2%** Divide the part ($1,100) by the base ($46,000) to find the rate. $1,100 ÷ $46,000 = 0.0239, rounded to the nearest percent is 2%.

12. **60%** Divide the part ($570) by the base ($950) to find the rate. $570 ÷ $950 = 0.6 = 60%

PAGE 194

1. **(2) grams of fat and number of calories in a gram of fat**

grams of fat $\times \dfrac{9 \text{ calories}}{\text{g fat}}$ = calories from fat

2. **(3)** $\dfrac{9}{200} = \dfrac{?}{100}$

$$\frac{\text{part}}{\text{base}} = \frac{\text{rate}}{100}$$

$$\frac{9}{200} = \frac{?}{100}$$

3. **(2) 4.5%** Solve the proportion.

$$\frac{9}{200} = \frac{?}{100} \quad \begin{array}{l} 9 \times 100 = 900 \\ 900 ÷ 200 = 4.5 \end{array}$$

4. **(2)** $\dfrac{18}{120} = \dfrac{?}{100}$

5. **(1) 15%** Solve the proportion.

$$\frac{18}{120} = \frac{?}{100} \quad \begin{array}{l} 18 \times 100 = 1,800 \\ 1,800 ÷ 120 = 15 \end{array}$$

6. **Andrew doesn't need to avoid either food. For spaghetti, only 4.5% of the calories come from fat, and in oat cereal 15% of the calories come from fat. Each of these percents is less than 30%.**

PAGE 196

1. **(4) $1,600** Add the amounts on the graph to find the total.

$$\begin{array}{r} \$600 \\ 300 \\ 250 \\ 350 \\ +\ 100 \\ \hline \$1,600 \end{array}$$

2. **(3) 38%** Divide the part ($600) by the base ($1,600; see the explanation for Item 1) to find the rate.

$$\overset{0.375}{\$1,600\overline{)\$600.000}} \quad \begin{array}{l} 0.375 \text{ rounds to } 0.38. \\ 0.38 = 38\% \end{array}$$

3. **(3) 28%** Add the amounts budgeted for personal expenses ($350) and other ($100). $350 + $100 = $450. Then divide the part ($450) by the base ($1,600; see the explanation for Item 1) to find the rate.

$$\overset{0.281}{\$1,600\overline{)\$450.000}} \quad \begin{array}{l} 0.281 \text{ rounds to } 0.28. \\ 0.28 = 28\% \end{array}$$

4. **(5) $200,000** Add the amounts on the graph to find the total. Then multiply by 1,000 because the key says $ *in thousands.*

$$\begin{array}{r} \$50 \\ 85 \\ 30 \\ 20 \\ +\ 15 \\ \hline \$200 \end{array} \qquad \begin{array}{r} \$200 \\ \times\ 1,000 \\ \hline \$200,000 \end{array}$$

5. **(5) 68%** Add the sales from T-shirts ($50,000) and from sweatshirts ($85,000).
$50,000 + $85,000 = $135,000
Then divide the part ($135,000) by the base ($200,000; see the explanation for Item 4) to find the rate.

$$\overset{0.675}{\$200\overline{)\$135.000}} \quad \begin{array}{l} 0.675 \text{ rounds to } 0.68. \\ 0.68 = 68\% \end{array}$$

6. **Subtract the sales from jackets ($20,000) from the sales from caps ($30,000).**
$30,000 – $20,000 = $10,000
Then divide the part ($10,000) by the base ($200,000; see the explanation for Item 4) to find the rate.

$$\overset{0.05}{\$200\overline{)\$10.00}} \quad 0.05 = 5\%$$

PAGES 197–198

1. **160%** $1.6 = 160\%$

$$\overset{1.6}{10\overline{)16.0}}$$

2. **75%** $0.75 = 75\%$

$$\overset{0.75}{72\overline{)54.00}}$$

3. **4%** $0.04 = 4\%$

$$\overset{0.04}{300\overline{)12.00}}$$

4. **500%** $5 = 500\%$

$$\overset{5}{15\overline{)75}}$$

5. **40%** Divide the part (240) by the base (600) to find the rate.

$$\overset{0.4}{600\overline{)240.0}} \quad 0.4 = 40\%$$

6. **83%** Divide the part ($80) by the base ($96) to find the rate.

$$\overset{0.833}{\$96\overline{)\$80.000}} \quad \begin{array}{l} 0.833 \text{ rounds to } 0.83. \\ 0.83 = 83\% \end{array}$$

7. **57%** Divide the part (125) by the base (220) to find the rate.

$$\overset{0.568}{220\overline{)125.000}} \quad \begin{array}{l} 0.568 \text{ rounds to } 0.57. \\ 0.57 = 57\% \end{array}$$

8. **30%** Divide the part (12) by the base (40) to find the rate.

$$\overset{0.3}{40\overline{)12.0}} \quad 0.3 = 30\%$$

9. **20,000** Add the amounts on the graph to find the total. Then multiply by 1,000.
$7 + 2 + 5 + 4 + 2 = 20$
$20 \times 1,000 = 20,000$

10. **25%** Divide the part (5,000) by the base (20,000; see the explanation for Item 9) to find the rate.

$$\overset{0.25}{20,000\overline{)5,000.00}} \quad 0.25 = 25\%$$

11. **45%** Add the amounts for coffee (7,000) and tea (2,000) to find the part.
$7,000 + 2,000 = 9,000$
Then divide the part by the base (20,000; see the explanation for Item 9) to find the rate.

$$\overset{0.45}{20,000\overline{)9,000.00}} \quad 0.45 = 45\%$$

12. **(3) number of calories in a gram of fat**
$$5 \text{ grams of fat} \times \frac{9 \text{ calories}}{\text{g fat}} =$$
45 calories from fat

13. **(4)** $\dfrac{45}{180} = \dfrac{?}{100}$

14. **(2) 25%** Solve the proportion.
$$\frac{45}{180} = \frac{?}{100} \qquad 45 \times 100 = 4,500$$
$$4,500 \div 180 = 25$$

15. **(2) $25.00 ÷ $500.00** Divide the part deducted for the credit union account ($25.00) by the base ($500.00).

16. **(3) 14%** Add the deductions for federal ($63.40) and state ($6.60) income taxes to find the part.

$$\begin{array}{r} \$63.40 \\ + \ \ 6.60 \\ \hline \$70.00 \end{array}$$

Then divide the part by the base ($500.00) to find the rate.

$$\begin{array}{r} 0.14 \\ \$500\overline{)\$70.00} \end{array} \qquad 0.14 = 14\%$$

17. **(2) $20.00** Multiply the base ($500.00) times the rate (4%).

$$4\% = 0.04 \qquad \begin{array}{r} \$500.00 \\ \times \ \ \ \ 0.04 \\ \hline \$20.0000 \end{array}$$

SECTION 15

PAGE 200

1. **(2) $17.52 ÷ 0.06** Divide the part ($17.52) by the rate (0.06) to find the base (gross pay).

2. **(2) $270 ÷ 0.75** Divide the part ($270) by the rate (0.75) to find the base (number of people who came into the store).

3. **(3) 22 ÷ 0.88** Divide the part (22) by the rate (0.88) to find the base (number of questions on the test).

4. **(2) 18 ÷ 0.45** Divide the part (18) by the rate (0.45) to find the base (hours Curtis worked).

5. **(4) 85 ÷ 0.20** Divide the part (85) by the rate (0.20) to find the base (number of sales).

6. **(5) Not enough information is given.** You do not know the rate to answer the question.

7. **(4) $13.60 ÷ 0.40** Divide the part ($13.60) by the rate (0.40) to find the base (regular price).

8. **part = base × rate, divide both sides by the rate**
   part ÷ rate = base × rate ÷ rate
   part ÷ rate = base

PAGE 202

1. **(2) $7.20 ÷ 30%** Divide the part ($7.20) by the rate (30%) to find the base (regular price).

2. **(3) $54.00 ÷ 45%** Divide the part ($54.00) by the rate (45%) to find the base (regular price).

3. **(5) $36.00 − $12.60** Subtract the amount you save ($12.60) from the regular price ($36.00) to find the sale price.

4. **(2) $5.60 ÷ 40%** Divide the part ($5.60) by the rate (40%) to find the base (regular price).

5. **(2) $64.00 × 25%** Multiply the base ($64.00) times the rate (25%) to find the part (savings).

6. **Divide the part ($6.50) by the base ($32.50) to find the discount rate (20%). To find the sale price of the skirt, subtract the amount saved ($6.50) from the regular price of the skirt ($32.50).**

PAGE 204

2. 90 $\qquad$ 40% = 0.4 $\qquad \begin{array}{r} 90. \\ 0.4\overline{)36.0.} \end{array}$

3. $12 $\qquad$ 75% = 0.75 $\qquad \begin{array}{r} \$12. \\ 0.75\overline{)\$9.00.} \end{array}$

4. 19 $\qquad$ 300% = 3 $\qquad \begin{array}{r} 19 \\ 3\overline{)57} \end{array}$

5. 500 $\qquad$ 8% = 0.08 $\qquad \begin{array}{r} 500. \\ 0.08\overline{)40.00.} \end{array}$

6. 20 $\qquad$ 175% = 1.75 $\qquad \begin{array}{r} 20. \\ 1.75\overline{)35.00.} \end{array}$

7. **$75** Divide the part ($22.50) by the rate (30%) to find the base.

$$30\% = 0.3 \qquad \begin{array}{r} \$75. \\ 0.3\overline{)\$22.5.0} \end{array}$$

8. **20** Divide the part (14) by the rate (70%) to find the base.

70% = 0.7   0.7⟌14.0. = 20.

9. **$1,400** Divide the part ($490) by the rate (35%) to find the base.

35% = 0.35   0.35⟌$490.00. = $1400.

10. **$685** Divide the part ($34.25) by the rate (5%) to find the base.

5% = 0.05   0.05⟌$34.25. = $685.

11. **14,830** 16,313 ÷ 1.1 = 14,830

12. **$7,600** $2,356 ÷ 0.31 = $7,600

PAGE 206

1. **(3) $316.80** Multiply the base ($1,320.00) times the rate (24%) to find the part.

24% = 0.24
$$\begin{array}{r} \$1,320 \\ \times\ \ 0.24 \\ \hline 52\ 80 \\ +264\ 0 \\ \hline \$316.80 \end{array}$$

2. **(4) $2,125.00** Divide the part ($425.00) by the rate (20%) to find the base.

20% = .20   0.20⟌$425.00. = $2125.

3. **(1) 22%** Divide the part ($290.40) by the base ($1,320.00) to find the rate.

$1,320⟌$290.40 = 0.22   0.22 = 22%

4. **(3) $134.60** Subtract the lowest commission ($290.40) from the highest commission ($425.00) to find the difference.

5. **(2) 20%** Divide the part ($870.00) by the base ($4,350.00) to find the rate.

$4,350⟌$870.0 = 0.2   0.2 = 20%

6. **(3) $7,564.00** Divide the part ($1,891.00) by the rate (25%) to find the base.

25% = 0.25   0.25⟌$1,891.00. = $7564.

7. **(2) $1,248.21** Multiply the base ($5,427.00) times the rate (23%) to find the part.

23% = 0.23
$$\begin{array}{r} \$5,427 \\ \times\ \ 0.23 \\ \hline \$1,248.21 \end{array}$$

8. **Find the commission for the couch. $630 × 0.06 = $37.80
Find the commission for the dining room set. $820 × 0.05 = $41.00
Subtract to find the difference in commissions. $41.00 − $37.80 = $3.20**

PAGE 208

2. **25%** $8.80
$$\begin{array}{r} \$8.80 \\ -\ 6.60 \\ \hline \$2.20 \end{array}$$
$8.80⟌$2.20.00 = 0.25   0.25 = 25%

3. **40%**
$$\begin{array}{r} 25 \\ -\ 15 \\ \hline 10 \end{array}$$
25⟌10.0 = 0.4   0.4 = 40%

4. **12%** $\frac{576}{4,800} = \frac{?}{100}$ 576 × 100 = 57,600
57,600 ÷ 4,800 = 12
12 ÷ 100 = 12%

5. **10%** $\frac{7}{70} = \frac{?}{100}$ 7 × 100 = 700
700 ÷ 70 = 10
10 ÷ 100 = 10%

6. **6%** $\frac{\$1.44}{\$24.00} = \frac{?}{100}$ $1.44 × 100 = $144.00
$144.00 ÷ 24.00 = 6
6 ÷ 100 = 6%

7. **50%** Subtract the original premium ($750.00) from the new premium ($1,125.00). Then divide by the original premium.
$1,125.00 − $750.00 = $375.00
$375.00 ÷ $750.00 = 0.50
0.50 = 50%

8. **15%** Subtract the new fare ($23.80) from the original fare ($28.00). Then divide by the original fare.
$28.00 − $23.80 = $4.20
$4.20 ÷ $28.00 = 0.15
0.15 = 15%

PAGE 210

1. **(3) 30** The colored bar for dinner in September reaches halfway between the marks for *140* and *160*. So, the number of customers in September was 150. The number of dinner customers in August was 120. Subtract to find the difference. $150 - 120 = 30$

2. **(3) 25%** Divide the change in the number of dinner customers (30; see the explanation for Item 1) by the number of dinner customers in August (120).

$$\begin{array}{r} 0.25 \\ 120\overline{)30.00} \end{array} \qquad 0.25 = 25\%$$

3. **(2) 18%** Add the number of customers for breakfast (60), lunch (130), and dinner (150) in September. Then divide the part (number of breakfast customers, 60) by the base (total customers) to find the rate.

$$\begin{array}{r} 60 \\ 130 \\ +150 \\ \hline 340 \end{array} \qquad \begin{array}{r} 0.176 \\ 340\overline{)60.000} \end{array} \qquad \begin{array}{l} 0.176 \text{ rounds to } 0.18. \\ 0.18 = 18\% \end{array}$$

4. **(3) 20%** Subtract the number of customers for breakfast in August (50) from the number of customers for breakfast in September (60). Then divide by the number of customers in August.

$$\begin{array}{r} 60 \\ -50 \\ \hline 10 \end{array} \qquad \begin{array}{r} 0.2 \\ 50\overline{)10.0} \end{array} \qquad 0.2 = 20\%$$

5. **(1) 7%** Subtract the number of customers for lunch in September (130) from the number of customers for lunch in August (140). Then divide by the number in August.

$$\begin{array}{r} 140 \\ -130 \\ \hline 10 \end{array} \qquad \begin{array}{r} 0.071 \\ 140\overline{)10.000} \end{array} \qquad \begin{array}{l} 0.071 \text{ rounds to } 0.07. \\ 0.07 = 7\% \end{array}$$

6. **(4) 650** Find the number of customers for September: $60 + 130 + 150 = 340$ (see the explanation for Item 3). Find the number of customers for August: $50 + 140 + 120 = 310$. Add the two totals: $340 + 310 = 650$.

7. Add the number of customers for breakfast (50), lunch (140), and dinner (120) in August. Then add the number of customers for breakfast (60), lunch (130), and dinner (150) in September. Subtract the total for August (310) from the total for September (340). Then divide by the total for August.

PAGES 211–212

1. 90 $30\% = 0.3$ $\begin{array}{r} 90. \\ 0.3.\overline{)27.0.} \end{array}$

2. 2,000 $3\% = 0.03$ $\begin{array}{r} 2000. \\ 0.03.\overline{)60.00.} \end{array}$

3. $60 $68\% = 0.68$ $\begin{array}{r} \$60. \\ 0.68.\overline{)\$40.80.} \end{array}$

4. 20 $20\% = 0.2$ $\begin{array}{r} 20. \\ 0.2.\overline{)4.0.} \end{array}$

5. $28 $125\% = 1.25$ $\begin{array}{r} \$28. \\ 1.25.\overline{)\$35.00.} \end{array}$

6. 48 $75\% = 0.75$ $\begin{array}{r} 48. \\ 0.75.\overline{)36.00.} \end{array}$

7. **(5) $120 ÷ 80%** Divide the part ($120) by the rate (80%) to find the base.

8. **(3) 9,610 ÷ 62%** Divide the part (9,610) by the rate (62%) to find the base.

9. **(2) 8%** Subtract the original premium ($216.00) from the new premium ($233.28). Then divide by the original premium.

10. **(3) 10%** Subtract Stan's current wage ($6.50) from his new wage ($7.15). Then divide by his current wage.

$$\begin{array}{r} {}^{6}\ {}^{11} \\ \$7.\cancel{1}5 \\ -\ 6.50 \\ \hline \$0.65 \end{array} \qquad \begin{array}{r} 0.1 \\ \$6.50.\overline{)\$0.65.0} \end{array} \qquad 0.1 = 10\%$$

11. **$14.40** Divide the part ($3.60) by the rate (25%) to find the base.

$$25\% = 0.25 \qquad \begin{array}{r} \$14.40 \\ 0.25.\overline{)3.60.00} \end{array}$$

12. **$10.80** Subtract the amount you save ($3.60) from the regular price (see the explanation for Item 11).

$$
\begin{array}{r}
{\overset{3\ \ 14}{\$1\cancel{4}.\cancel{4}0}} \\
-\ \ \ 3.60 \\
\hline
\$10.80
\end{array}
$$

13. **89%** Subtract the amount of July's gas bill ($10) from January's gas bill ($95). Then divide by the amount of January's gas bill.

$$
\begin{array}{r}
\$95 \\
-\ 10 \\
\hline
\$85
\end{array}
\qquad
\begin{array}{r}
0.894 \\
\$95\overline{)\$85.000}
\end{array}
\qquad
\begin{array}{l}
0.894 \text{ rounds to } 0.89. \\
0.89 = 89\%
\end{array}
$$

14. **88%** Subtract the amount of January's electric bill ($40) from July's electric bill ($75). Then divide by the amount of January's electric bill.

$$
\begin{array}{r}
\$75 \\
-\ 40 \\
\hline
\$35
\end{array}
\qquad
\begin{array}{r}
0.875 \\
\$40\overline{)\$35.000}
\end{array}
\qquad
\begin{array}{l}
0.875 \text{ rounds to } 0.88. \\
0.88 = 88\%
\end{array}
$$

15. **$1,397.44** Multiply the base ($6,352.00) times the rate (22%) to find the part.

$$
22\% = 0.22
\qquad
\begin{array}{r}
\$6,352 \\
\times\ \ \ \ 0.22 \\
\hline
\$1,397.44
\end{array}
$$

16. **$4,163** Divide the part ($874.23) by the rate (21%) to find the base.

$$
21\% = 0.21
\qquad
\begin{array}{r}
\$4163. \\
0.21\overline{)\$874.23.}
\end{array}
$$

17. **25%** Divide the part ($1,310.00) by the base ($5,240.00) to find the rate.

$$
\begin{array}{r}
0.25 \\
\$5,240\overline{)\$1,310.00}
\end{array}
\qquad
0.25 = 25\%
$$

18. a. **$770.40**   base × rate = part
     $4,280 × .18 = $770.40

    b. **$1,321.56** $7,342 × .18 = $1,321.56

    c. **$1,082.70** $6,015 × .18 = $1,082.70

SECTION 16

PAGE 214

1. **(4) 135** Multiply the base (500) times the rate (27%) to find the part.
   $500 \times 0.27 = 135$

2. **(2) $\frac{2}{5}$ or 40%** The probability is the sum of the probabilities of running or jogging and walking.
   $27\% + 13\% = 40\%$ which equals $\frac{2}{5}$.

3. **(2) 20%** Divide the part (12) by the base (60) to find the rate.
   $12 \div 60 = 0.2 = 20\%$

4. **(4) 67%** Divide the part (28 + 12 = 40) by the base (60) to find the rate.
   $40 \div 60 - 0.666$, rounds to 67%

5. **0.2% of the population** $60 \div 30,000 = 0.002$; $0.002 = 0.2\%$; **Answers may vary: The size of the sample surveyed is probably too small to predict probability accurately. A larger sample of the population would make the survey a better predictor of probability.**

PAGE 215

2. **1 or 100%** A box contains 15 blue marbles. Each of the 15 outcomes is equally likely to occur. The probability is 1 or 100% that a blue marble will be picked from the box. There is no other possibility.

3. **0 or 0%** None of the pairs of socks in the drawer is brown, so there is no possibility that the pair of socks picked will be brown.

4. **0 or 0%** There is 100% chance of rain tomorrow. Since there is 100% chance of rain, then there is no other possibility. Therefore, the probability that it will *not* rain is 0%.

PAGE 216

2. **0 or 0%** The numbers on the number cube are from 1 to 6, so you cannot roll a 9. The probability of rolling a 9 is 0 or 0%.

3. **1 or 100%** All of the possible rolls of the number cube are numbers less than 7, so the probability of rolling a number less than 7 is 1 or 100%.

4. **$\frac{1}{2}$ or 50%**

$$
\frac{\text{number of ways to roll an even number}}{\text{number of possible outcomes}} = \frac{3}{6}
$$

$$
= \frac{1}{2} \text{ or } 50\%
$$

PAGE 218

2. $\frac{1}{2}$ **or 50%** There are 5 even numbers on the wheel.

$$\frac{\text{number of sections with even numbers}}{\text{number of sections}} = \frac{5}{10} = \frac{1}{2} \text{ or } 50\%$$

3. $\frac{3}{5}$ **or 60%** The numbers on the wheel that are greater than 4 are 5, 6, 7, 8, 9, and 10.

$$\frac{\text{number of sections greater than 4}}{\text{number of sections}} = \frac{6}{10} = \frac{3}{5} \text{ or } 60\%$$

4. **1 or 100%** All of the numbers on the wheel are less than 12. So the probability of the wheel stopping on a number less than 12 is 1 or 100%.

5. $\frac{3}{8}$ **or 37.5%**

$$\frac{\text{number of gray marbles}}{\text{number of marbles}} = \frac{3}{8} \text{ or } 37.5\%$$

6. $\frac{3}{4}$ **or 75%**

$$\frac{\text{number of marbles that are black or gray}}{\text{number of marbles}} = \frac{6}{8} = \frac{3}{4} \text{ or } 75\%$$

7. $\frac{5}{8}$ **or 62.5%**

$$\frac{\text{number of marbles that are not black}}{\text{number of marbles}} = \frac{5}{8} \text{ or } 62.5\%$$

$$5 \div 8 = 0.625 = 62.5\%$$

8. $\frac{1}{5}$ **or 20%**

$$\frac{\text{number of white marbles}}{\text{number of marbles}} = \frac{4}{20} = \frac{1}{5} \text{ or } 20\%$$

9. $\frac{7}{100}$ **or 7%**

$$\frac{\text{number of defective jeans}}{\text{number of jeans}} = \frac{35}{500} = \frac{35 \div 5}{500 \div 5} = \frac{7}{100}$$
$$= 7\%$$

PAGE 220

2. $\frac{1}{4}$ **or 25%** After a blank card is drawn, there are 4 cards left, *Drive* and 3 blanks.

$$\frac{\text{number of cards with } Drive}{\text{number of cards left}} = \frac{1}{4} \text{ or } 25\%$$

3. $\frac{2}{3}$ **or $66\frac{2}{3}$%** After two blank cards are drawn, there are 3 cards left, *Drive* and 2 blanks.

$$\frac{\text{number of blank cards}}{\text{number of cards left}} = \frac{2}{3} \text{ or } 66\frac{2}{3}\%$$

4. $\frac{1}{3}$ **or $33\frac{1}{3}$%**

$$\frac{\text{number of \$100 bills}}{\text{number of bills}} = \frac{3}{9} = \frac{1}{3} \text{ or } 33\frac{1}{3}\%$$

5. $\frac{1}{4}$ **or 25%** After a \$100 bill is drawn, there are 8 bills left, 6 \$20 bills and 2 \$100 bills.

$$\frac{\text{number of \$100 bills}}{\text{number of bills}} = \frac{2}{8} = \frac{1}{4} \text{ or } 25\%$$

6. $\frac{2}{7}$ **or 28.5%** After a \$20 bill and \$100 bill are drawn, there are 7 bills left, 5 \$20 bills and 2 \$100 bills.

$$\frac{\text{number of \$100 bills}}{\text{number of bills}} = \frac{2}{7} \text{ or } 28.5\%$$

$$2 \div 7 = 0.285 = 28.5\%$$

7. $\frac{1}{25}$ **or 4%**

$$\frac{\text{Amy's ticket}}{\text{number of tickets}} = \frac{1}{25} \text{ or } 4\%$$

8. $\frac{8}{22}$ **or 36%** After three tickets are drawn, there are 22 tickets left.

$$\frac{\text{number of men}}{\text{number of tickets}} = \frac{8}{22} \text{ or } 36\%$$

PAGE 222

1. **(3) 25** Multiply the radius (4 ft.) by 2 to find the diameter (4 ft. $\times$ 2 = 8 ft.).
   $$\begin{aligned} C &= \pi d \\ &= 3.14 \times 8 \\ &= 25.12 \text{ feet} \end{aligned}$$
   25.12 feet rounds to 25 feet.

2. **(4) 75** $\begin{aligned} C &= \pi d \\ &= 3.14 \times 24 \\ &= 75.36 \text{ feet} \end{aligned}$
   75.36 feet rounds to 75 feet.

3. **(2) 38** $\begin{aligned} C &= \pi d \\ &= 3.14 \times 12 \\ &= 37.68 \text{ meters} \end{aligned}$
   37.68 meters rounds to 38 meters.

4. **(2) 63** Multiply the radius (10 in.) by 2 to find the diameter (10 $\times$ 2 = 20 in.).
   $$\begin{aligned} C &= \pi d \\ &= 3.14 \times 20 \\ &= 62.8 \text{ inches} \end{aligned}$$
   62.8 inches rounds to 63 inches.

5. **(2) 3.14 × 4.5**
$$C = \pi d$$
$$= 3.14 \times 4.5$$

6. **(4) 201** Multiply the radius (32 in.) by 2 to find the diameter ($32 \times 2 = 64$ in.).
$$C = \pi d$$
$$= 3.14 \times 64$$
$$= 200.96 \text{ inches}$$
200.96 inches rounds to 201 inches.

7. **(3) 78.5 ft.**
$$C = \pi d$$
$$= 3.14 \times 25$$
$$= 78.5 \text{ ft.}$$

8. **The ratio of the circumference to the diameter is the same. This ratio is π, or about 3.14.**

PAGE 224

1. **(4) 201**
$$A = \pi r^2$$
$$= 3.14 \times 8 \times 8$$
$$= 3.14 \times 64$$
$$= 200.96 \text{ square feet}$$
200.96 square feet rounds to 201 square feet.

2. **(5) 254** Divide the diameter (18 ft.) by 2 to find the radius ($18 \div 2 = 9$).
$$A = \pi r^2$$
$$= 3.14 \times 9 \times 9$$
$$= 3.14 \times 81$$
$$= 254.34 \text{ square feet}$$
254.34 square feet rounds to 254 square feet.

3. **(4) 201** Divide the diameter (16 in.) by 2 to find the radius ($16 \div 2 = 8$).
$$A = \pi r^2$$
$$= 3.14 \times 8 \times 8$$
$$= 3.14 \times 64$$
$$= 200.96 \text{ square inches}$$
200.96 square inches rounds to 201 square inches.

4. **(3) 3.14 × 12 × 12**
$$A = \pi r^2$$
$$= 3.14 \times 12 \times 12$$

5. **(3) 154**
$$A = \pi r^2$$
$$= 3.14 \times 7 \times 7$$
$$= 3.14 \times 49$$
$$= 153.86 \text{ square feet}$$
153.86 square feet rounds to 154 square feet.

6. **(5) Not enough information is given.** You do not know the diameter, radius, or area of the circle.

7. **Find the area of the large circle: 3.14 × 12 × 12. Find the area of the small circle: 3.14 × 3 × 3. Find the difference of the two areas to find the area of the colored region.**

PAGES 225–227

1. **100** Add the numbers from each section: $34 + 20 + 10 + 12 + 24 = 100$

2. **20%** Divide the part (20) by the base (100) to find the rate.

3. **$\frac{1}{10}$ or 10%** Divide the part (10) by the base (100) to find the rate.

4. **1 or 100%** The drawer contains only multi-colored scarves. The probability is 1 or 100% that a multi-colored scarf will be taken from the drawer. There is no other possibility.

5. **0 or 0%** There is 100% chance of snow. Since there is 100% chance of snow, there is no other possibility. Therefore, the probability that it will *not* snow is 0%.

6. **$\frac{1}{5}$ or 20%**
$$\frac{\text{number of ways you can choose a green marble}}{\text{number of marbles}} = \frac{1}{5} \text{ or } 20\%$$

7. **$\frac{4}{5}$ or 80%**
$$\frac{\text{number of ways you can choose a marble that is not blue}}{\text{number of marbles}} = \frac{4}{5} \text{ or } 80\%$$

8. **$\frac{2}{5}$ or 40%**
$$\frac{\text{number of ways you can choose an orange or yellow marble}}{\text{number of marbles}} = \frac{2}{5} \text{ or } 40\%$$

9. **0 or 0%**
$$\frac{\text{number of ways you can choose a black marble}}{\text{number of marbles}} = \frac{0}{5} = 0 \text{ or } 0\%$$

10. **(2) $\frac{3}{5}$ or 60%**
$$\frac{\text{number of black marbles}}{\text{number of marbles}} = \frac{6}{10}$$
$$= \frac{3}{5} \text{ or } 60\%$$

11. **(1) 0 or 0%** There are no green marbles in the box, so there is no possibility of choosing one.

12. **(2) $\frac{1}{10}$ or 10%**

$$\frac{\text{number of gray marbles}}{\text{number of marbles}} = \frac{1}{10} \text{ or } 10\%$$

13. **(5) $\frac{7}{10}$ or 70%**

$$\frac{\text{number of marbles}}{\text{that are not white}} = \frac{7}{10} \text{ or } 70\%$$

14. **$\frac{1}{3}$ or $33\frac{1}{3}$%** After the gray marble is drawn, there are 9 marbles left: 6 black and 3 white.

$$\frac{\text{number of white marbles}}{\text{number of marbles}} = \frac{3}{9}$$
$$= \frac{1}{3} \text{ or } 33\frac{1}{3}\%$$

15. **$\frac{1}{4}$ or 25%** After the gray and white marbles are drawn, there are 8 marbles left: 6 black and 2 white.

$$\frac{\text{number of white marbles}}{\text{number of marbles}} = \frac{2}{8} = \frac{1}{4} \text{ or } 25\%$$

16. **$\frac{5}{7}$ or 71%** After 1 gray, 1 white, and 1 black marble are drawn, there are 7 marbles left: 5 black and 2 white.

$$\frac{\text{number of black marbles}}{\text{number of marbles}} = \frac{5}{7} \text{ or } 71\%$$

17. **$\frac{1}{3}$ or $33\frac{1}{3}$%** After 1 gray, 1 white, and 2 black marbles are drawn, there are 6 marbles left: 4 black and 2 white.

$$\frac{\text{number of white marbles}}{\text{number of marbles}} = \frac{2}{6}$$
$$= \frac{1}{3} \text{ or } 33\frac{1}{3}\%$$

18. **50 feet; 201 square feet**

$$\begin{aligned} C &= \pi d \\ &= 3.14 \times 16 \\ &= 50.24 \text{ feet} \end{aligned}$$

50.24 feet rounds to 50 feet.
Divide the diameter (16 ft.) by 2 to find the radius (16 ÷ 2 = 8).

$$\begin{aligned} A &= \pi r^2 \\ &= 3.14 \times 8 \times 8 \\ &= 3.14 \times 64 \\ &= 200.96 \text{ square feet} \end{aligned}$$

200.96 square feet rounds to 201 square feet.

19. **63 yards; 314 square yards**
Multiply the radius (10 yd.) by 2 to find the diameter (10 × 2 = 20).

$$\begin{aligned} C &= \pi d \\ &= 3.14 \times 20 \\ &= 62.8 \text{ yards} \end{aligned}$$

62.8 yards rounds to 63 yards.

$$\begin{aligned} A &= \pi r^2 \\ &= 3.14 \times 10 \times 10 \\ &= 3.14 \times 100 \\ &= 314 \text{ square yards} \end{aligned}$$

20. **44 inches; 154 square inches** Multiply the radius (7 in.) by 2 to find the diameter (7 × 2 = 14).

$$\begin{aligned} C &= \pi d \\ &= 3.14 \times 14 \\ &= 43.96 \text{ inches} \end{aligned}$$

43.96 inches rounds to 44 inches

$$\begin{aligned} A &= \pi r^2 \\ &= 3.14 \times 7 \times 7 \\ &= 3.14 \times 49 \\ &= 153.86 \text{ square inches} \end{aligned}$$

153.86 square inches rounds to 154 square inches.

21. **(2) 3.14 × 32**

$$\begin{aligned} C &= \pi d \\ &= 3.14 \times 32 \end{aligned}$$

22. **(3) 3.14 × 3 × 3** Divide the length of the diameter by 2 to find the radius (6 ÷ 2 = 3).

$$\begin{aligned} A &= \pi r^2 \\ &= 3.14 \times 3 \times 3 \end{aligned}$$

23. **82 feet** Multiply the radius (13 ft.) by 2 to find the diameter (13 × 2 = 26).

$$\begin{aligned} C &= \pi d \\ &= 3.14 \times 26 \\ &= 81.64 \text{ feet} \end{aligned}$$

81.64 feet rounds to 82 feet.

24. **254 square inches**

$$\begin{aligned} A &= \pi r^2 \\ &= 3.14 \times 9 \times 9 \\ &= 3.14 \times 81 \\ &= 254.34 \text{ square inches} \end{aligned}$$

254.34 square inches rounds to 254 square inches.

MATH AT WORK
PAGE 229

1. **(4) 3:1**   3 cloves : 1 cup cheese

2. **(1)** $\frac{6}{8} = \frac{4}{x}$ $\dfrac{\text{6 people—orig. recipe}}{\text{8 people to be served}} =$

   $\dfrac{\text{4 peppers—orig. recipe}}{?}$

3. **(4) 50%**   3 cloves ÷ 6 people = .5 = 50%

UNIT 4 REVIEW
PAGES 230–231

1. **14** $8 \times 35 = 280; \; 280 \div 20 = 14$

2. **324**   450% = 4.5

   $$\begin{array}{r} 72 \\ \times\ 4.5 \\ \hline 324.0 \end{array}$$

3. **2.5%** $\begin{array}{r} 0.025 \\ 320\overline{)8.000} \end{array}$ 0.025 = 2.5%

4. **$45** $\begin{array}{r} \$45. \\ 0.09.\overline{)\$4.05.} \end{array}$ 9% = 0.09

5. **40%** $\begin{array}{r} \$17.50 \\ -\ 12.50 \\ \hline \$5.00 \end{array}$ $\begin{array}{r} 0.4 \\ \$12.50.\overline{)\$5.00.0} \end{array}$

   0.4 = 40%

6. **67%** $\begin{array}{r} 45 \\ -15 \\ \hline 30 \end{array}$ $\begin{array}{r} 0.666 \\ 45\overline{)30.000} \\ -27\ 0 \\ \hline 3\ 00 \\ -2\ 70 \\ \hline 300 \end{array}$

   0.666 rounds to 67%

7. **17%** $\begin{array}{r} 60 \\ -50 \\ \hline 10 \end{array}$ $\begin{array}{r} 0.1666 \\ 60\overline{)10.0000} \\ -6\ 0 \\ \hline 4\ 00 \\ -3\ 60 \\ \hline 400 \\ -360 \\ \hline 400 \end{array}$

   0.1666 rounds to 17%

8. **$151.20** Subtract the down payment from the cost to find the principal.
   $520 − $100 = $420.
   Use the formula to find the interest.
   $i \ = \ p \times r \times t$
   $= \ \$420 \times 0.18 \times 2$
   $= \ \$151.20$

9. **$23.80** Add the principal and interest (see the explanation for Item 8) to find the amount to be paid back.
   $420 + $151.20 = $571.20
   Divide the amount to be paid back by the number of monthly payments.
   (2 years = 24 months)
   $571.20 ÷ 24 = $23.80

10. **$7.60**
    $\dfrac{3}{\$2.85} = \dfrac{8}{?}$ $2.85 × 8 = $22.80
    $22.80 ÷ 3 = $7.60

11. **$995.94** Multiply the profit times the rate to find the commission earned.
    $4,527 × 0.22 = $995.94

12. **(3)** $\frac{3}{5}$ **or 60%**
    $\dfrac{\text{number of nickels}}{\text{number of coins}} = \dfrac{3}{5} = 60\%$

13. **(3)** $\frac{1}{2}$ **or 50%** After a nickel has been chosen, there are 4 coins left: 2 nickels and 2 dimes.
    $\dfrac{\text{number of nickels}}{\text{number of coins}} = \dfrac{2}{4} = \dfrac{1}{2}$ or 50%

14. **(3) 108°** Multiply the base (360° in a circle) times the rate (30%) to find the part.

    30% = 0.3 $\begin{array}{r} 360 \\ \times\ 0.3 \\ \hline 108.0 \end{array}$

15. **(2) 188 inches** Multiply the radius (30) times 2 to find the diameter (30 × 2 = 60).
    $C \ = \ \pi d$
    $= \ 3.14 \times 60$
    $= \ 188.4$ inches
    188.4 inches rounds to 188 inches.

16. **28 square feet** Divide the diameter (6 feet) by 2 to find the radius (6 ÷ 2 = 3). Use the area formula.
    $A \ = \ \pi r^2$
    $= \ 3.14 \times 3 \times 3$
    $= \ 28.26$ square feet
    28.26 square feet rounds to 28 square feet.

17. **44%** Divide the part (600 + 500 = 1,100) by the base (2,500) to find the probability.
    $1,100 \div 2,500 = \frac{11}{25}$ or 44%

18. **28%** To find the probability, divide the part (700) by the base (2,500).
    $700 \div 2,500 = \frac{7}{25}$ or 28%

MATH CONNECTION: RATIOS, PROPORTIONS, PERCENTS AND EARTH SCIENCE

PAGE 233

1. **(1)** 10%
2. **(3)** at a certain depth, the ice in the glacier appears blue.
3. **(5)** 6:1 Ice Age: Today = 32% + 30%: 10% = 62%: 10% = 6: 1
4. **(2)** 3 meters
5. **(4)** Glaciers are not found near the Equator.
6. **Answers may vary.** Sample answer below. I believe scientists are concerned that glaciers are receding at a faster rate because it shows that our planet is getting warmer and warmer. Since glaciers hold most of our fresh water, that could be a problem for living things that need fresh water to grow.

UNIT 4 CUMULATIVE REVIEW

PAGES 234–235

1. six million, thirty thousand, seventeen
2. seven and ninety-six thousandths
3. $54,2\underline{7}7 > 54,2\underline{2}7$ 7 is greater than 2.
4. $0.083 < 0.83$
   Add a zero: $0.\underline{0}83$
   $0.\underline{8}30$
   Then compare: 0 is less than 8.
5. $5.10 = 5.1$
6. $9\frac{5}{12}$ $\qquad 5\frac{2}{3} = 5\frac{8}{12}$
   $\qquad\qquad +3\frac{3}{4} = 3\frac{9}{12}$
   $\qquad\qquad\quad 8\frac{17}{12} = 8 + 1\frac{5}{12} = 9\frac{5}{12}$
7. $2\frac{11}{30}$ $\qquad 7\frac{1}{6} = 7\frac{5}{30} = 7\frac{\overset{6\ \overset{35}{\not{5}}}{\not{5}}}{30}$
   $\qquad\qquad -4\frac{4}{5} = 4\frac{24}{30} = 4\frac{24}{30}$
   $\qquad\qquad\qquad\qquad\qquad 2\frac{11}{30}$
8. 21 $\qquad 5 \times 4 - 2 + 9 \div 3$
   $\qquad\qquad 20 - 2 + 9 \div 3$
   $\qquad\qquad\quad 20 - 2 + 3$
   $\qquad\qquad\qquad\quad 18 + 3$
   $\qquad\qquad\qquad\qquad 21$
9. 5 $5 \times 5 = 25$, so $\sqrt{25} = 5$
10. 81 $9^2 = 9 \times 9 = 81$
11. $3.04
12. 4,560
13. 0.00485
14. 1,108
15. $5.52
16. $11\frac{1}{4}$ $3\frac{3}{8} \times 3\frac{1}{3} = \frac{\overset{9}{\not{27}}}{\not{8}} \times \frac{\overset{5}{\not{10}}}{\not{3}} = \frac{45}{4} = 11\frac{1}{4}$
17. $\frac{10}{13}$ $2\frac{1}{2} \div 3\frac{1}{4} = \frac{5}{2} \div \frac{13}{4} = \frac{5}{\not{2}} \times \frac{\overset{2}{\not{4}}}{13} = \frac{10}{13}$
18. 59.048
19. 4.674
20. **(5)** 4.3 + 4.3 + 2.5 + 2.5 To find the perimeter, add the lengths of all four sides.
21. **(4)** 15% 2 grams × 9 = 18
   $\frac{\text{part}}{\text{base}} = \frac{?}{100}$
   $\frac{18}{120} = \frac{?}{100} \qquad 18 \times 100 = 1,800$
   $\qquad\qquad\qquad\quad 1,800 \div 120 = 15$
   $\frac{15}{100} = 15\%$
22. **(4)** 3° Start at −5 and count 8 units to the right.
23. **(4)** $62.80 Multiply the cost of a 2-day pass for children ($11.95) times the number of children (3). Then add the cost of one 2-day pass for adults.
   $\qquad$ $11.95 $\qquad$ $35.85
   $\qquad \underline{\times \qquad 3}$ $\underline{+ \ 26.95}$
   $\qquad$ $35.85 $\qquad$ $62.80
24. **41** To find the mean, first find the total of the numbers of the set of data.
   46 + 39.5 + 41.5 + 38.5 + 39 + 41.5 = 246
   Then divide the total by the number of items in the set. 246 ÷ 6 = 41
25. **40.5** To find the median, arrange the numbers in order. Find the middle numbers.
   38.5, 39, <u>39.5</u>, <u>41.5</u>, 41.5, 46
   Then find the mean of the middle numbers.
   39.5 + 41.5 = 81; 81 ÷ 2 = 40.5
26. **150** The bar for Friday is the tallest. It reaches the 4 and represents 400 deliveries. The bar for Tuesday is the smallest. It ends between 2 and 3 and represents 250 deliveries. Subtract to find the difference. 400 − 250 = 150
27. **1,650** Read the bars for each day. Then add the deliveries.
   Mon. $\qquad$ 350
   Tues. $\qquad$ 250
   Wed. $\qquad$ 300
   Thurs. $\qquad$ 350
   Fri. $\qquad \underline{+400}$
   $\qquad\qquad$ 1,650

POSTTEST

PAGES 236–241

1. **two million**
2. **five hundredths**
3. **25,7̲44 > 25,4̲47**
   7 is greater than 4.
4. **0.83 > 0.083**
   Add a zero: 0.8̲30   0.0̲83 Then compare.
   830 is greater than 83.
5. **421,000**
   42⓪,536 The number to the right of 0 is
   5; add 1 to the circled digit.
6. **7.8**
   7.⑧39 The number to the right of 8 is less
   than 5. Do not change the circled digit.
7. **$1\frac{5}{6}$** Each figure is divided into 6 equal parts.
   One figure is completely shaded, and 5 of
   the 6 parts of the last figure are shaded.
8. **$3.17**
9. **4.646**
10. **292,878**
11. **208 r21** $38\overline{)7,925}$
    $$\begin{array}{r} 208\ \text{r}21 \\ 38\overline{)7,925} \\ \underline{-7\ 6} \\ 325 \\ \underline{-304} \\ 21 \end{array}$$
12. **$9\frac{11}{20}$**
    $$\begin{array}{r} 3\frac{3}{4} = 3\frac{15}{20} \\ +5\frac{4}{5} = 5\frac{16}{20} \\ \hline 8\frac{31}{20} = 8 + 1\frac{11}{20} = 9\frac{11}{20} \end{array}$$
13. **$3\frac{19}{24}$**
    $$\begin{array}{r} 7\frac{1}{6} = 7\frac{4}{24} = 7\frac{\overset{6\ \ 28}{4}}{24} \\ -3\frac{3}{8} = 3\frac{9}{24} = 3\frac{9}{24} \\ \hline 3\frac{19}{24} \end{array}$$
14. **14**
    $$\begin{array}{c} 8 + 2 \times 9 \div (7 - 4) \\ 8 + 2 \times 9 \div \quad 3 \\ 8 + \quad 18 \quad \div \quad 3 \\ 8 + \quad\quad 6 \\ 14 \end{array}$$
15. **563**
16. **40.196**
17. **384**
18. **0.00536**
19. **$5\frac{3}{5}$** $2\frac{2}{3} \times 2\frac{1}{10} = \frac{\overset{4}{\cancel{8}}}{\underset{1}{\cancel{3}}} \times \frac{\overset{7}{\cancel{21}}}{\underset{5}{\cancel{10}}} = \frac{28}{5} = 5\frac{3}{5}$
20. **$3.46**

21. **62.5**
22. **$1\frac{1}{5}$** $6\frac{2}{5} \div 5\frac{1}{3} = \frac{32}{5} \div \frac{16}{3} = \frac{\overset{2}{\cancel{32}}}{5} \times \frac{3}{\underset{1}{\cancel{16}}} = \frac{6}{5} = 1\frac{1}{5}$
23. **216** $6^3 = 6 \times 6 \times 6 = 216$
24. **0.035** $3\frac{1}{2}\% = 3.5\% = .0\underset{\frown}{3}5 = 0.035$
25. **650%** $6.5 = 6.\underset{\frown}{50}. = 650\%$
26. **$\frac{3}{4}$** $75\% = \frac{75}{100} = \frac{75 \div 25}{100 \div 25} = \frac{3}{4}$
27. **60%** $3.0 \div 5 = 0.6$ $0.6 = 60\%$
28. **$14.40** $40\% = 0.4$ $\$36 \times 0.4 = \$14.40$
29. **125%** $20 \div 16 = 1.25$ $1.25 = 125\%$
30. **400** $2\% = 0.02$ $8.00 \div 0.02 = 400$
31. **30** $40 \times 9 = 360; \ 360 \div 12 = 30$
32. **(4) 144** To find the area, multiply the
    length (18 feet) times the width (8 feet).
    $18 \times 8 = 144$
33. **(3) 60** Compare the number of cash
    customers (63) with the total number of
    customers (121). Round the numbers to
    the nearest ten and subtract to find the
    difference.
    $$\begin{array}{rcr} 121 & \text{rounds to} & 120 \\ -\ 63 & \text{rounds to} & -\ 60 \\ \hline & & 60 \end{array}$$
34. **(2) $\frac{2}{5}$ or 40%**
    $\dfrac{\text{number of white marbles}}{\text{number of marbles}} = \frac{4}{10} = \frac{2}{5}$ or 40%
35. **(4) $2\frac{1}{2} \times 2$** To double the recipe, multiply
    the amount of milk by 2.
36. **(4) 200 ÷ 10** The area of the rectangle
    and the width are given. Divide the area by
    the width to find the length.
37. **(2) 2** Multiply the base (8) times the rate
    (25%) $8 \times 0.25 = 2.00$
38. **(5) $25.98 ÷ 3** To find the unit cost,
    divide the total cost ($25.98) by the
    number of shirts (3).
39. **(2) $75.40 ÷ $1,200.00** Divide the part
    ($75.40) by the base ($1,200.00) to find
    the rate.
40. **(4) $17.00** Divide the part ($6.80) by the
    rate (40%) to find the base.
    $\$6.80 \div 0.40 = \$17.00$
41. **(4) $\frac{6}{\$4.98} = \frac{10}{?}$** Both ratios are in the same
    order. The first ratio means that 6 pairs
    cost $4.98. The second ratio means that 10
    pairs cost an unknown amount.

42. **(4) ($8.60 − $8.30) ÷ $8.60** Subtract the new wage ($8.30) from Marge's current wage ($8.60) to find the amount of change. Then divide by the original amount, her current wage.

43. **2,400 cubic inches** $V = l \times w \times h$
$$= 20 \times 10 \times 12$$
$$= 2,400$$

44. **7 miles** Round each distance and add to find the total.

$2\frac{1}{5}$	rounds to	2
$3\frac{9}{10}$	rounds to	4
$+1\frac{3}{10}$	rounds to	$+1$
		7

45. **63 tables** Divide the number of reservations for Saturday in the ballroom (504) by the number that can be seated at each table (8).
$$504 \div 8 = 63$$

46. **154 square feet** $A = \pi r^2$
$$= 3.14 \times 7 \times 7$$
$$= 3.14 \times 49$$
$$= 153.86$$
153.86 square feet rounds to 154 square feet.

47. **$5.02** Add the cost of the paint and the brush to find the subtotal. Add the tax to find the total. Then subtract the total from the amount of cash the customer gave George.

$\overset{1\ 1}{\$11.37}$	$\$14.26$	$\$\overset{1\ 9\ 9\ 10}{\cancel{2}\cancel{0}.\cancel{0}\cancel{0}}$
$+\ \ 2.89$	$+\ \ 0.72$	$-\ \ 14.98$
$\$14.26$	$\$14.98$	$\$5.02$

48. **3.5 inches** Add the number of inches for August (0.9) and the number of inches for September (2.6).

$$\begin{array}{r} \overset{1}{0.9} \\ +2.6 \\ \hline 3.5 \end{array}$$

49. **126 inches** Multiply the radius (20 in.) by 2 to find the diameter (20 × 2 = 40 in.)
$C = \pi d$
$$= 3.14 \times 40$$
$$= 125.6$$
125.6 inches rounds to 126 inches.

50. **32** Find the total of the numbers in the set.
$$42 + 36 + 19 + 31 = 128$$
Then divide by the number of items in the set.
$$128 \div 4 = 32$$

51. $\frac{3}{2}$
$$\frac{18 \text{ warehouse employees}}{12 \text{ drivers}} = \frac{18 \div 6}{12 \div 6} = \frac{3}{2}$$

52. **15%** Add the amounts on the graph to find the total. Multiply by 100.
$6 + $4 + $4 + $3 + $3 = $20;
$20 × 100 = $2,000
Divide the part ($300) by the base ($2,000) to find the rate.
$300 ÷ $2000 = 0.15   0.15 = 15%

53. **+3 degrees** Start at -3 on the number line. Count 6 units to the right.

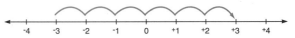

CALCULATOR HANDBOOK

PAGE 243

1. **$628** Add the amounts for the cost of the oven, tax, and delivery:
$499 + $29 + $25 + $75 = $628.

2. **$1,820** Add the amounts:
$780 + $345 + $290 + $85 + $320 = $1,820.

PAGE 244

1. **8 points** Subtract the Bluebirds' first-half score from the Redwings' first-half score:
47 − 39 = 8.

2. **27 points** Subtract the Bluebirds' final score from the Redwings' final score:
116 − 89 = 27.

3. **931** Subtract the paid attendance from the number of people who attended the game:
14,899 − 13,968 = 931.

PAGE 245
1. **15,914 square meters** Multiply the length times the width: $218 \times 73 = 15,914$.
2. **768 cubic inches** Multiply the length times the width times the height: $12 \times 8 \times 8 = 768$.

PAGE 246
1. **$2,500,000** Divide the total lottery winnings by the number of winners: $\$32,500,000 \div 13 = \$2,500,000$.
2. **$15,500** Divide the state grant by the number of counties in the joint project: $\$93,000 \div 6 = \$15,500$.

PAGE 247
1. **3,125** $5 \times 5 \times 5 \times 5 \times 5 = 3,125$
2. **6.25** $2.5 \times 2.5 = 6.25$
3. **25** $\sqrt{625} = 25$
4. **17** $\sqrt{289} = 17$

PAGE 248
1. **20,132.8 miles** Add the number of miles at which the oil was last changed to the number of miles after which the oil should be changed: $16,632.8 + 3,500 = 20,132.8$.
2. **$10.73** Add the cost of the parts: $\$5.88 + \$4.25 + \$0.60 = \$10.73$.

PAGE 249
1. **350.9 miles** Subtract the May 10 odometer reading from the May 15 odometer reading: $21,734.6 - 21,383.7 = 350.9$.
2. **$5.85** Subtract the amounts from $20: $\$20.00 - \$11.75 - \$1.39 - \$1.01 = \$5.85$

PAGE 250
1. **256.68 square meters** Multiply the length times the width: $20.7 \times 12.4 = 256.68$.
2. **38.465** Multiply the value for π (3.14) times the value of the radius squared: $3.14 \times 3.5 \times 3.5 = 38.465$.

PAGE 251
1. **2.625 feet** Divide the length of the board by the number of pieces: $10.5 \div 4 = 2.625$.
2. **$7.50** Divide the total by the number of people splitting the cost: $\$22.50 \div 3 = \$7.50$.

PAGE 252

Fraction	Decimal	Percent
$\frac{1}{10}$	.1	10%
$\frac{1}{8}$	.125	12.5%
$\frac{1}{5}$	.2	20%
$\frac{1}{4}$	.25	25%
$\frac{1}{3}$	.33	$33\frac{1}{3}\%$
$\frac{2}{5}$	.4	40%
$\frac{1}{2}$	.5	50%
$\frac{3}{5}$	.6	60%
$\frac{2}{3}$	.67	$66\frac{2}{3}\%$
$\frac{3}{4}$	.75	75%
$\frac{4}{5}$	.8	80%
$\frac{9}{10}$	.9	90%
1	1	100%

PAGE 253
1. **$2.19** Multiply the purchase price by the tax. $\$35 \times 6.25\% = 2.1875$. 2.1875 rounds to 2.19.
2. **1%** Divide the number of defective parts (the part) by the number of parts tested (the whole): $5 \div 500 = .01 = 1\%$.

PAGE 254
1. **Subtotal = $215.93; Deposit = $145.93** Add the two checks to find the subtotal: $\$165.43 + \$50.50 = \$215.93$. Subtract the amount of cash received from the subtotal to find the deposit: $\$215.93 - \$70.00 = \$145.93$.
2. **Subtotal = $221.69; Deposit = $166.69** Add the three checks to find the subtotal: $\$85.25 + \$70.89 + \$65.55 = \$221.69$. Subtract the amount of cash received from the subtotal to find the deposit: $\$221.69 - \$55.00 = \$166.69$.

Glossary

addition combining numbers to find a total or sum

area the amount of surface something takes up, measured in square units; for a square or rectangle, the product of the length times the width; for a circle, the product of *pi* (π) times the radius squared

average the usual, or typical, value that can be used to represent a group of values; the mean, the sum of a group of numbers divided by the number of numbers used to find the sum

axis line the vertical and horizontal lines that create a reference point for a graph

bar graph a graph that displays data using horizontal or vertical bars to compare numbers

base (*b*) the whole amount in a percent problem; the base represents 100%; one of the terms in the percent formula; the quotient of the part divided by the rate

base the amount being multiplied by itself when working with exponents

canceling finding values that will divide evenly into the numerators and denominators of fractions to convert them into smaller fractions that are easier to work with

central angle one of the sections of a circle graph, measured in degrees. The central angles of a circle add up to 360°.

chance outcome; the probability that something will happen

circle a closed figure with a curved edge. All points on the edge of a circle are the same distance from the circle's center.

circle graph a graph that represents data as parts of a whole using a circle divided into segments

circumference (*C*) the distance around the edge of a circle; the product of the diameter times *pi* (π)

column information arranged in a table so that it can be read from top to bottom

commission the amount a salesperson earns on a sale; a percent of the sale

common denominator a denominator that is the same for two or more fractions

convert to change from one form to another

cross-multiply multiplying the numerator of one fraction by the denominator of another fraction; used to solve a proportion and to determine if two fractions are equivalent

cross-product the result of cross-multiplication. The cross-products of equivalent fractions are equal.

cube a six-sided figure in which each side is the same-sized square

cube a number multiplied by itself three times, represented by the exponent 3; raised to the third power

cubic foot a box shape that measures one foot on each side (the length, width, and height)

data information that has been gathered to be studied and analyzed

decimal a value that shows part of a whole number; a number containing a decimal point; a fraction that uses the place value system based on the number 10

denominator the bottom number in a fraction that shows the number of parts a whole is divided into

dependent probability the chance of something happening that is affected by another outcome

diameter (*d*) the distance from the edge of a circle, through the center, to the other edge of the circle; twice the length of the radius

difference the answer to a subtraction problem; the result of subtraction

digit one of ten numbers on which all other numbers are based. Our number system uses ten digits: 0, 1, 2, 3, 4, 5, 6, 7, 8, and 9.

digital scale a tool that measures weight using decimal amounts

distance formula the distance traveled in a specific amount of time at a specific speed; the product of the rate times the time

division finding out how many times one number goes into or divides another number

double bar graph a graph that uses two bars in each category as a means of comparison

equation a statement formed by comparing two expressions using an equals sign

equivalent equal

equivalent fractions a fraction that has the same value as another fraction; fractions that are equal

estimation a method for finding an approximate answer or value close to the actual answer or value

exponent the number that shows how many times a number (the base) is multiplied by itself

formula an algebraic equation that shows the constant relationship between certain variables

fraction a value that represents part of a whole

graph a way to represent some mathematical relationships as a picture

higher terms finding an equivalent fraction with a greater numerator and denominator

horizontal axis the part of the graph with values that are read across

improper fraction a fraction with a value equal to or greater than one; a fraction in which the numerator is greater than the denominator

integer any positive or negative whole number or zero

interest (*i*) a charge paid to borrow or use someone else's money; the product of the principal times the rate times the time period

interest rate (*r*) the percent used to figure out how much interest a borrower has to pay

invert to switch the numerator and denominator of the fraction being divided by; to turn a fraction upside down

label a word or name used to identify something

least common denominator the smallest number that the denominators of two or more fractions will divide evenly into

like fractions fractions that have the same, or a common, denominator

like quantities amounts that are expressed in the same, or common, units

line graph a graph that uses a line to show changes over time

lowest terms a fraction with a numerator and denominator that can only be divided evenly by the number one

mathematical expression numbers and symbols representing variables and the mathematical operations used to represent a problem

mean the average; the sum of a group of numbers divided by the number of numbers used to find the sum

median the middle number of a group or set of numbers arranged in order

mixed number a number that contains a whole number and a proper fraction

multiplication the mathematical operation that is the same as adding the same number many times

negative number an integer that is less than zero; a number to the left of zero on a number line

number line a visual that shows the relationship between positive and negative numbers and zero. On a number line, positive numbers are to the right of zero, and negative numbers are to the left of zero.

numerator the top number in a fraction that shows the number of parts of a whole being discussed

order of operations an accepted rule stating the sequence in which to perform mathematical operations such as addition, subtraction, multiplication, and division

outcome the result; the probability that something will happen

part (p) a portion of the whole amount; one of the terms in the percent formula; the product of the base times the rate

percent a way of representing part of a whole equal to 100; means "for every 100" or "out of 100"

percent formula mathematical equation that shows the relationship between the base (b) or whole, rate (r), and part (p)

percent of decrease the rate of change by which an amount went down; the quotient of the amount of change (the original amount minus the new amount) divided by the original amount

percent of increase the rate of change by which an amount went up; the quotient of the amount of change (the new amount minus the original amount) divided by the original amount

perimeter the distance around the outside of a shape; the sum of the measurements of all the sides of an object

pi (π) for any circle, the ratio of its circumference to its diameter; a constant equal to about 22/7 or 3.14

place value the value of a digit that depends on its location in the number

population in statistics, an entire group from which information is gathered

positive number an integer that is greater than zero; a number to the right of zero on a number line

powers of ten any number that is evenly divisible by 10, such as 10, 100, 1,000, and so on

principal (p) the amount of money borrowed on which interest is paid; a factor in the interest formula

probability the study of the chance of something happening

proper fraction a fraction in which the numerator is less than the denominator

proportion an equation comparing two equal ratios or fractions; expressed as a number from 0 to 1 or as a percent; the quotient of the number of ways an outcome can occur divided by the total number of possible outcomes

radius (r) the distance from the center of a circle to any point on the circle's edge; equal to half the length of the diameter

rate a ratio comparing two unlike quantities (amounts that are expressed in different units); a fraction usually with a denominator of one

rate (r) one of the terms (the percent) in the percent formula; always followed by the percent sign (%) or the word *percent*; the quotient of the part divided by the base

rate of commission the percent of the profit earned by a salesperson on the sale of an item

ratio a way of comparing two like quantities (amounts expressed in the same units)

reducing finding an equivalent fraction with a smaller numerator and denominator

remainder an amount left over in a division problem

rounded number an approximate, or estimated, amount

row information arranged on a table so that it can be read across, from left to right

sample in statistics, a smaller group that represents a population

scale a key that gives values to help read a graph or map

square a rectangle that has four sides of equal length; to multiply a number times itself; the exponent 2

square root the number which, when multiplied by itself, gives a specified number

square yard a square that measures one yard on each side

statistics data that is organized and analyzed

subtotal the sum of part of a group of numbers

subtraction finding the difference between amounts; making a comparison between two numbers

sum the total; the answer to an addition problem

survey a way of gathering data

table a grid that organizes information using rows and columns

tax additional money paid based on the amount of a purchase

term one of the numbers in a proportion; the numerator or denominator of a fraction

time (t) the period or duration of something; used to compute distance and interest; often the denominator of a rate

total the sum; the complete cost of something

total cost the amount found by multiplying the number of units by the unit cost

unit cost the price of one item

unlike fractions fractions that have different denominators

unlike quantities amounts that are expressed in different units

variable a letter or symbol used to represent an unknown value

vertical axis the part of the graph with values that are read from bottom to top

volume the amount of space inside a solid or three-dimensional object

whole number an amount made up of one or more digits

Index

A

adding, 24, 44, 243, 248
  calculator skills, 16, 24, 128, 243, 248
  decimals, 126, 128, 248
  fractions, 82, 86
  money, 126
  whole numbers, 22, 24, 243
addition sign, *see* plus sign
algebra topics
  using formulas, 58
  order of operations, 130
  percent formula, 187, 199
  writing and solving equations, 42
amount of change, 207–208
angles, 182
area
  of circle, 223
  of rectangle, 48, 52, 130
  of square, 52
average, 46
axis lines
  of bar graph, 30
  of line graph, 132

B

bar graphs, 30
  double, 209
base, 54, 176–209, 247
  exponent, 54, 247
  percent and, 176, 191
  solving for, 176, 199, 203, 205
base number of squares and cubes, 54, 247

C

calculator handbook, 242–254
calculator skills, 16, 17, 24, 26, 36, 39, 44, 46, 54, 128, 140
canceling, 98, 100
central angle, 182
chance, 213, 215, 217, 219
circle, 221, 223
  area of, 223
  circumference of, 221
  diameter of, 221
  radius of, 221, 223
circle graph, 182, 195, 213

circumference, 221
commission, 205
common denominator, 84, 86, 88
communicate, 23, 29, 35, 117, 165, 171, 177, 194, 210
comparing
  decimals, 120
  fractions, 79
  like quantities, 79
  unlike quantities, 158
  whole numbers, 14
connect, 59, 91, 143, 190, 222
cross-multiplying, 76, 77, 78, 79, 162, 164, 166, 193
cross products, 162
cube, 54, 56, 247
cubic foot, 56
cubic unit, 56

D

data, 46
  in bar graph, 30, 209
data analysis
  budgets and circle graphs, 195
  finding averages mean and median, 46
  making circle graphs, 182, 195
  probability, 213
  reading circle graphs, 182, 195
  reading tables, 82
  using bar graphs, 30
  using double bar graphs, 209
  using line graphs, 132
  using tables, 30, 82
decimals, 114–155, 248–252
  adding, 126, 128, 248
  changing to fractions, 119, 252
  changing percents to, 173, 252
  changing to percents, 173
  comparing, 120
  converting with a calculator, 252
  dividing, 136, 140, 203, 251
  multiplying, 136, 138, 140, 250
  percents and, 170, 172, 173, 203
  placeholders in, 118, 119, 138, 173
  place value and, 116, 118, 119

powers of ten, 142
  reading, 118
  reading decimal scales, 139
  rounding, 120
  subtracting, 128, 249
  writing, 118
  zeros in, 118, 119, 120, 128, 173
denominator, 72
  common, 84, 86, 88
dependent probability, 219
diagram, using, 102
diameter, 221
difference, 26, 30
digits, 12, 13
distance, finding, 58
dividing, 34, 38, 44, 246, 251
  calculator skills 17, 39, 140, 246, 251
  decimals, 136, 140, 251
  fractions, 96, 100
  whole numbers, 34, 38, 246
  money, 38
division bracket, 38
division sign, 38
double bar graph, 209

E

equal or equivalent fractions, 76, 79, 172
equals sign, 14, 42
equation, 42; *see also* formulas
estimate, 18, 19
exponent, 54, 247
expression, 28, 42, 130
extend, 123
extra or extraneous information, 28

F

formulas, 22, 48, 52, 56, 58, 176, 178, 180, 187–193, 199–208
fractions, 68–113, 156–175, 193, 252
  adding, 82, 86
  canceling, 98, 100
  changing decimals to, 119, 252
  changing percents to, 174, 252
  changing to decimals, 252
  changing to percents, 174, 252

Steck-Vaughn
Connections Mathematics
Correlation to Mathematics Skill Books

Section	Skill	Mathematics Skill Book: Pages
1	Whole Number Theory	5500: pp. 2–4, 44
2	Add/Subtract Whole Numbers	5500: pp. 5–21, 44–45 8800: pp. 2–11, 31, 33, 44–45
3	Multiply/Divide Whole Numbers	5500: pp. 22–45 8800: pp. 2–11, 28, 30, 44–45
4	Squares, Cubes, and Square Roots	8800: pp. 34–35, 37, 44–45
5	Fraction Theory	6600: pp. 2–6, 16, 44
6	Add/Subtract Fractions	6600: pp. 7–29, 44–45 8800: pp. 12, 14–22, 44–45
7	Multiply/Divide Fractions	6600: pp. 30–43, 45 8800: pp. 13–22
8	Add/Subtract Decimals	7700: pp. 2–12, 20–21, 44 8800: pp. 2–11, 28–29
9	Multiply/Divide Decimals	7700: pp. 13–21, 44 8800: pp. 2–11, 40
10	Ratio and Proportion	8800: pp. 41–42, 44–45
11	Percent Theory	7700: pp. 22–27, 44
12	Solving for the Part	7700: pp. 28–30, 40–45 8800: pp. 20–27, 31, 44–45
13	Solving for the Rate	7700: pp. 31–33, 42–45 8800: pp. 20–27, 31, 44–45
14	Solving for the Base	7700: pp. 34–38, 42–45 8800: pp. 20–27, 44–45
15	Percent of Change	7700: pp. 34–38, 42–45 8800: pp. 20–27, 44–45
16	Probability Theory	
17	Triangles and Circles	8800: pp. 33–35, 44–45
18	Integers	Measurement and Geometry: pp. 42–43, 45 Algebra: pp. 2–3, 42–43, 45